The best book on CB

AF770801

by Evan Herbert

Publisher: Bill Fisher
Editors: Bill Fisher
Marcia Redding
Carl Shipman
Cover Design: Josh Young
Design & Assembly: Don Burton
Typography: Frances Ruiz
Drawings: Erwin Acuntius
Chris Crosson

ABOUT THE COVERS: The happy CBer on the front cover is using the remote-control mike of Hy-Gain's 2679-A. At press time, this 40-channel set awaits FCC type-approval prior to being offered for sale.

On the back cover, the marine scene is Hy-Gain. CBers hunting, farming and trucking are courtesy of Pace/Pathcom. The RV and base-station photos are from E.F. Johnson.

ISBN: 0-912656-52-2, Library of Congress Catalog Card No. 76-41094, H. P. Book No. 52, © 1976 Printed in U.S.A. 9-76
H. P. Books, P. O. Box 5367, Tucson, AZ 85703 602/888-2150

CHAPTER 1

CATCHING UP WITH CB TODAY

Howboutcha, good buddy! Got your ears on? **CB two-way radio has become everybody's portable telephone booth.**

People don't talk to each other the way they used to. They aren't friendly to strangers any more. Right?

Wrong! They talk to each other more than ever, using personal two-way radio stations. Millions of strangers talk to each other and help each other with Citizens Band radios. CB has suddenly become a way of life.

You can talk to other vehicles from your car. Or call ahead to home, office, hotel or restaurant. You can back-pack through the woods and keep in touch with your camp or other hikers. Lots of outdoorsmen have gotten "unlost" with portable CB sets. Boaters let each other know where the fish are biting; signal docks and marinas for supplies and service. Farmers coordinate work in the fields by CB. Boy meets girl. There have even been weddings on CB with the guests tuned in!

CB sounds like a great electronic three-ring circus. It's at least that and for millions of CBers, it's more fun. As the world's greatest party line, it is enormously useful. People shop by CB in unfamiliar territory—just by picking up the microphone and asking for local info. You may be looking for a good buy in Indian blankets. You are heard by half a dozen CBers who tell you just the

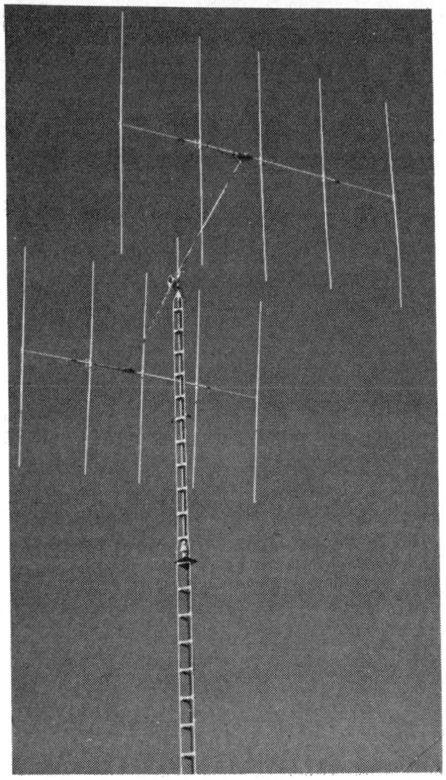

How far can you talk? At first, a CBer is happy to contact anybody who wants to chat. After a while, CBitis sets in and you start thinking of a dream beam to zap your signal over far horizons.

New generation sets handle your CB social life. This rig samples the channels so you can pick an interesting conversation. It will also search for an empty channel so you can arrange to rendezvous with a friend. At press-time this Tennelec 300 was not yet FCC type-approved and not offered for sale.

place. If you are lost, give your location and someone will direct you.

CB simplifies life and makes it safer, too. Working mothers check on children home from school. Hitchhikers with portable sets tell approaching vehicles where they'd like to have a ride—then call their home base station and report the license number of the car that stopped, just in case. If you're in trouble, people are listening on CB.

You can enjoy the benefits and advantages of CB almost instantly. All you need is a CB set, antenna, license to operate and simple instructions. That gets you on the air, but you won't stop at that level because CBitis is catching.

Having your own personal broadcasting station arouses the dormant tinkerer in almost everybody. It won't be long before you'll want to improve the performance of your equipment. Doing that doesn't require technical knowledge beyond what is in this book. You can add accessories or trade up your equipment. Knowing *what to do* is important even if you have somebody else do it. Knowing *how to do it* can save you money and bring you the pride of a job well done.

This book covers the entire field of CB—from making a start up to achieving maximum performance. It helps you make a graceful entrance into on-the-air social circles by teaching you how to talk like a pro. It makes you a knowledgeable member of the CB community.

Besides the boom in add-on CB sets for vehicles, Detroit cars for 1977 offer CB from the factory or CB installed at car dealers. Either way, your CB-equipped vehicle is ready for two-way radio communications as soon as you get your license.

NEW CHANNELS

At least a half-million people are joining the fun each month. The national party line on CB is getting crowded in some areas, so the Federal Communications Commission (FCC) expanded the band from 23 to 40 channels, effective January 1, 1977. A channel on CB is like a channel on TV—a location on the tuning dial. This chapter lists the official tuning frequencies for the complete citizens band and the channel numbers you'll find on many new sets sold in 1977.

THE IMPACT OF 40-CHANNEL CB

Even though the FCC has authorized more channels, you may not need them. In densely populated areas the original 23 channels are so full at times it's hard to find a clear channel to use. People there can use 40 channels provided the other party in each communication also has 40 channels.Otherwise a 40-channel set has no advantage over a 23-channel set.

In much of the U.S., the original 23 channels aren't crowded. Because all sets sold during previous years were 23-channel units, there are a lot more people to talk to on channel 23, or lower, than on the new channels.

In July, 1976, the FCC announced the new channels, effective January 1, 1977. But no 40-channel sets can be sold until they are tested and accepted—called type accepted—by the FCC. This left prospective CB buyers with a problem. Get a 23-channel set immediately or wait for type-accepted 40-channel sets which will cost more.

NEW 40-CHANNEL CITIZENS BAND

Effective January 1, 1977, the Class D Citizens Radio Service includes these frequencies. The FCC does not refer to channel numbers in its new rules. For the sake of tuning convenience and simplicity, CB sets will continue to mark channel numbers on selector switches. Some of the new channels are between the frequencies for channels 22 and 23.

MHz	MHz
1 – 26.965	21 – 27.215
2 – 26.975	22 – 27.225
3 – 26.985	23 – 27.255
4 – 27.005	24 – 27.235
5 – 27.015	25 – 27.245
6 – 27.025	26 – 27.265
7 – 27.035	27 – 27.275
8 – 27.055	28 – 27.285
9 – 27.065	29 – 27.295
10 – 27.075	30 – 27.305
11 – 27.085	31 – 27.315
12 – 27.105	32 – 27.325
13 – 27.115	33 – 27.335
14 – 27.125	34 – 27.345
15 – 27.135	35 – 27.355
16 – 27.155	36 – 27.365
17 – 27.165	37 – 27.375
18 – 27.175	38 – 27.385
19 27.185	39 – 27.395
20 – 27.205	40 – 27.405

Neighborhood roofs are sprouting base-station antennas. Install your own just by following simple, step-by-step instructions in this book.

CB set makers made it easier to decide. They offer to re-manufacture *certain models* of 23-channel sets into type-accepted 40-channel sets as soon as the FCC allows it. The cost of this re-manufacturing is comparable to the additional cost of the new 40-channel rigs, so the total is about the same either way. Purchasers of 23-channel sets covered by a CB company's 40-channel re-manufacturing policy have three advantages over those who wait:

Immediate use of CB.

The option not to spend more money for 40 channels if 23 is enough for the area and use of a particular set.

The option to acquire 40-channel capability as soon as it is legal to do so.

Re-manufacturing of a transceiver for 40 channels is the only way it can be done. The FCC forbids kits or owner modifications, but nearly everybody would rather have the factory do it anyway.

If you are considering a CB set between July 1976 and the time new 40-channel sets are type-accepted in 1977, there's no need to wait. You can have the fun and convenience of CB immediately. If you think you may need 40 channels later, check your dealer on the sets covered by the re-manufacturing offer. Select one of those, install and use it as described in this book, and you're in for a whole new world of fun!

HOW TO GET AN INSTANT LICENSE

You don't have to wait to buy a CB set and you no longer have to wait for your license to operate your CB station.

Today, you issue yourself a temporary permit instantly, mail your application to the FCC for a permanent station license, familiarize yourself with the essential rules and regulations, and turn on the set. The forms you need for doing this are in Chapter 20. So are the latest FCC Rules and Regulations the law says you must have.

In Chapter 2 I'll tell you how to fill out the forms correctly so your permit and license application won't be declared defective. If you're going to visit Canada and would like to use your CB on the air in the equivalent Candian General Radio Service, you'll find the forms for a temporary Canadian permit in Chapter 20, *The CBer's Scene.*

With those formalities done, the rest of CB is delightfully informal, like a grand costume party where everybody dresses as characters they'd like to be. On CB you cloak yourself in colorful language and use a nickname—*a handle*—such as Bald Eagle or The Red Baron. Once you've given your legally-required station call sign, you can call yourself whatever you like.

CBers often use 10-codes in place of time-consuming sentences. They also love rythmic slang like *put the pedal to the metal.*

There's a right way and a wrong way. Learn to talk on the air like a pro—see Chapter 3.

IN EMERGENCIES, CB BRINGS HELP FAST

When there are emergencies—anything from highway accidents to natural disasters—CB radio may be the most immediate and possibly the only means of communication. Today you can get help faster by using your CB than any other means of communication. CB is now officially recognized by most police departments—they monitor emergency channel 9 from headquarters, patrol cars, helicopters and harbor patrols. Moreover, CBers themselves have created volunteer emergency monitoring services everywhere. Find out *how to handle an emergency*, your own or somebody else's, in Chapter 4 which includes a handy chart to

Boaters put CB aboard even when they have marine radios. This salty skipper is an architect who likes to keep in touch with other week-end sailors along coastal waterways.

post at your CB station so you can keep your cool when handling emergencies.

CHOOSING AMONG THE NEW SETS

A CB set is called a *transceiver* because it is both a transmitter and a receiver. The swirl of new technology brings new models in all shapes and sizes which offer only 1 channel or up to 40 channels.

Features of CB transceivers are often stated in technical jargon which won't mean a thing unless you know how to interpret it.

Chapter 6 takes the mystery out of specifications by telling you which words and numbers are important and what they mean to CB communicating and to your wallet.

When you understand specs and features, you should make a good choice of a mobile or marine transceiver, or a base-station rig, or a hand-held portable. The most popular type is the mobile set. There are so many models it would be impossible for you to compare them all even if you shopped for months. Even then, you couldn't compare them under actual operating conditions—which are what really counts. The way a set performs in a crowded city is no indication of how well it will do in the suburbs or open country. You can't get that kind of information from specs or laboratory tests. We have a team of professional mobile-communication engineers who field test popular models under actual real-world conditions. They go out to play in the traffic and drive in the country to get performance ratings on popular models. You'll see the test results in the *Consumer's Performance Guide* of Chapter 8.

INSTALL YOUR CB STATION RIGHT—THE FIRST TIME!

It's easy to select and install transceivers if you know how. Chapter 7 sorts out mobile mounting locations and procedures so you'll know where it's best to put a CB rig, how to make it convenient to use, easy to install but difficult to steal. In later chapters I show you how to select and install base stations, portable and marine rigs. It's not a fine art, but you really have to pay attention to the *practical details* which I've emphasized in words, photos and drawings that anybody can understand and follow.

HOW TO TALK FARTHER

There is a legal limit to the transmitter power you can use. If you want to get the most communication range, your best approach is to pick the right antenna. You'll find what you *really* need to know about antennas in Chapter 12, followed by big chapters on how to install them properly. These chapters cover cars, trucks, RV's, motorcycles, boats, buildings and towers.

AVOID RIP-OFFS

CB theft has become a crime wave. Chapter 15 tells what you can do to keep your set from being stolen and about insurance coverage if you lose the game.

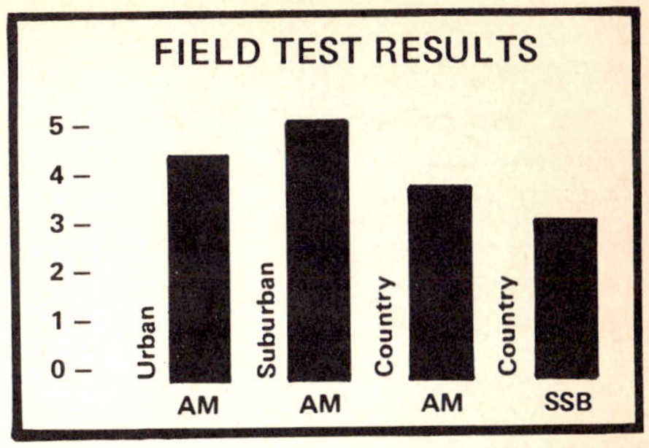

Comparing CB sets in a store is like comparing race horses in a paddock. You never know what one can do until it counts. Our professional mobile communication engineers compared leading brands under typical conditions where CBers operate. Their ratings are in the Consumers Performance Guide of Chapter 8.

HOLD DOWN THE NOISE AND INTERFERENCE

CB-land is not a quiet place. Electrical noise is a major problem. If the built-in noise controls on your mobile set don't do the job, your own car is probably making too much electrical noise. Chapter 18 gives you a handy trouble-shooting chart and shows you how to fix your car so it stops interfering with your CB fun. CB sets also interfere with TV sets—your neighbor's and your own. There is also information about how to solve these problems whether you are the cause of TV interference or the victim.

UPGRADING YOUR CB STATION

Make the most of your equipment by periodic maintenance and check-ups you can do yourself. Chapter 17 shows you how and also covers ways to make significant improvements through new equipment and accessories. Chapter 19 is almost a catalog of accessories you may need. Most of the supplies and accessories you need are readily available but a few are hard to find. I've listed names and addresses of manufacturers of the hard-to-find stuff in Chapter 20. When you get that far and are looking for the place to buy a super-gadget so you can have more fun with CB, be listening for my call. *Have a good trip. KEJ 6386, standing by.*

CHAPTER 2

HOW TO GET YOUR LICENSE

Getting on the air immediately is easy to do. Until mid-1976, you had to wait until you received a regular license from the FCC before you could legally operate a CB radio. You've probably heard agonizing stories: People waiting for months while the FCC struggled under a mountain of new applications every week.

Today you can get an *instant permit.* You fill it out and issue it to yourself. It's a two-step procedure.

Here's how it works. Obtain copies of FCC Form 505 and Form 555-B called *Temporary Permit for a Class D Citizens Radio Station.* Copies of these forms are in Chapter 20. You can also get them at your local CB stores, local or county police departments if they are on the ball, from the FCC if you are in no particular hurry, and packed with new CB equipment manufactured after April, 1976.

Don't take a shortcut. You *must* fill out both forms to be legal. Operating without a license or temporary permit makes you subject to *criminal* prosecution by the Department of Justice.

HOW TO FILL OUT FORM 505

Your application can bounce back if you have the right information in the wrong places. Item 2 asks for your birthdate, not the date of application, which goes near your signature at the bottom. You as an individual must be 18 or over to hold a license, but you can operate a CB with permission of a licensed member of your immediate family in the same household, or if asked to do so by your employer in the course of your duties. Of course, the date of birth applies only to application by an individual.

If the application is not for an individual person, but for an organization or a business, here are the rules about who can apply:

PART OF FORM 505

5. If you gave a P.O. Box No., RFD No., or General Delivery in Item 4A, you must also answer items 5A, 5B, and 5C

5A. NUMBER AND STREET WHERE YOU OR YOUR PRINCIPLE STATION CAN BE FOUND
(If your location can not be described by number and street, give other description, such as, on RT. 2, 3 mi., north of York.)

5B. CITY

5C. STATE

(See reverse side of this form for filling in Item 5C.)

Just below your mailing address—if it's a box number or RFD—you must show the actual location of your principal station. The government spells *principal* wrong but wants you to describe the station location accurately.

PART OF FORM 505

7. This application is for

Corporation

l Entity

New License

Renewal

Increase in Number of Transmitters

IMPORTANT
Give Official FCC Call Sign

9. Indicate number of transmitters applicant will operate during the five year license period (Check Only One Box)

1 to 5

6 to 15

16 or more (Specify No. and attach statement justifying need.)

resentative thereof.

of Part 95 of the Commission's
erse side for ordering informa-

compliance with the applicable
n will not be used for any pur-
greater power than authorized.

tory power of the United States
use of the medium of transmis-
s use, whether licensed or un-

THIS APPLICATION WILL NOT BE PROCESSED UNLESS SIGNED AND DATED.

WILLFUL FALSE STATEMENTS MADE ON THIS FORM OR ATTACHMENTS ARE PUNISHABLE BY FINE AND IMPRISONMENT. U.S. CODE, TITLE 18, SECTION 1001.

12. DATE

authorized person on behalf of a
tion or association

Here's where to anticipate CBitis. Five transceivers may not be enough for multi-vehicle families. Within a year I'll bet you have a base station and at least a pair of portables. Add up the probabilities and check the right box.

You and your business *partners* may apply for a license in the name of the partnership provided the communications you will engage in relate to the business of the partnership.

Members of an unincorporated *association* can apply for a license in behalf of the organization provided the communications relate to its business. For example, a monitoring team formed by a CB club may obtain a license to carry on its business of handling emergency messages and providing assistance. However, if the purpose of the license is to provide radio communication among the members of the association, you must get specific prior approval of the FCC. An officer of the association must sign the application.

A *corporation* may apply for a license to provide radio communication among its own facilities, or among facilities that include parent or subsidiary corporations. The license application must be signed by an officer of the corporation. If the radio facilities costs are shared among the several related corporations, it must be done so on a nonprofit basis spelled out in a contract between the licensee and the other parties. Records of the costs and the sharing basis must be maintained by the licensee. A corporation must get prior specific approval of the FCC for its particular use of CB. What's more, employees of the corporation may use its stations only when specifically authorized by the corporation and only to conduct business. In short, you can't grab the mike and give a shout to your home base station to put up dinner.

The *sole proprietor* of a business or someone doing business under a company name may apply for a license and authorize employees to use station facilities in the conduct of the business as required by their employment.

Government entities, state, territorial or local, may apply for a license. It must be signed by a duly elected or appointed official. The FCC warns that the Citizens Band is shared without distinction among licensees–government or anyone else.

Form 505 has different item numbers for the same information, depending on when the form was printed. Check item 7 or 12 carefully because if you are applying for a license renewal or for an increase in the *number of transmitters,* you'll need to give your current call sign. Item 9 or 14 gives you the option of specifying the number of transmitters you'll be operating during the license period of 5 years. One to five *seems* like enough when you start out. You'll probably start out with one mobile. Then a base station. With more than one vehicle in the family, CBitis may spread. So you add one or two more units. Now you're almost up to your limit, which makes it hard to add a pair of portables without applying for an increase in the number of transmitters. You can specify 6 to 15 radio sets without explaining why you need them, so why not pick enough to be sure? If you check 16 transmitters or more, you have to attach a statement justifying your need.

Your *address*, item 4, has a catch to it which could make your application defective. You must show the *actual location* of your principal base station or the address where your original license will be displayed for inspection.

If your address given in item 4 is a PO Box or RFD number, you must go on to fill out items 5, 5A, 5B, 5C or items 8, 9 and 10. This tells precisely how to find your base station or the place where you will make your license available for inspection. You can give mileage from a particular road intersection, map coordinates or even latitude and longitude as stated on a surveyor's map.

Your *certification* in items 10 or 15 should not be taken lightly. It is your part of the bargain in being authorized to operate a CB station. Your signature on this form says you accept certain responsibilities.

First, you state that you are a US citizen, or the organization is not a representative of a foreign government. Rejoice! Only a few countries in the world allow private citizens to own and operate two-way radio communications.

Second, you state that you have in your possession or have ordered from the Government Printing Office a *current* copy of Part 95 of the FCC Rules and Regulations. If you own this book, Part 95 is in the back–current as of the indicated date. You can also get Part 95 by ordering "FCC Part 95, Stock Number 004-000-00324-1."

UNITED STATES OF AMERICA
FEDERAL COMMUNICATIONS COMMISSION
FCC Form 452-C
(July 1972)
TRANSMITTER IDENTIFICATION CARD

1. Station call sign:

2. Name and Address of Permittee or Licensee:

Attach a Transmitter Identification card like this, or something equivalent, to each transceiver. You don't have to use the official form but you must show your temporary or permanent call sign. Include your name and the address where the original license is available for inspection by the FCC.

Include payment by money order or check for $1.50 payable to the Superintendent of Documents and mail to:

Superintendent of Documents
Government Printing Office
Washington, DC 20402

Third, you agree to operate your CB in full compliance with the current applicable Rules and Regulations of the FCC. You also agree not to use the station for any purpose contrary to Federal, state or local law. This paragraph is really the hooker. I've heard CBers argue that you have to *own* Part 95, but nothing requires you to *read* it. If you are found in violation of these regulations, you cannot plead ignorance. The FCC is no longer reducing fines. It is getting tough and is under pressure by Congress and other government agencies to get even tougher.

Also, you agree that even if you can't find a clear CB channel you can use, you won't sue the FCC. Your license gives you the right to use and *share* the designated channels whenever they are available. The best you can do is to encourage radio discipline among fellow CBers and practice it yourself.

Some versions of Form 505 have a space near the top for your name and mailing address to speed return of your application if it is defective. At the bottom is a space for your call sign under the temporary permit I'll discuss next.

Mail your application to the FCC with $4 by check or money order payable to the FCC; *don't send cash.* Make note of the date. Here's a *special address* just for CB licenses:

FCC
Gettysburg, PA 19326

FCC forms are going out of date so quickly you're likely to find an application with a PO Box Number and different Zip Code packed in the carton with your new CB set. The latest address above is intended to take care of the mailbag pile-up. It's no coincidence that the now obsolete PO Box 1010, which used to get you a license, has an appropriate meaning in ten-code: 10-10 means *transmission completed, standing by.*

THEN FILL OUT FORM 555-B

While you are standing by, waiting for your regular license, you can operate on a Temporary Permit that you issue to yourself, using Form 555-B, after you have mailed in the regular application. You don't mail in the Temporary Permit form, just fill it out. It authorizes up to 5 transmitters and no more, even if your license application, Form 505, requests permission for more than 5.

Read the guidelines for using your CB station on the back. This is not a waste of time even under a temporary permit because you should know the most important rules the FCC wants you to observe.

Sign your name, print your name and address, and fill in the date you mailed Form 505 to the FCC. Your permit is valid for 60 days from the date you mailed *Form 505,* the application for a regular license.

Now give yourself a temporary call sign. Part 3 of the Temporary Permit has blocks to be filled in. The first block already has the letter K. In the next 2 blocks fill in the initials of your first and last name. Then put your Zip Code in the last 5 blocks. Here's how my temporary call sign would come out: |K|E|H|0|7|7|2|4|.

Do not mail your permit. Keep it in a safe place; the original becomes a legal document even if it is not certified or notarized. Post photocopies of it at all fixed station locations, indicating on the copy where the original permit is kept.

Attach a card with your name, address and temporary call sign to each CB radio you are responsible for. You do not have to use FCC Transmitter Identification Card Form 452, but it's a neat, useful reminder that you are operating under FCC authority so you tend to think about operating legally and being one of the good guys in CB.

If your Temporary Permit runs out before you get your license; call:

202/632-7175

Going to Canada with your CB? The lower 22 channels there are the same as the US Citizens Radio Service. You won't be able to operate on channels 23-40. Be sure to get your Canadian temporary permit.

HOW TO IDENTIFY YOUR STATION PROPERLY IN ANOTHER COUNTRY

Proper station identification is particularly important when you are operating your CB on a temporary permit issued for your stay in another country. US citizens visiting Canada and Canadian citizens visiting the USA should give the nearest location along with their call signs. Here's the correct procedure:

This is KEJ 6386 at Thunder Bay, Ontario.

A likely result of your call may be the issuance of a 90-day Special Temporary Authority while the problem is corrected. *Do not write* instead of telephoning; even the opening of mail-bags is backlogged at Gettysburg because of the phenomenal number of license applications.

OPERATING YOUR CB IN MEXICO

US borders are easy to cross. It would be nice to take your CB mobile or portable with you. *Mexico no longer permits tourists to bring in CB.* Transceivers have become so useful to smugglers—of narcotics and wetbacks—that the Mexican Government won't trust anyone.

OPERATING IN CANADA

Canada has a reciprocal agreement with the USA permitting American citizens to operate CB in Canada and Canadian citizens to operate in the USA. You must already hold a valid license in your own country before applying for a permit to operate in the other country. Here's how to apply:

US citizens wishing to operate a CB in Canada as a tourist or visitor may apply to the Regional Superintendent, Telecommunications Regulation, Department of Communication. There are several addresses and it doesn't matter which one you contact, even if you plan to go to a different part of Canada. The application form is included in Chapter 20. Although the Canadian application for a temporary permit does not specifically ask for a copy of your US authority to operate a CB, it's a good idea to attach a *copy* of your US Station License or Temporary Permit.

If you are going to Canada with your CB and can't wait the 30 days or so for paper to shuffle back and forth across the border, telephone the Regional Superintendent or any one of the main Regional Offices. You'll find the Canadians delightfully obliging good neighbors, willing to take your call sign and other information over the phone to expedite your permit. Or they will make arrangements for you to visit a district Regional Office on arrival so you can walk out with an official operating authorization in hand.

You will receive a permit for the

WHERE TO APPLY FOR A TEMPORARY PERMIT TO OPERATE CB WHEN VISITING CANADA

US Citizens licensed by the FCC may register with the Canadian Government for a temporary permit to operate your CB during a visit. Your license must be current when you enter Canada. Your equipment must be FCC Type Approved.

Here are the five Regional Service Offices of the Canadian Department of Communications. Any one of them will process your application by mail. Or you may call to arrange for a visit to one of the many District Offices in each Province to get your permit in person; your equipment will be sealed by Customs at the border unless you have a valid temporary permit.

Atlantic Region
Regional Superintendent
Department of Communications
7th Floor
Terminal Plaza Building
PO Box 1290
MONCTON, N.B.
ELC 8P9
CANADA

Quebec Region
Regional Superintendent
Department of Communications
20th Floor
2085 Union Street
MONTREAL, Que.
H3A 2C3
CANADA

Ontario Region
Regional Superintendent
Department of Communications
9th Floor
55 St. Clair Avenue East
TORONTO, Ont.
M4T 1M2
CANADA

Central Region
Regional Superintendent
Department of Communications
2300-One Lombard Place
WINNIPEG, Man.
R3B 2Z8
CANADA

Pacific Region
Regional Superintendent
Department of Communications
325 Granville Street
Room 300
VANCOUVER, B.C.
V6C 1S5
CANADA

VISITOR WITH AN EMERGENCY?

In both the USA and Canada, channel 9 is the emergency channel. Both countries have monitoring teams listening, but CB communication range is short so you may not be heard in the unpopulated areas.

duration of your visit, but not longer than one year. You may apply for renewal not to exceed one year, provided the Canadian Government has not received any complaints about your station operation.

In Canada, the channels are the same as channels 1 through 22 in the USA. Photocopy your Canadian permit and keep a copy with each transceiver you take into the country.

Canadian regulations are about the same as current Citizens Band Regulations of the FCC. There is an important exception. In the US, the FCC acknowledges the hobby aspects of CB and you can chat about your rig or your station's performance and the like. You cannot indulge in hobby talk on the air in Canada.

You aren't formally required to own and read the Regulations of the Canadian General Radio Service but it's a good idea to be familiar with them. Pertinent extracts from the rules affecting operation in Canada are in Chapter 20.

CANADIANS VISITING THE USA

Canadian citizens desiring to operate their General Radio Service stations when visiting the USA must apply for a permit on FCC Form 410. You can write to The Federal Communications Commission. Washington, DC 20554 for copies of Form 410. The FCC asks for a separate application for each station *or* transmitter, so request enough forms. Form 410 is also included in the back of this book. Mail it at least 60 days in advance to: FCC, Gettysburg, PA 17325. Don't use the box number you saw earlier in this chapter; it is strictly for CB applications from the US.

You must have a valid Canadian license for General Radio Service (GRS). A permit to operate in the USA is good for one year, or less if your Canadian GRS license expires before that date. Each Canadian permittee, operating under subpart E, shall have read and understood Part 95, Citizens Radio Service. Part 95 is reproduced in full in the back of this book, including subpart E on operation of CB stations in the US by Canadians.

GOING TO CANADA?
Parlez-vous GRS?

Officially, Canada has two languages—English and French. There are parts of Canada where French is predominant; you'll encounter it on all channels of the General Radio Service (GRS). It's a good idea to know some conventional expressions in an emergency—yours or those of an *opérateur du service radio général.*

Calls to "all stations" are not permitted in the Canadian General Radio Service, except in an emergency you can call "any GRS station" in a specific location. For example, in Windsor, Ontario, you can say:

Any GRS station in Windsor, this is KEJ 6386.

Or, in Quebec:

N'importe quelle station du service radio general a Quebec ici KEJ six trois huit six.

Here are a few of the most important words used in message handling in Canada. They sound a lot more formal than the lingo rattling around the United States.

The Canadian Department of Communications feels strongly about Canadian operators behaving like professional radio operators.

ENGLISH	FRENCH
Acknowledge	Veuillez accuser réception
Affirmative	Affirmatif
Break	Séparatif
Channel	Voie
Confirm	Confirmez
Correction	Rectification
Go Ahead	Continuez
How Do You Read	Comment recevez-vous
I Say Again	Je dis de nouveau
Negative	Négatif
Over	Répondez
Out	Terminé
Read Back	Relisez
Standby	Attendez
That Is Correct	C'est exact
Verify	Vérifiez
Wilco	Wilco
Words Twice	Chaque mot deux fois

CHAPTER 3

HOW TO TALK ON THE AIR LIKE A PRO

You can press that microphone button and be on the air. Heard by millions or maybe even dozens. What should you say?

Mike fright happens to almost everyone the first few times the transmit button is pressed. Here are the most likely reasons. First, you feel funny about addressing an audience of strangers. Second, you're not sure you can speak the language. You're reluctant to break into the conversation even though there is a standard way of doing it. *Breaker 1-9,* you say, or *Break 1-9,* meaning "May I break into the conversation on channel 19?" The answer is always yes whether spoken directly or not.

Relax. CB language is a way of going back to being just plain folks. On CB you will be speaking on the national party line where every stranger is immediately a *good buddy.* Strangers are coming on the line at a rate approaching a million a month. Talking on the air is like the first day of school, or summer camp, or joining a club. There are lots of newcomers and plenty of old-timers, glad to help you because it gives them a chance to play expert—and to become your friend.

You don't have to talk CB lingo and *ten-codes* to get on the air. After a few hours of listening, you'll be talking like most CBers, which is good—and bad. This chapter will speed up your learning and alert you to bad communication habits you might pick up by imitation.

Talking CB lingo is fun, for the most part. But it has its place, for there are times involving road information and emergencies when it is critical that you be precise and use only clear language—not even ten-codes will do.

Remember there is no *official* CB language. Only a few FCC rules govern the way you speak or what you say on the air. I'll cover these shortly. Ten-codes—those mysterious numbers—are a way of compressing common messages into short transmissions. Two pros talking ten-code are a pleasure to hear and those crackling businesslike codes reduce their air time so you can use the channel too. Some

A real pro communicates on the air in language another person is sure to understand: Brief and precise. That's a Browning LTD mobile being used to talk to pilots at Lake Winnipesaukee, New Hampshire.

ten-codes mean different things to law-enforcement agencies than to CBers. Compare the two 10-codes given here and you'll see why private citizens shouldn't attempt 10-codes in an emergency. You may do more harm than good.

Don't be bamboozled by slang or bugged by style. I included CB slang in Chapter 20 at the back of this book, though no such dictionary can ever be complete. It's so much fun to make up new words or phrases that CB lingo keeps changing as people criss-cross the country swapping the latest colorful phrases. Regional language rapidly becomes part of the national lingo of CB communication and "in" phrases rapidly fall out of use.

A real CB expert considers who is listening and how little or how much the other person may understand. It's a mark of pride to say it so you never have to repeat or explain.

EVERYBODY HAS A HANDLE

Handles—nicknames—are universally used. You can choose and use a permanent handle, or call yourself whatever you feel like at the time.

Many CBers simply pick a name that rolls right off the tongue such as Steamin' Demon or Double Trouble. Handles that conjure images of World War fighter squadrons are popular—The Red Baron, Blue Max, and others. Comic strip characters are all over the CB band; you'll overhear L'il Abner talking to Daisy Mae; Blondie rapping with Charlie Brown.

Your handle can tell a little about yourself. An IRS agent uses The Revenooer. While writing this book, my handle is The Happy Booker.

Handles that make CB playful are those chosen for additional effect in typical exchanges:

You've got The Tombstone here, passin' Exit 2.

TEN CODE USED BY LAW ENFORCEMENT AGENCIES

10-0	Caution	10-50	Accident (F, PI, PD)
10-1	Unable copy—change location	10-51	Wrecker needed
10-2	Signal good	10-52	Ambulance needed
10-3	Stop transmitting	10-53	Road blocked at . . .
10-4	Acknowledgement (OK)	*10-54	Livestock on highway
10-5	Relay	*10-55	Intoxicated driver
10-6	Busy—stand by unless urgent	*10-56	Intoxicated pedestrian
10-7	Out of service	*10-57	Hit and run (F, PI, PD)
10-8	In service	*10-58	Direct traffic
10-9	Repeat	10-59	Convoy or escort
10-10	Fight in progress	*10-60	Squad in vicinity
10-11	Dog case	10-61	Personnel in area
10-12	Stand by (Stop)	10-62	Reply to message
10-13	Weather—road report	10-63	Prepare make written copy
*10-14	Prowler report	10-64	Message for local delivery
*10-15	Civil disturbance	10-65	Net message assignment
10-16	Domestic problem	10-66	Message cancellation
10-17	Meet complainant	10-67	Clear for net message
10-18	Complete assignment quickly	*10-68	Dispatch information
10-19	Return to . . .	*10-69	Message received
10-20	Location	10-70	Fire alarm
10-21	Call . . . by telephone	*10-71	Advise nature of fire
10-22	Disregard	*10-72	Report progress on fire
10-23	Arrived at scene	10-73	Smoke report
10-24	Assignment completed	*10-74	Negative
10-25	Report in person (meet) . . .	10-75	In contact with
10-26	Detaining subject, expedite	*10-76	En route
10-27	(Drivers) license information	10-77	ETA (Estimated Time Arrival)
10-28	Vehicle registration information	*10-78	Need assistance
10-29	Check record for wanted	*10-79	Notify coroner
10-30	Illegal use of radio	*10-80	Chase in progress
10-31	Crime in progress	*10-81	Breatherlizer report
10-32	Man with gun	*10-82	Reserve lodging
10-33	EMERGENCY	10-83	Work school xing at . . .
10-34	Riot	*10-84	If meeting . . . advise time
*10-35	Major crime alert	10-85	Delayed due to . . .
10-36	Correct time	*10-86	Office/operator on duty
*10-37	(Investigate) suspicious vehicle	*10-87	Pickup/distribute checks
*10-38	Stopping suspicious vehicle	10-88	Advise present telephone number of . . .
10-39	Urgent—use light, siren	10-89	Bomb threat
*10-40	Silent run—no light, siren	*10-90	Bank alarm at . . .
*10-41	Beginning tour of duty	10-91	Pick up prisoner/subject
*10-42	Ending tour of duty	10-92	Improperly parked vehicle
*10-43	Information	10-93	Blockade
10-44	Request permission to leave patrol . . . for . . .	10-94	Drag racing
10-45	Animal carcass in . . . lane at . . .	*10-95	Prisoner/subject in custody
10-46	Assist motorist	*10-96	Mental subject
*10-47	Emergency road repairs needed	*10-97	Check (test) signal
*10-48	Traffic standard needs repairs	*10-98	Prison/jail break
*10-49	Traffic light out at . . .	10-99	Records indicate wanted or stolen

*CB code has no equivalent number.

UNOFFICIAL "OFFICIAL" CB TEN-CODE

10-1	Receiving Poorly
10-2	Receiving Well
10-3	Stop Transmitting
10-4	Okay, Acknowledged
10-5	Relay Message
10-6	Busy, Stand By
10-7	Out of Service
10-8	In Service
10-9	Repeat Message
10-10	Transmission Completed, Standing By
10-11	Speak More Slowly
10-12	Visitors Present
10-13	Weather/Report Conditions
10-16	Make Pickup At
10-17	Urgent Business
10-18	Anything For Us?
10-19	Return To Base
10-20	Location
10-21	Call By Telephone
10-22	Report In Person To
10-23	Stand By
10-24	Assignment Completed
10-25	Can You Contact?
10-26	Disregard Last Information
10-27	I Am Moving To Channel
10-28	Identify Your Station
10-29	I Am Leaving This Location
10-30	Does Not Conform to FCC Rules
10-31	Crime In Progress
10-32	Radio Check
10-33	Emergency Traffic
10-34	Trouble Here, Help Needed
10-36	Correct Time
10-39	Your Message Delivered
10-44	I Have a Message For
10-45	All Units Please Report
10-46	Assist Motorist
10-50	Accident
10-51	Wrecker Needed
10-52	Ambulance Needed
10-53	Road Blocked
10-59	Convoy or Escort
10-60	What is Next Message Number?
10-62	Unable to Copy, Use Phone
10-63	Prepare To Make Written Copy
10-64	Net Directed To
10-65	Net Clear
10-66	Message Cancellation
10-67	All Units Comply
10-70	Fire At
10-73	Speed Trap At
10-75	You Are Causing Interference
10-77	E.T.A. (Estimated Time Of Arrival)
10-82	Reserve Lodging
10-84	My Telephone Number Is
10-85	My Address Is
10-88	Advise Telephone Number Of
10-89	Radio Repairman Needed
10-90	I Have TVI
10-91	Talk Closer To Mike
10-92	Have Your Transmitter Checked
10-93	Check My Frequency
10-94	Give Me A Long Count
10-95	Transmit 5 Second Dead Carrier
10-99	Mission Completed, Units Secure
10-100	Personal Reasons
10-200	Police Needed At

YOU CAN TELL YOUR PROFESSION BY YOUR HANDLE

Loan Arranger — Bank officer
Diesel Doctor — Truck mechanic
Liquid Plumber — Pipefitter
Green Doc — Groundskeeper
Mother Hubbard — Kindergarten teacher
Human Erase — Mortician
Wood Butcher — Carpenter
Penny Pincher — Investment banker
Sour Dough — Pawn broker
Electric Bill — Electrician
Pop Eye — Optician
Piece Maker — Gunsmith
The Happy Booker — This author
First Momma — The President's wife
Tooth Fairy — Dental assistant

10-4, Tombstone. You've got The Devil looking over your shoulder.

I like simple handles that command attention:

Thanks for that Smokey report, breaker. What's your handle?

You've got Trouble.

Is your handle your exclusive property? No more than your given name. You'll see many advertisements in CB magazines offering to "officially" register your handle for several dollars. Don't waste your money.

Handles are fun, but they also serve a useful purpose you hadn't considered. They are easier to understand and remember than call signs' combinations of letters and numbers. Choose your handle with care; it may be heard through poor-quality loudspeakers in noisy environments so your handle may not be clearly understood or heard correctly. I'm having second thoughts about calling myself The Happy Booker.

The law says you can use a handle, but it is not to be a substitute for a call sign in properly identifying your station. A call sign is the legal requirement.

CALL SIGNS COME FIRST

Perhaps you haven't heard call signs used very often on CB. There has been a lot of abuse of the FCC regulation which says: "All transmission from each unit of a citizens radio station shall be identified by the transmission of its assigned call sign at the beginning and end of each transmission or series of transmissions, but at least at intervals not to exceed ten (10) minutes." That means give your call when you come on and when you sign off.

Failure to do it can cost your license or a fine of $50 to $100, but that's not the important reason to comply. After you have been into CB for a while, you will hear plenty of people with sick minds, abusing the air and everybody listening with boorishness, profanity and outright antagonism toward the world. These weak sisters don't identify themselves of course, except sometimes with an offensive handle. "So what," they say, "nobody else identifies himself either."

Unfortunately they are more nearly right than wrong about people identifying themselves. But if those of us who don't mind being good guys will follow the law and the rules of decency, maybe the weirdos will get a message and go back under their rocks.

Playing it straight doesn't even slow you down. It goes like this:

This is KEJ 6386, The Happy Booker, lookin' for a northbounder on Route 30.

After you exchange Smokey reports, form a convoy, or find out the good places to eat, make your call sign an automatic part of the sign-off:

KEJ 6386, The Happy Booker. Thanks for the info good buddy, have a safe trip.

You can use your handle and you can call the other station's operator by his handle; it is not necessary to use the other station's call sign, just your own.

Units of the same station have the same license and call sign. You may be at your base, calling employees or members of your family who are out in vehicles or hiking with portable units.

KEJ 6386 base to mobile or
KEJ 6386 base to unit 1 or
KEJ 6386 unit 1 to unit 2

A typical response from the other station using your call sign would be:

KEJ 6386 unit 1 to base. Go ahead.

The other station *must* use the call sign as if it were fully independent. It is not legal simply to say "unit 1." You are responsible for the proper operation of all stations using your call sign, so be sure their language conforms to the rules. The FCC notice of violation will have *your* name on it.

If you're talking to another station with a different license, you have only 5 minutes to use the channel before you must remain silent for 1 minute. If you're talking to another unit of your own station, the only restriction on transmission time is "the minimum practicable." Nevertheless, the FCC says you have to identify your station at intervals not to exceed ten minutes.

There are various conditions you may wish to declare about your sign-off. Here are two of the most useful:

Clear and on the side, or *10-10* means "Transmission completed, standing by."

Down or *gone, 10-7* means "Out of service."

Many CBers make it a practice *not* to carry on communications with people who don't give their official call signs. Actually it is illegal to communicate with an unlicensed station. You may say:

10-28—Please identify your station. or *10-30—You are not conforming to FCC Regulations.* Lots of CB club members monitor the channels and request call signs if they are not properly given. They remind recalcitrant CBers that their transmission is illegal and promise to report it to the FCC if the practice continues. I know how that sounds, but I'd rather give my call letters than pay higher taxes or license fees to support more enforcement by the FCC.

Some of the tradition of amateur radio operators is rubbing off on serious CB operators; they are proud of the snappy way they respond to calls or send messages. While they use handles, they sign on and off with their call sign because it *is* unique—a brand name exclusively yours and registered

HANDLES THAT REVEAL PERSONALITY

Happy Critter
The Sour Kraut
Sassie Lassie
Plain Jane
Double Trouble
Happy Pappy
Dum Dum
Eight Ball
Super Elf
Deputy Dog
Great Body
Jolly Joker
Cadillac Cowboy
Spell Binder
Melancholy Monkey
Society Slim
The Wild Card

10-4, 88's, AND OTHER SECRET CODES

Here's a kinky little bit of the law. The FCC says you must use *plain language* on the air, though it does not say the language must be English. That means no secret codes —not even abbreviations—unless you keep a list of all such abbreviations and their meaning in your station records and show it to any FCC person who asks. Howbout that!

If you are using ten-codes and other recognized abbreviations, to be strictly legal you must keep a copy of these. Of course you can use the list in this book for that purpose. If you are not a flagrant violator of other FCC rules, you're not likely to get in trouble for saying, "10-4." But if you are caught doing a lot of things wrong don't be surprised if they nail you for everything possible. "Throwing the book" at one violator tends to keep a lot of others in line, and it's an effective law-enforcement technique. It costs $100 per violation up to a maximum of 5.

with the United States Government.

ACTION ON THE CHANNELS

Listen around on various channels and you'll discover CB largely concerns information, services and home channel topics. If you're learning to talk CB, the channels are your best classrooms for picking up the fine points of handling various conversations. Here's where the action usually is. Certain channel uses may vary by locale:

Channels 10 and 19—Vehicles on the move; chatter about finding your way, convoys, avoiding traffic jams, spotting Smokey, staying out of trouble with the law, reporting vehicular and road obstructions, weather conditions ahead, local information on places to eat or stay or shop, noting pretty girls in cars and toll booths, reporting hitchhikers in need of a ride. Conversations are usually short so the channel can be shared; ratchet jaws agree to switch to another channel, and are often asked to do so.

Channel 11—This used to be the calling channel for establishing communications. As of January 1, 1977, it is an ordinary channel used for ordinary purposes.

Channels 13 and 3—Marine CB operators tend to gather here in areas where there is boating activity. They use their boat names as handles, call marinas with base stations. A great channel to find out where the fish are biting.

Channel 16—The sidebanders have adopted this channel. They love to talk about super antennas, new CB transceivers and accessories. Many CBers here are also radio amateurs—*hams*—and tend to use Q-signals instead of ten-codes. A list of useful Q-signals is in the back of this book.

Other channels—These are used by local custom. In my area of the North Jersey Coast, teenagers gather after school on channel 22. Housewives are on channels 1 and 2; if they're working mothers, they call their children—mobile to base—to be sure the kids are eating a good lunch. Channel 7 is the singles channel—I've never heard 10-4 said with such passion. Business communication is heard on various channels.

> **CHILDHOOD HANDLES**
>
> *The Cookie Monster*
> *Rubber Ducky*
> *L'il Abner*
> *Woody Woodpecker*
> *Charlie Brown*
> *Soda Pop Kid*
> *Orphan Annie*
> *Bugs Bunny*
> *Crackerjack Kid*
> *Goldielocks*

COMMUNICATING ON THE HIGHWAY

If the lingo puzzles you, just ask your question in plain English. A sophisticated CBer will judge your knowledge by the way you ask your question; if he's a good knight of the road, he'll reply in language he assumes you'll understand.

The road ahead—Here, minus most break and call signs, are two versions of typical questions about the road ahead followed by three versions of the same reply:

How about a road report on Interstate 95?

How about a 10-13 on Interstate 9-5 northbound above mile marker 3-oh?

The second version is more into CB language and pronunciation. 10-13 is a request for road and weather conditions. Numerals are spoken separately and distinctly. Though the message seems longer, it will require less air time for reply because the caller has defined the area he wants to know about. Otherwise he would be deep into a conversation about his location and direction of travel. Lots of people don't know the ten-codes except 10-4, so 10-13 will only confuse; while asking in words for a road report will get you an answer of sorts:

> **"Q" IS YOUR CUE TO SIDEBANDERS**
>
> Many CBers using sideband come from the ranks of radio amateurs. They've brought with them International "Q" Signals—code for many standard transmissions about radio operation. Sidebanders love to flip Q-signals at each other; that's fine because it saves air time so other stations may use the channel, too.
>
> Here are typical examples of Q-signal brevity:
>
> QRB How far are you from my station?
> QRK What is the readability of my signals?
> QSY Shall I change to transmission on another frequency?
> QTH What is your position in latitude and longitude?
> QTX Will you keep your station open for further communication with me until further notice?
>
> Additional Q-signals that may be of use are given in Chapter 20.

10-4, breaker. It's clean and green to mile marker 4-2.

Ah, he thought you simply wanted a Smokey report. No police in sight.

Traffic is light and doing the double-nickel.

That's a different answer, though worthy of your suspicion. It's hard to conceive of light traffic doing only 55 MPH unless there's a Smokey in sight.

Traffic rollin' 5-5 into Eskimo cotton. Hold 'er down, it's slippery.

That's helpful, though not everybody listening would recognize Eskimo cotton as snow. In critical communications, tell it like it is—plainly.

As you see, asking a general question gets you a variety of answers. I prefer to ask specifically for road and weather conditions in a given area, then get the location of the station replying to cross-check the freshness of the report.

Be Specific—In wording your replies to questions, remember you may be talking to people unfamiliar with roads and with slang. Here are typical problems to avoid:

Breaker to all you northbounders on this green stamp. Look out for a four-wheeler fixing bubble trouble on the side.

Which toll road? Sometimes there are two or more within radio range? Will new CBers know *four-wheeler* as a car? Will everybody know *bubble trouble* as a flat tire? Which shoulder of the road? *On the side* isn't very specific, and it also means transmission completed but standing by. A transmission like this makes the channel even busier when people call for clarification.

Better to say: *Breaker to northbounders on the New Jersey Turnpike approaching mile marker 8. Make room in the right lane for a disabled auto changing a tire on the right shoulder.*

One day I heard this urgent transmission repeated several times:

Northbounders on the green stamp take the eastern extension. Avoid the western extension. Head for the hole in the wall, not the GWB.

Again, which toll road? Where? Eastern and western extensions? All the toll roads in the area run north and south. Why detour? What's a hole in the wall? And what is the GWB?

Better to say: *Northbounders on the New Jersey Turnpike above Exit 14 take the eastern extension toward Lincoln Tunnel. The western extension toward George Washington Bridge is backed up because of an accident.*

When you broadcast road information, remember your listeners are on the move. During an exchange of transmissions requiring only one minute a listener may travel a mile closer to trouble. Talk like a pro so listeners can immediately alter course instead of calling you back.

Be brief but specific. Use clear language, avoiding slang or ten-codes requiring time to translate. Tell what to do in terms of actual road signs your listener will see. If it's urgent information, broadcast it on all mobile channels.

TRUE CONFESSION HANDLES

Virgin Territory
Eager Beaver
The Rabbit Habit
Flim Flam Man
Jail Bait
Heart Breaker
Modified Rabbit
Steamin' Demon
Weird Harold
Filthy Phil
Gadabout
Jeannie Without a Bottle

I've said that channels 10 and 19 have been adopted by CBers for vehicular use. Most likely you'll be on 19. Channel 10, the old mobile channel is still an alternate when 19 is busy, though experienced CBers prefer not to risk bleeding over into the adjacent 9—the official emergency channel. Still, 19 and 10 are no longer the only mobile channels to notify.

Where two super slabs intersect or run parallel, CBers frequently traveling one of them tend to pick a regular alternate channel to avoid radio collisions. For example, you'll hear this in Northern New Jersey where the Garden State Parkway crosses the New Jersey Turnpike near Woodbridge:

Breaker 1-9 to northbounders and southbounders on the Garden State. Drop down to 1-5 for mobiles the next ten miles.

Sure enough, channel 15 is not quite as busy, though you sometimes get the impression you've driven into somebody's living room when a teenager with a base station protests you're on his home channel. In your area, 15 may not be an alternate mobile channel to avoid congestion; listen carefully in the vicinity of multiple major highways. And be courteous to the so-called home channels—you're just passing by.

If You're Lost—First try to pick out mile markers, exit signs or landmarks to pin-point your location—then call.

Breaker 1-9 for some local information. I'm near the Carlton Theatre on Spokane Avenue and looking for the Valley Shopping Center.

You'll get directions. Sometimes more than one CBer will chime in. There have been cases where CBers ask for your vehicle identification and then come by to lead you to the destination. Chivalry is not dead—it has simply gone electronic.

When You're Hungry—You can break for local information on anything from hundred-mile coffee to the best restaurant around. Some fast-food establishments have base stations; they'll accept take-out orders on CB while you're driving toward them. Give your location when calling so they'll know when to expect you. To find such places, including names, calls signs and channel monitored, ask other CBers for local information.

If You Need Overnight Accommodation—Hotels, motels and campgrounds are beginning to install and monitor base stations. The trouble is finding the channel monitored. Some RV travel directories such as *Woodall's* list campgrounds with CB stations. Big hotel chains like Hilton may not have CB channel information in their central reservation services, although some chain hotels and inns monitor CB for room requests 24 hours a day. So break for local information on the mobile channels in the vicinity.

Shopping—CBers entering unfamiliar territory often use the channels like Yellow Pages in the 'phone book. First check for a reasonably free channel. Then break for local information on the mobile channel and mention what you're after. Somebody will be back to you. Then agree to continue the conversation on another channel so you don't clutter the air. Other helpful

SEDUCTIVE HANDLES

Passion Fruit
Cherry Lips
Sexy Lady
The Italian Stallion
Captain Fantastic
Loose Goose
Electric Lips
Love Bear
Easy Lee

CBers may overhear the request and join the conversation on the new channel. I know of one CBer who got best buys in Indian jewelry by CB shopping throughout the Southwest. But I worried about another CBer heading my way—he was shopping for a gun.

EMERGENCY LANGUAGE

Each person's definition of an emergency is different. Often it depends on whether it's your emergency or one confronting somebody else. To sort things out, check the pertinent parts of the FCC emergency regulations governing your use of CB in Chapter 20. Read 95.85, "Emergency and Assistance to Motorist Use."

95.85 Emergency And Assistance To Motorist Use

All citizens radio stations shall give priority to the emergency communications of other stations which involve the immediate safety of life of individuals or the immediate protection of property.

That's very clear. *All* stations must give priority. Note there is no restriction as to channel. While channel 9 is the official emergency channel and may be used *only* for emergencies, you may communicate about emergencies on other channels where there is likelihood of more listeners. Try channel 9 first; it's usually monitored by police and volunteer organizations. Channel 13 may be monitored by volunteer organizations who will report marine emergencies to the Coast Guard.

Here's the 95.41 official word on the way you may use channel 9:

The frequency 27.065 MHz (Channel 9) shall be used solely for:

Emergency communications involving the immediate safety of life of individuals or the immediate protection of property or

Communications necessary to render assistance to a motorist.

To be an emergency communication, the message must have some direct relation to the immediate safety of life or immediate protection of property. If no immediate action is required, it is not an emergency. What may not be an emergency under one set of circumstances may be an emergency under different circumstances. There are many worthwhile public service communications that do not qualify as emergency commmunications. In the case of motorist assistance, the message must be necessary to assist a particular motorist and not, except in a valid emergency, motorists in general. If the communications are to be lengthy, the exchange should be shifted to another channel, if feasible, after contact is established. No nonemergency or nonmotorist assistance communications are permitted on Channel 9 even for the limited purpose of calling a licensee monitoring a channel to ask him to switch to another channel. Although Channel 9 may be used for marine emergencies, it should not be considered a substitute for the authorized marine distress system. The Coast Guard has stated it will not "participate directly in the Citizens Radio Service by fitting with and/or providing a watch on any Citizens Band Channel."

The following are examples of permitted and prohibited types of communications. They are guidelines and are not intended to be all inclusive.

Permitted	*Example message*
Yes	A tornado sighted six miles north of town.
No	This is observation post number 10. No tornados sighted.
Yes	I am out of gas on Interstate 95.
No	I am out of gas in my driveway.
Yes	There is a four-car collision at Exit 10 on the Beltway, send police and ambulance.
No	Traffic is moving smoothly on the Beltway.
Yes	Base to Unit 1, the Weather Bureau has just issued a thunderstorm warning. Bring the sailboat into port.
No	Attention all motorists. The Weather Bureau advises that the snow tomorrow will accumulate 4 to 6 inches.
Yes	There is a fire in the building on the corner of 6th and Main Streets.
No	This is Halloween patrol unit number 3. Everything is quiet here.

Breaker 9 for a 10-33—If you don't know the ten-code don't worry. Say *emergency.* Do *not* use an international distress signal like *Mayday* except in the most desperate situations. It is strictly forbidden to use international distress signals unless the CB station transmitting them is *on* a vehicle, ship or aircraft threatened by grave and imminent danger. The distress signal must be a call by you for immediate assistance to you.

In an emergency, you can talk as long as is necessary. What's more, if you are unable to operate the CB, somebody else can operate it for you without further specific authorization. Channel 9 can save a life—see the next chapter.

CB TALK

I want to talk on this channel:	*Break 10* (or whatever channel number) *Break 1-oh or Breaker 1-oh*
Finding somebody on the channel:	*Howboutcha Big Ben?* (or whatever handle) *How about that River Boat?* *I'm looking for the one Daffy Duck?* *Answer your phone, Blue Flash.*
Go ahead and transmit:	*Come on* *Go, breaker* *Bring it on*
Did you receive me?	*Got a copy?* *Do you read?*
No:	*Negatory* *Negative*
You had interference:	*You got stepped on* *You were walked on*
Repeat:	*Bring it back* *Say again* *Crank the handle* (repeat your handle)
Listen in:	*Copy the mail*
OK:	*10-4* *4-roger* *40-roger* *That's a 4* *That's a big 10-4* *4-10* (sometimes a question—OK? or for emphasis)
Ending conversation:	*Clear* *10-10* *On the side* *We thank you for the break* *We down* *We gone* *Have a good day today and a better day tomorrow* *Catch you on the flip-flop* (return trip) *3's and 8's, 73 and 88, stack them 8's* *Put the good numbers on you*

LANGUAGE STYLE

Once CBers get practice slinging the language, they like to develop style. By style, I don't mean high-speed ratchet jawing punctuated by "hey!"

You got the one Wild Wolf Whistle, hey, drivin' the Funny Farm Express, hey, keep your nose between the ditches and the Smokeys out of your britches, hey! Howboutcha, good buddy, hey? Got peanut butter in your ears?

That's patter, not communication. Such channel clutter sounds like somebody has mastered an elocution lesson. You'll hear it more than once, word-for-word, because the dimwit has memorized it and can do no better. I'll translate:

"My handle is the Wild Wolf Whistle, driving a frozen food express. Drive carefully and stay out of speed traps. Why don't you answer me when I call you like this?

Why indeed? Hey?

Style is the use of certain words or phrases which are a generally accepted way of conversation. Shortly I'll come to even more polished style. Here are some often-used phrases followed by a brief discussion of their fine points:

"Those good numbers," mean anything from best regards to love and kisses.

Perhaps you noted the elimination of a verb—*we down, we gone.* I'm not sure whether this is a deliberate attempt to reach a common denominator by sinking toward the illiterate, or a commendable approach to shortening messages the way you eliminate expensive words in telegrams. You wonder why verbose CBers drop the verb but add "em" as in "pick 'em up truck" or "rest 'em up stop." And why place-names are made longer by adding the word *town* to everything—Tulsa town, New York town. Sometimes even "that ole Tulsa town" or "that ole Boston town."

The worst of this practice is to make a location obscure for other CBers. *T-town* can be Topeka, Tulsa or Tucson; only a local could guess. Another variation is "Bean town" for Boston, "Bingo town" for Binghamton, "Shakey town" for Los Angeles, "Steel town" for Gary, Indiana (or Pittsburgh, or wherever), "Cactus town" for Phoenix. When style makes messages clumsy or obscure, it's better to be crisp and direct.

Pro Style With Polish—Short, precise transmissions and scrupulous politeness are hallmarks of the pros on CB. Many have learned good manners and radio discipline in other communication-based fields like the military services, aviation and amateur radio. They shy away from jargon.

Break 1-4. KEJ 6386 for a radio check from mile marker 1-oh-9. Here's a count. 1-2-3-4-5. How do you read? Over.

4-roger. This is KEH 07724 from mile marker 1-oh-2. You read S-8. Over.

Thank you for the break. KEJ 6386 clear. Good day.

When transmissions ripple right along, they're like the short, businesslike and courteous conversations between cockpit and busy control tower.

Clear language: *What is your location?*

Beginning CBer: *What's your 10-20?*

Run of the mill CBer: *What's your 20?*

Real pro: *10-20?*

There's More Than One Ten-code—Two ten-codes were listed earlier in this chapter. One is the CBer's widely used but not official code—because there's nobody to make it official. The FCC does not publish a ten-code for CB use. Whatever you hear on the air is by agreement or happenstance among the early users of CB.

The other ten-code is the APCO code used on the job by most law-enforcement agencies. It is only for agencies such as your local sheriff and police departments who have adopted it officially. Agencies in some areas have not moved toward the general standard and use special codes of their own. So you run the risk of having some ten-codes misunderstood depending on what you mean and who is listening.

For example, you signal 10-34 and give your 10-20. As a CBer you are saying help is needed at a certain location. A law-enforcement officer overhearing your transmission might conclude that a riot is underway at the location you gave. If so, you might get more help than you bargained for.

Another touchy double-use of the same code is 10-89 which in CB language means "Radio repairman needed," but in law-enforcement code means "Bomb threat."

Imagine what could happen if you broadcast 10-89 and gave your location.

Fortunately these and similar troublesome ten-codes with dual meanings are seldom used by either CB people or law officers. If you restrict your use of ten-codes to the ones in common use, you'll benefit two ways. CBers will understand you and law-enforcement people won't misunderstand you.

Fun Style—I admire the pro style and urge you to use it when the channels are crowded. You are obligated to share channels. But CB style can be easy-going fun, too. The secret is so simple you can develop the fun-style right away. It's so smooth in its own way that you also sound like a pro.

Look for words that have similar rythms and pronunciations, though they still must convey meaning, too.

It's a challenge to sling the the lingo with the lilt and it takes the boredom out of over-the-road driving. But if you haven't been in CB very long and find this difficult, don't let it worry you. The more you're on the air, the more CB language becomes part of you. In fact, it's hard to get back to straight conversational English when you get home.

You: "Howboutcha, Big Mama?"

The YL (young lady): "Hi! Say, would you call the dog in?"

"Rover, Rover. What's your 20? 4-roger. He's comin' with the hammer down. Break for some local information. Where's the Colorado Kool-Aid?"

"We're out of Coors. Please speak English."

"All right. Want to go to the movie?"

"10-4."

HOW RELIABLE IS SLANG?

In ordinary conversational use, everybody expects slang phrases to come into style, fade away, and even change meaning or have different meanings to different people.

CB slang captured widespread interest and several books have been published as dictionaries of the new slang. This gives CB slang more dignity and stature than street-corner lingo. People even debate the correct meaning of CB slang words and check these "reference books" to cite authority.

Like any other slang, CB slang sometimes means what the speaker intends it to—never mind what the hearer thinks he hears. Don't use slang in emergencies or anytime it's important that you are clearly understood.

Here's an example: *Runnin' barefoot* came into CB language in my neighborhood as a way of saying, "I am operating illegally," usually without a license. But in some parts of the country it means to operate without an amplifier to boost the transmitter power higher than the law allows. Depending on who says it, *Runnin' barefoot* means "I am operating illegally," or it means "I am operating legally."

When used in this book, it has the first meaning—illegal operation—except in Chapter 20 where I give both definitions.

CHAPTER 4

EMERGENCY! CHANNEL 9 CAN SAVE A LIFE!

You say you buy CB for fun, or highway information, or keeping in touch with tractors working the south 40. But when you plunk down your money for the rig you think, "Someday this could save my life."

CB has saved countless lives by advance warning of natural and man-made disasters, by summoning emergency vehicles, by coordinating search and rescue efforts, and even by simply clearing the way for speeding ambulances.

When a tornado hit Xenia, Ohio, telephone wires went down along with everything else. There was no way to warn survivors wandering through the wreckage that a second tornado had been sighted until CB-equipped teams of REACT—Radio Emergency Associated Citizens Teams—began moving systematically through the area. Quickly establishing an emergency radio network in cooperation with police, Red Cross and Civil Defense units, they directed people to shelter in churches and schools that were still intact. They accompanied medical personnel through the stricken area in a search for the injured. Using their transceiver public-address features they called to people who might be lying unseen in the rubble; then reported to base via CB.

WHEN YOU NEED HELP ON THE HIGHWAY, CALL ON CITIZENS RADIO EMERGENCY CHANNEL 9

Here's how a monitoring organization like REACT works. A disabled motorist, a passing CBer or a mobile REACT unit calls for help on channel 9. The monitor acknowledges the call by CB and dispatches proper aid by telephone, keeping the emergency channel clear for other important messages.

Spectacular disasters like tornados and earthquakes may not be what you had in mind. You just want to be able to shout "Help!" into the microphone and know for certain somebody will respond. With CBers being licensed at a rate of 750,000 a month, chances that you will be heard keep improving, even if most channels are busy.

When a car careened off the road and hit a tree, the injured motorist was alone in the wreckage and invisible to passing vehicles. Miraculously his transceiver still worked. An emergency call on channel 9 brought his rescue.

A woman in labor was being rushed to a Dallas hospital. Commuter traffic was so heavy at that hour the husband thought he would lose the race. As he picked his way through traffic, he radioed his plight to a REACT monitor on channel 9. The monitor notified

CBers get involved in search and rescue operations. This news photo from the *Arizona Republic* shows the rescue of an injured climber from Camelback Mountain by Superstition React Unit # 8 in cooperation with police and mountain rescue teams. The officer at right is also a member of REACT.

police and the hospital. Then he called for a break on channel 19 where the mobiles would be listening. He gave a description of his speeding stork-wagon and its plight, plus location and travel route to the hospital. Vehicles gave way and the frantic driver didn't become a frantic father until shortly after he reached the hospital.

Sometimes CB brings you more than emergency communications services. A woman rushing her stricken father to a hospital was unsure of directions. When she called for assistance on the car's transceiver, a REACT monitor listening on channel 9 took her location and told her to stand by. Then he switched to channel 19 and found some mobile CBers nearby who escorted her to the hospital.

Two CB operators monitoring channel 9 for Garden State REACT in Brick Township, New Jersey, helped guide the Coast Guard to the rescue of five people aboard a disabled cabin cruiser drifting in ten-foot swells 20 miles off shore. The vessel, prophetically named "Out of Bounds," was equipped with CB but not a marine radio.

HOW TO HANDLE EMERGENCY MESSAGES

You don't have to be a member of a formal monitoring service to use your CB for emergencies. The FCC rules tell you what to do. A good guideline for how to do it is the proven procedure used by

INTERNATIONAL PHONETIC ALPHABET

(A) ALPHA
(B) BRAVO
(C) CHARLIE
(D) DELTA
(E) ECHO
(F) FOXTROT
(G) GOLF
(H) HOTEL
(I) INDIA
(J) JULIETTE
(K) KILO
(L) LIMA
(M) MIKE
(N) NOVEMBER
(O) OSCAR
(P) PAPA
(Q) QUEBEC
(R) ROMEO
(S) SIERRA
(T) TANGO
(U) UNIFORM
(V) VICTOR
(W) WHISKEY
(X) XRAY
(Y) YANKEE
(Z) ZULU

REACT. It is based on this monitoring organization's experience in handling 55-million emergency calls since its founding in 1962. Most monitoring services use a similar procedure. When you call for help, messages get handled faster when procedures are standard, no matter who calls and who answers.

Before you ever have troubles of your own, it's useful to listen to channel 9 for a while to get a feel for the exchanges of emergency messages. It's fascinating listening and you might even find you're needed as a relay between two stations almost out of range of each other. REACT's excellent advice to its monitors applies to all CBers involved in an emergency. Let me paraphrase it for you:

Be calm, courteous, correct and concise.

Remain courteous at all times. If it's not your personal emergency, lots of other CBers may be trying to help, too. Another station may be in a better position to handle the emergency message traffic.

Keep your communications correct. Use a phonetic alphabet when it is necessary to spell anything. Any attempt at phonetics is better than a garbled message. The pros at handling emergency messages use the International Phonetic Alphabet shown here. Repeat a message you receive where appropriate to be sure you get names, locations and call signs accurately. If you are relaying a message, write down the information for further reference. Report emergencies to the correct authority. In some areas you may quickly be able to raise the police on channel 9 or 19. Don't waste time if you happen to be at your base station or anywhere else near a phone. Call on a land line (telephone) for the police, fire department, hospital, Coast Guard, gas, electric or water company as needed.

Make your message concise so channel 9 won't be tied up if there is more emergency traffic. Arrange to communicate on another channel if subsequent messages don't fit emergency priorities.

To REACT's advice let me add:

Speak in clear language. The other person may not understand jargon or CB slang.

Declare an emergency ended so there will not be massive over-response to your calls.

When information about an emergency can prevent additional trouble, broadcast it on other channels. For example, having reported an accident on channel 9, be sure to warn other motorists not to speed headlong into the wreckage; transmit on mobile action channels 19 and 10 plus whatever other channels you know mobiles are using locally.

Know and follow these priorities for use of channel 9:

Priority 1: Communications relating to an existing situation dangerous to life or property—fire, automobile accident, similar hazards.

Priority 2: Communications relating to a potentially hazardous situation—car stalled in a dangerous place, lost child, boat out of fuel.

Priority 3: Road assistance to a disabled vehicle on the highway or street.

Priority 4: Road and street directions.

You must be engaged in assisting a *particular* motorist to broadcast on channel 9. You can't broadcast on this channel to motorists in general except in a valid emergency where your message will be vital to all motorists. Most of them will be on the mobile channels anyway. Strictly speaking, you can't even broadcast a request to another person known to be monitoring channel 9 to meet you on another channel so you can talk.

PREPARE FOR SPECIFIC EMERGENCIES

Now that you know what you can and should do about emergency communications in general, you can prepare for what actually may happen so you know exactly what to do. It's a bit like stocking a first-aid kit, though in this case the things you need at hand are procedures and important phone numbers in your locality.

This chapter includes tables and procedures which help you organize in several ways. For example, Table 4-1 lists a horrifying assortment of emergencies in alphabetical order—from Aircraft Accident to Unconscious Person. For each type of emergency, the table shows what information you will need so you can make a satisfactory report to the authorities.

The table has blank spaces for you to fill in. Do it now—the life you save may be your own. Look up and fill in phone numbers in your area for police, fire, gas company and the other agencies indicated on the table. If you live inland, there may be no nearby Coast Guard for boating emergencies and of course a long-distance call won't help. For any

Cooperation with other volunteer organizations makes CB monitoring of channel 9 especially effective. In Kansas, the Saline County REACT Organization provides two-way radio mobile units and a mobile command post when working with the Civil Air Patrol and Red Cross.

emergency that doesn't fit the table or your ability to handle it, contact your local law-enforcement agency.

Once you have the phone numbers filled in, you have a remarkably useful document which almost automatically makes you operate in a cool and confident manner even if the emergency is your own. You may want several copies—one for your car, one for your home or base station, and perhaps additional copies for relatives or friends just because being prepared is being prepared.

This table is a slightly modified version of the standard instructions given by REACT. It has been developed through years of experience.

Many kinds of emergencies require you to call the police. If you live in an area protected by a sheriff, use that number. All blanks for phone numbers should be completely filled in even though some are the same number used for other emergencies. The idea is to get ready to work fast and smoothly and if you have all the information where you can get it without cross-checking or fumbling, you'll do a better job.

What To Do—Table 4-1 applies to virtually any emergency except marine emergencies in areas served by the Coast Guard. More information on that in a minute.

If you are at a base station, monitoring channel 9, it's your job to get the needed information from the person communicating with you by radio. That person may be flustered, excited, frightened or dazed so you must be calm and reassuring. Ask simple questions with simple answers and move steadily toward getting the complete story. Check table 4-1 again, it tells you what you need and what to do.

Marine Emergencies—This section assumes you are in an area served by the Coast Guard. In that case, procedures are well defined. If you are near other bodies of water, rivers and inland lakes, the rescue agency will vary depending on who controls that water. If you don't know, call your local police. They will know.

The preferred radio channel for emergency marine communications is not on the CB band. It is VHF Marine Channel 16. If you have it, use it!

When CB is used on a boat in trouble, or by another party on behalf of the boat in trouble, it

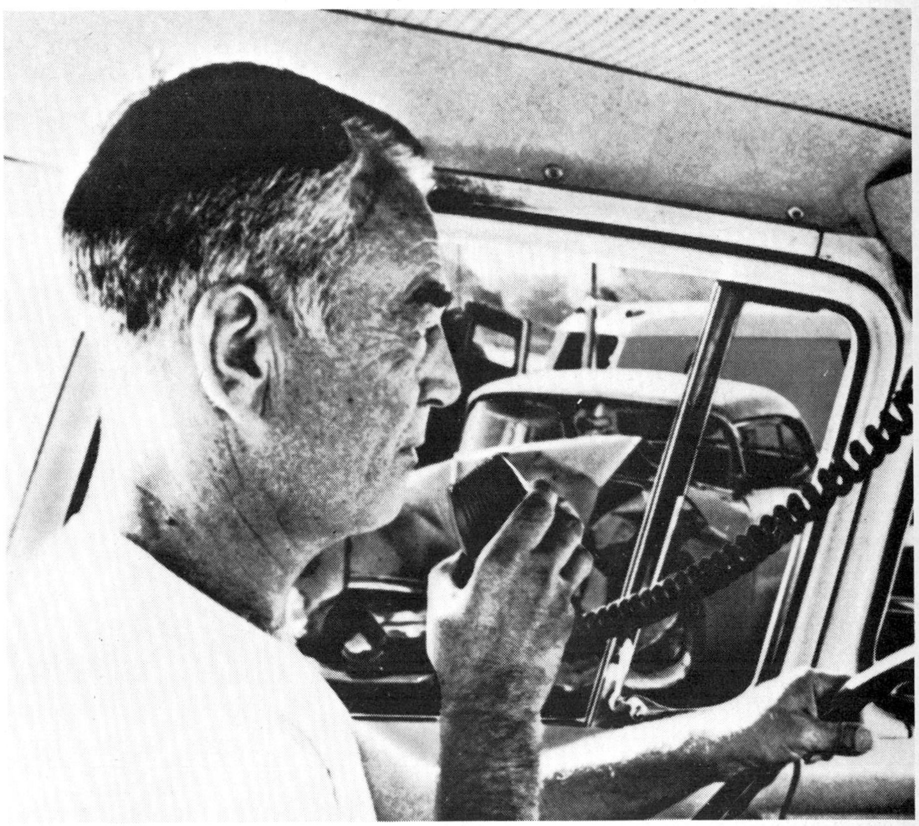

Emergencies are often reported first by CBers, especially truckers, because so many vehicles on the road are CB-equipped. If you see an accident get on channel 9 first, then switch to the local mobile channels to advise other drivers.

TABLE 4-1
HOW TO HANDLE EMERGENCY CALLS

EMERGENCY	INFORMATION REQUIRED	PHONE REPORT TO:		SPECIAL INSTRUCTIONS
Aircraft Accident	Description: Commercial, private, military. Location. Possible injuries, fatalities. Passengers, crew, people on ground.	Police		Reporter should remain on scene at safe distance until help arrives.
Auto Accident	Location. Number of vehicles. Injuries, fatalities. Is traffic blocked?	Police		Reporter should remain on scene until help arrives. Don't administer first aid unless qualified.
Auto Repairs Needed	Location. Auto description, license number. Problem. Is traffic blocked?	Police, if traffic blocked.		Acknowledge help on way to reporter.
		Service station or auto club per motorist's request.		
Boating Emergency	Location. Description of boat. Problem.	Police		Ask for instructions.
		or Coast Guard		
Civil Disturbance	Location. Number of people.	Police		Reporter should keep out of way.
Crime Report	Location. Description. Number of people.	Police		Reporter should seek safety. Do not pursue criminals.
Explosion	Location. Description of damage. Injuries, fatalities. Fire?	Police		Reporter should stay at safe distance until help arrives.
		Fire		
Fire	Location, Extent of fire. People in building?	Fire		Reporter should help evacuation and stay on scene until help arrives.
Floods	Location. Are lives threatened? Height of water.	Police		Reporter should remain on scene and look for people. Assist in evacuation.
Hazardous Driving Conditions	Get extent of area covered, period covered. Description. (Snow, icy roads, obstruction, etc.)	Police		Reporter should proceed with caution if possible.
Leaking Gas	Location. Description. Indoors, outdoors.	Gas Company		Reporter should stay at safe distance until help arrives.
		Police		
Lost Child	Location. Thorough description. How long missing.	Police		Reporter should initiate search by others. Await arrival of help.
Medical Emergency	Location. Description.	Police		Reporter should await help.
Miscellaneous (not otherwise listed)	Location. Description. Urgency. Injuries.	Police		Reporter should await arrival of help.

EMERGENCY	INFORMATION REQUIRED	PHONE REPORT TO:	SPECIAL INSTRUCTIONS
Natural Disaster	Location. Description. Threat to life and property.	Police	Reporter should seek safety.
Power Failure	Location. Extent of area concerned.	Power Company Police	Reporter should seek safety.
Road Directions	Destination. Location.	No report required.	Give clear direction.
Road Hazard	Location. Description.	Police	Proceed with caution.
Search & Rescue Call	Location. Purpose. Number of people requested. Person in charge. Have authorities been notified?	Police (if they haven't been called in).	Request reporter to stand by for reply.
Severe Weather Warning	Description. Location. Duration.	Check with Weather Bureau.	Tell reporter to listen for verification.
Signals Out	Location. Condition of traffic.	Police	Reporter to proceed with caution.
Stalled Car	Location. Description. License number. Blocking traffic?	Police	Reporter to proceed.
Street Lights Out	Location. Extent (number of lights, area out). Results?	Police	Reporter to proceed.
Suspicious Behavior	Location. Description.	Police	Reporter to proceed with caution.
Tornado	Location. Damage. Direction. Speed	Police Weather Bureau	Reporter to seek safety.
Unconscious Person	Location. Description.	Police	Reporter to remain until help arrives. Don't administer first aid unless qualified.

gets complicated because CB transmitting range is limited. Even over water it's a lot shorter than the distances some vessels travel away from shore. Also, the Coast Guard does not officially monitor CB channel 9 although somebody is listening just in case. Nevertheless, you don't have any guarantee that your call to the Coast Guard on CB channel 9 will be received. Because of short CB range, some relays may be involved or the reporting link may reach to the shore and then to the Coast Guard by land line.

In preparing this book, I checked all Coast Guard Districts about marine use of CB radio. Officially their replies followed current Coast Guard policy: The marine bands are more reliable than Citizens Band. Therefore emergency watch will be maintained on the International Distress, Safety and Calling Frequency, and on the National Distress, Safety and Calling Frequency—neither is on CB.

Unofficially, the Coast Guard concedes that the frailties of Citizen Band communication will not deter boaters from using it while hoping there is never an emergency. In actual emergencies, the Coast Guard will consider information from any source likely to prove useful for search and rescue, including individual CB operators and CB-monitoring groups. I was strongly, sincerely urged by several Coast Guard District Commanders to outline to CBers precisely the kind of information needed to deal with a maritime emergency.

Not only are marine emergencies more complicated from a communications standpoint when using CB, the emergencies themselves

TABLE 4-2
HOW TO HANDLE MARINE EMERGENCY AND DISTRESS MESSAGES

Use CB channel 9 *only* if you do not have communication on VHF Marine channel 16. Speak Slowly and Clearly.

Call:

If you are in distress (i.e. when threatened by grave and imminent danger) transmit the International Distress call, "MAYDAY" "MAYDAY" "MAYDAY" "THIS IS (Your vessel's name and call sign repeated three times)."

If calling from a vessel in trouble—give:

1. WHO you are (Your vessel's call and name).
2. WHERE you are (Your vessel's position in latitude/longitude or true bearing and distance in nautical miles from a known geographical point—local names known only in the immediate vicinity are confusing).
3. WHAT is wrong (Nature of distress or difficulty, if not in distress).
4. Kind of assistance desired.
5. Number of persons aboard and the condition of any injured.
6. Present seaworthiness of your vessel.
7. Description of your vessel (length, type cabin, masts, power, color of hull, superstructure and trim).
8. Your listening frequency and schedule.

If calling while observing another vessel in difficulty—give:

1. Your position and the bearing and distance of the vessel in difficulty.
2. Nature of distress or difficulty, if not in distress.
3. Description of the vessel in distress or difficulty, (See item 7 above).
4. Your intentions, course and speed, etc.
5. Your radio call sign, name of your vessel, listening frequency and schedule.

If calling by land line phone [ ]

Notify the Coast Guard promptly as soon as the emergency terminates.

require much more detailed and specific information than problems on land.

Table 4-2 is a checklist of things to communicate if you are on a vessel in trouble, or observing another vessel in an emergency. In either case, when you make the initial report, stand by to give more information as requested.

EMERGENCY MONITORING SERVICES

In an emergency, it doesn't matter who is monitoring channel 9 as long as someone competent is listening and prepared to act. The number of monitoring services is mushrooming as more CBers coming on the air want to give assistance. Competitive spirit and differences within local organizations may give rise to more than one monitoring service in the same area.

Teams often take Red Cross training. They own mobile vans with rescue equipment as well as superb CB stations. Base stations not only have detailed maps and directories, but may even have direct lines to official emergency services and to key utility offices. In addition to emergency services, many teams render community assistance at parades, sporting events and fairs, using CB to help direct traffic, re-unite lost children with their parents and keep track of scarce parking places. Some teams are even headquartered in rehabilitation hospitals where this useful activity is welcomed by the patients; they participate in 24-hour emergency monitoring.

REACT Radio Emergency Associated Citizens Teams, is an independent non-profit organization established in 1962. It consists of 1500 teams and 70,000 participants active in all 50 states of the USA, Puerto Rico, 7 Canadian provinces, Mexico and West Germany. Teams must agree to establish a goal of 24-hour coverage of channel 9. A formal cooperative understanding exists between the American National Red Cross and REACT.

In some states REACT research programs are conducted in cooperation with the highway patrol. For information about joining a team, write: REACT International, Inc., 111 E. Wacker Drive, Chicago, IL. 60601.

ALERT, Affiliated League of Emergency Radio Teams, accepts individual members who may organize into teams. Anyone may join who is over 18 years of age. Parents may join and include their children even if under 18.

Present membership is about 14,000 with chapters in all 50 states of the USA. Approximately 25% of the teams maintain a 24-hour watch; about 60% maintain 12-hour monitor service. For information write: ALERT, National Press Building, Washington, DC 20004.

HAM, Highway Assistance Modulators, is composed of individuals and teams throughout the USA and Canada. Team charters require a minimum of five supporting members in a team. Teams and members work independently and are free to set up their own procedures. The national director suggests that only sincere CB operators write: Highway Assistance Modulators, 5221 Creekwood Drive, Harrisburg, PA 17109.

REST, Radio Emergency Safety Teams, is sometimes known as REST Marine. Members must be at least 18 years of age and hold a valid CB license or a Marine Radio/Telephone Station license with an R/T Operator permit. Individual and team memberships are accepted. Teams are usually along seacoasts and inland waterways. REST monitors act as interface stations, receiving emergency messages from mariners on CB and relaying them to the Coast Guard by land line or VHF/FM. For information write: Radio Emergency Safety Teams, Inc., 1039

S. 26th St., Arlington, VA 22202.

Highway Radio Patrol International is a Canadian-based private organization performing a community watch against crime and rendering assistance to motorists. It aids public safety and law-enforcement agencies on request. Membership is limited to 18 years of age or above. Some teams have been formed in the USA. For information write: International Emergency Radio Patrol, Box 434, St. John, New Brunswick, Canada.

Community Radio Watch members are screened by local police and work closely with them in observing and reporting street crimes, fires, accidents, suspicious acts and unusual occurrences. A centralized base station is maintained under strict security measures. Standard operating procedures are maintained with a high degree of radio discipline. For information write: Community Radio Watch, National Program Coordinator, Carl M. Oldberg, 1301 E. Algonquin Road, Schaumberg, IL 60172.

DON'T GET CARRIED AWAY

Let's assume you've caught the spirit of this chapter and see the importance of volunteers helping other people with CB radio. You want to participate, so you have studied the rules and regulations. You have filled out the tables and are all set with full instructions, emergency phone numbers, and all the info you need. Perhaps you have joined one of the organizations listed here or some similar group in your area.

Don't get carried away with your own authority, *because you don't have any.* Don't try to do things you aren't trained for or you may be carried away in the next departing ambulance.

Here are some tips derived from the Highway Assistance Modulators operating procedures and other sources. These are intended to protect you while you are helping others.

If you hear about an accident on your CB, don't go there to help unless the people handling the emergency ask you to. If somebody is already on the job, you aren't needed.

Don't play traffic cop unless you are qualified. Drivers don't pay much attention to a person in plain clothes out there waving his arms. Remember, you have no authority and you are responsible for yourself.

If you come upon an accident, try to protect yourself and other traffic by placing flares. When the police arrive, they are usually glad to replace what you have used. Just ask.

At the scene of an accident, try to park ahead of the cars involved, not behind. Always park out of traffic lanes.

A big advantage of working with a monitoring team is the fact that somebody knows where you are and what you are doing. If you stop to assist a motorist or investigate a suspicious vehicle, be sure to let a monitor know you are out of your car. If you don't come back on the air in a reasonable period of time, the monitor can get you assistance. Always report back after you make a stop so the monitor knows you are clear.

Do not move an injured person unless there is real danger of loss of life if the person is not moved. For example, the vehicle is on fire and people are still inside.

Don't try to be a hero.

Don't attempt to stop the commission of a crime or violence of any kind.

Don't try to arrest anyone. A citizens' arrest may be legal but it is also very dangerous. Let the police do it.

When proper authorities arrive at the scene, offer information and further help but don't do anything unless they ask you to. Turn it over to people who are properly trained and legally authorized to act.

If you have been in communication about the emergency, report the arrival of authorities on the scene.

I have advised participation in an emergency only to the extent it is needed. The contribution you make may not seem like much, but consider these facts taken from the remarkably large log of CB assists to the Missouri State Highway Patrol:

You Bring Help Faster—A study of accident reporting to the Missouri Highway Patrol in a single month covered 6391 CB radio contacts. The average notification time by CB was 9 minutes. Telephone and other conventional notification averaged 15 minutes. In many cases, CBers were able to accurately describe the situation at the accident scene so proper emergency personnel and vehicles could be dispatched immediately.

You help maintain law and order—Of 9774 CB contacts related to law-breaking in a six-month period, Missouri Highway Patrol officers made 2434 arrests and issued 1836 warnings of violations. Sometimes the mere presence of your vehicle with its two-way radio antenna can be a deterrent.

CHAPTER 5

TALKIN' SMOKEY– YOU AND THE POLICE

Many state highway patrols have ears. When the budget doesn't permit, some officers furnish their own CB's. Troopers are usually amused or bored by what they hear on mobile channels but respond like lightning to emergencies reported on channel 9.

You're riding Radar Alley listening to Smokey reports:

Bear in the grass. Takin' pictures. Spy in the sky.

Tijuana taxi northbound with the hammer down. He's comin' up your back door at mile marker 109!

You haven't gotten an eyeball on the Smokey in the median strip, using radar to monitor traffic. It's impossible to see the police helicopter overhead. You glance coolly into your rear-view mirrors for that marked patrol car overtaking you at high speed. You're alert to the excitement—like a fighter pilot in enemy territory—ready for a dogfight—hero of a whole war.

If you're going to talk about Smokey, or even talk to him, you'd better know what you're doing. Is he friend or foe? Could you inadvertently be an outlaw yourself?

When you're on the road, most CB talk seems to concern Smokey Bear and his whereabouts. Like most CB terms, the name is apt and laced with wry humor. Many police officers—local yokels, county mounties and super troopers—wear ranger hats.

YOUR RIGHTS AND OBLIGATIONS AS A CBer

You paid for your CB rig. You are properly licensed to use it. You've looked at FCC Rules and Regulations, Part 95, which says how to use it. But Talkin' Smokey isn't that simple any more.

What you broadcast affects other people, for better or for worse. What you say into that microphone does not merely float away into space; it intersects with personal lives, constitutional rights, moral and emotional values of other people, fair play—meaning, *do unto others*—people in trouble, law enforcement, crime prevention, and a lot more.

It's your privilege to yak on the national party line. You can talk code as a member of a coast-to-coast society. But everyone uses the lingo and knows the code—including cops and crooks. Are you really one of the good guys?

HOW CONVOYS WORK

Is avoiding Smokey through CB radio a grand illusion? Can a CB-equipped vehicle really travel with the hammer down and speed all day? As it turns out, police in the USA and Canada indicate that CB doesn't make it much less likely you'll get caught speeding.

Smokey's taking pictures. Radar monitoring can lock on your vehicle in just 1/5th of a second. With radar backup, Sgt. Thorne of the Middletown, NJ police, hasn't lost a court case yet. He likes CB and his department has had it around for ten years. Smokey reports actually reduce speeding.

In Wisconsin, 71% of out-of-state vehicles caught speeding were equipped with CB radio.

What's happened is a reduction in speeding by CB-equipped vehicles, especially those traveling in convoys. If you think that's contradictory because convoys seem intended to outwit Smokey, consider how a convoy works.

Joining a Convoy—Channel 19 or 10—depending on where you are—is busy with transmissions. A convoy is moving up the highway—truckers, four-wheelers, or both.

You: *Breaker one-nine. This is KEH 07724 for a northbounder on this Interstate 8-oh.*

One-nine means channel 19. You're looking for anyone traveling in the same direction on Interstate 80. You give the number of the highway because there are lots of other highways within your transmission range and you might pick up northbounders on other roads.

Guy: *10-4, breaker. This is KEJ 6386, The Happy Booker. I'm at mile marker 105, northbound on the fast track of this green stamp. What's your 20? C'mon.*

OK says the guy. When you don't know somebody's handle, you call him guy, crude as that may seem. You gave your call letters, making your initial transmission legal. He gave his call letters also, either because he's a pro or he suspects you might work for Uncle Charlie—an FCC listening team. Now you know his handle or nickname and his location by mile post number on the express lane of the toll road. *Green stamp* once meant money; now it has been stretched to cover turnpikes that take your money. He asks your location—*your 20*— then asks you to answer. In some parts of the country the term is *go* rather than *c'mon.*

You: *Entering the fast triple-track from the piggybank at mile marker 101. Do you have a front door? C'mon.*

The Happy Booker: *Our front door is The Gooey Dipstick. He just reported clean and green from milemarker 112. You're behind The Italian Stallion at 103 so I guess you're the new back door and we're in the rockin' chair. Keep your eyes peeled over your shoulder for Smokeys on rubber. KEJ 6386 clear and 10-10 on the side.*

That exchange told of your entering one of the three express lanes from the toll plaza at mile post 101. You asked about a convoy leader—*a front door.* The Happy Booker gave you the handle of The Gooey Dipstick up front at mile post 112 and reported the road clear that far. You became the tail of the convoy while the others assumed the ease of "rocking chairs" in the middle of the caravan. You were warned to watch the rear for police on the move. The Happy Booker legally terminated his transmission and said he'd be listening.

Joining a convoy on the road gives you a nice feeling of togetherness, comradeship and mutual defense. Also somebody to talk to while you chase the white line. Just by looking, you can't tell which of these vehicles are teamed up to talk about Smokey. Maybe one of them is Smokey himself.

SMOKEY TALK

Advertising — A marked police car
Bear — Policeman, short for Smokey the Bear
Bear cave — Police station
Bear in the air — Airborne police
Bubblegum machine — Revolving flasher used on police and emergency vehicles
Camera — Police radar unit
Clean — No Smokeys around
County mounty — Sheriff or county highway patrol
Cop on top — Airborne police
Cub scouts — Sheriff's men
Feed the bears — Pay a fine to police
Houndmen — Policemen looking for CBers
Local yokel — Small town police officer
Picture taker — Radar speed trap
Plain wrapper — Unmarked police car; color may be indicated as in "plain green wrapper"
Smokey the Bear — Policeman, so-called because of the ranger hats worn by many troopers
Smokey on the ground — Trooper out of patrol car
Smokey on rubber — Trooper in vehicle
Smokey with ears — Policeman with CB rig
Spy in the sky — Police helicopter or aircraft
Tijuana taxi — Marked police car

A more complete glossary of CB lingo and technical terms is included at the back of this book.

What he said was correct in CB code, but redundant. 10-10 means transmission completed but subject to call. In CB lingo, *on the side* means you're hanging up your mike, receiving but not transmitting. Don't confuse it with a report that Smokey has a four-wheeler on the side which means a car pulled over and detained at the side of the road. Do you see why it's important to talk precisely and like a pro?

You: *10-4, good buddy. I won't be able to copy the front door because of these hills, so give me a shout if he lays eyes on anything. KEH 07724 clear and 10-59.*

OK. In CB everyone is your good buddy, you hope. You said you would not be able to receive messages directly from the head of the convoy because of the hilly terrain and the 11-mile distance. That's stretching a bit even in flat country; but in a convoy vehicles in the middle act as repeater stations. Also, the head of a convoy in one direction can contact the front door of a convoy passing in the opposite direction; by comparing notes on road conditions behind them, each convoy nearly doubles its road-monitoring capability.

Note that you cleared with the other station and said *10-59.* That means "in convoy" in both the law enforcement and CB codes but it isn't often used.

Highway patrols in Ohio and other states want you to know they're listening on CB and ready to help.

I've left out typical transmissions from other travelers in the convoy. It's common for people to speak up when a newcomer joins a convoy, or when a convoy forms. They exchange handles, destinations, reasons for travel, tell a little about themselves and generally get acquainted. If you find somebody who really wants to chatter, have him meet you on another channel so the rest of the convoy can get important transmissions. Don't do this if you're responsible for the front or back door.

Forming A Convoy—This isn't much different from joining one. Here's how you do it. Ask for a vehicle traveling in the same direction. Compare locations so you know who will be the leader. Agree to be a front or back door. Determine destinations so you know when you'll be leaving each other. Others will join you in the manner I've just described and take up appropriate positions in the convoy.

Watching The Road—Messages in a convoy are not confined to Smokey reports. Someone may spot a vehicle on the shoulder, its driver changing a tire. Back goes the call to give the disabled vehicle plenty of room and to flash directional signals for nearby vehicles that don't have ears (CB). The back door might spot an ambulance, moving up fast; out goes the word to give the emergency vehicle plenty of room to overtake and pass the convoy. Unlike military convoys, CB-linked travelers do not all use the same lane.

In Missouri, truckers heard a warning on CB that a state trooper was chasing a stolen car. Immediately all trucks on that route pulled off to the side and broadcast warnings ahead for all vehicles to do the same. Crash possibilities were minimized and the thief was captured.

One truck convoy saw a drunk driver weaving dangerously. They radioed his location to the police. Then they surrounded the drunk with their 18-wheelers. Maintaining coordination via CB, they skillfully slowed the formation and herded the dangerous vehicle to the shoulder, keeping him there until the police arrived. I don't advise trying this with a formation of four-wheelers. Nevertheless, watching out for drunks is a valuable safety service to all drivers. Having a CB to report the danger is safer than trying to apprehend the offender yourself. Smokeys monitor channels 9 and 19 when they are CB-equipped. Emergency monitoring organizations listen on channel 9 and will contact the police.

The police cannot be everywhere. It's astonishing how quickly CBers can switch from the us-against-the-Smokeys mood to taking command of a situation on behalf of the absent Smokeys.

As I write this, a heavy thunderstorm has broken over New Jersey at the height of rush hour on nearby highways. On my base station I am monitoring channel 19. Suddenly the Smokey reports cease and there are warnings of flooding on the roads; areas of heavy rain that reduce visibility. CBers are reporting the exact location of these problems, warning each other where to reduce speed.

An accident occurs and is described in detail by a CBer in a following car. Immediately he puts out a call to any Smokey with ears. He stops to assist and switches to channel 9 to describe the emergency.

One end of a disabled vehicle is hanging out in a traffic lane, gushing gasoline. The local emergency monitoring service—Garden State REACT—wastes no words. The monitor on duty asks pointedly about the cars, the drivers, and advises he has notified the police on a land line.

The CBer who observed and reported the accident has returned to channel 19, warning northbounders to hang far left and not toss any cigarettes. A minute and a half later he returns to channel 9 to report the first trooper on the scene.

And then: *Breaker one-nine for a Smokey report!*

Smokey's ears share a transmission-hump mount with police transceivers. Ohio Troopers have CB rigs in every patrol car and monitor any channel. Headquarters monitor only channel 9.

PUT THE PEDAL TO THE METAL

So far I've said little about speeding, which is what the popular songs say convoys are all about. You assume it's clear ahead until you hear from another traveler in the convoy reporting a Tijuana taxi or a plain wrapper—unmarked police car. Meanwhile, do you speed?

Maybe, but rarely for long. If you've got the front door, it's no fun beating the bushes—*shaking the leaves*, say the truckers—until a Smokey emerges. Mercy! The *gift certificate* will have your name on it, and while he's writing the ticket your good buddies in the convoy will roll on by—more slowly for sure. *Mercy* is an expletive you can use on the air—obscene language is illegal.

How about that clean and green report from a CBer heading in the opposite direction? All it buys you is a couple of minutes with the hammer down. Minutes—that's all. Remember you exchanged Smokey reports as you approached each other at 60 MPH, a speed often tolerated when conditions are safe. However, you are using up the road you both have under surveillance at a combined speed of 120 MPH—2 miles a minute. Because most convoys are about 4 miles in length, in 2 minutes you are passing their back door and you are out there all alone again.

Well, how about when you are the back door? Can't you speed as long as it's clean ahead. Mercy no, who knows whether a bear is following with his foot in the carburetor? So you just rake the leaves legally.

In between are all your good buddies of the CB caravan, riding the rockin' chair. Do they speed? Whoever does overtakes the front door and becomes exposed.

No wonder the average speed of convoys turns out to be surprisingly close to legal. Nor are these the only reasons why speeds stay down. Real ratchet jaws—inveterate talkers—drive more slowly with a mike at their mouth. In police vehicles, radar is often replaced or supplemented with VASCAR, a simple on-board computer which rapidly figures your speed as you pass any two check points observed by the officer even while his vehicle is in motion. Consequently a plain wrapper can easily infiltrate a convoy, identify speeders and radio ahead to a marked patrol on a frequency you can't hear.

You're really relying on the possibly unskilled judgements of an unseen person guarding your back door when you risk speeding. Listen to the convoy chatter. You'll often hear a back door getting chewed out for letting a plain wrapper overtake him without being reported. Worse, note the anxiety when somebody reports a plain wrapper and nobody can get an eyeball on him. Once I was in a convoy that became so frantic about an unseen infiltrator that it ran for the next 15 miles at 5 MPH *under* the speed limit.

Lots of CB purchases are justified by saying, "I wouldn't want my wife to be stranded on the road." When that emergency happens, stay in the car with the doors locked. If in doubt about whoever arrives to help, give the license number and description of the other vehicle to the police or channel monitor and wait for a real live Smokey.

Smokey sightings are inherently unreliable unless you are exercising your own eyeballs—which may be a a trifle too late. Here's why. On the highway, reports of Smokeys spread like wildfire among CBers. They become second, third or fourth-hand reports. Often such reports quickly grow stale, but they echo up and down the great superslab long after Smokey has moved to another location. In their wake is a stream of traffic scrupulously observing the speed limit.

SMOKEY'S STRATEGY

Clever Smokeys are aware that sightings announced on CB tend to make drivers very alert about all vehicles and extremely careful about speed. The Missouri State Highway Patrol *wants* its presence announced. That's why they wear uniforms and drive Tijuana taxis. Superintendent Samuel S. Smith says, "Resourceful troopers can change their locations frequently and produce reports of *wall-to-wall* Smokeys over considerable stretches of roadway."

CBers can make 15 patrol cars sound like 30 and warning each other about them does help to hold down speed in that area. As a taxpayer I'd rather support 15 patrols than 30. Delaware State Police have CB in all their patrol cars. Colonel Irvin B. Smith, Jr. comments that Delaware troopers listen to CB Smokey sightings so they'll know when to change location. As a result, they leave behind them ghost speed traps, still treated as real by CBers holding 'er down for miles around.

In most states, the police are instructed not to communicate directly with CBers except in an emergency. Most Smokeys don't. But there are always an anonymous few, irked by bad-mouthing or show-off CBers, who are not above reporting a highway *clean and green* to anyone who asks. First thing you know, they're right behind you with the bubble machine—police car light—flashing red and dread. *10-4. Gotcha for shore!*

Are you thus a victim of entrapment? No way. You talk CB slang to avoid specific statements that could be considered interference with an officer in the course of his duty. But you got an answer in slang, too, which is not specific enough for you to claim you were encouraged to break the law. So feed the bears—pay your fine.

Lots of mischievous kids have ears. To them, it's good fun to respond to any request for Smokey reports with clean and green, then watch speeding adults become bear bait.

Remember you are talking on a party line, shared more and more with Smokey. Lots of police departments haven't budgeted for CB, but their officers may be permitted to use their personal rigs on duty. They listen.

The driver of an 18-wheeler who called himself Mellow Yellow announced on CB that he was out to break the speed record between Los Angeles and San Francisco. Periodically he advised all good buddies on the channel that he intended to make it in record time and would appreciate their help. Asking for help in breaking the law is just plain dumb, like shouting in the ear of the California Highway Patrol. On one stretch of Interstate 5, Mellow Yellow had paper hung on him three times by the same bear.

Is it an obstruction of justice to use your CB set to warn of police cruisers in the area? There are cases on record, particularly in Texas and Wisconsin, of successful prosecution for such acts. These go back to 1975 and reflect early hostilities between CBers and Smokeys. At one time, several states asked the FCC for aid in catching CBers who used their radios to beat speed and weight limits.

The dividing line between proper and improper use of CB on the highway is not clear. Strictly speaking, advising others **not to speed** is an admonition by a good citizen not to break the law. Most Smokeys, unless they are saddled with a ticket quota to fill, would rather have drivers stay close to the legal limit and broadcasting their locations does a good job of slowing traffic. They know, too, that it's hard to hang paper on you for the way you use your CB set. But if you use it to break the law or encourage others to commit offenses they may do the following:

Track you and stop you for a check of your driver's license and vehicle registration.

Make a safety check of your vehicle and ticket you for violations.

Ticket you for minor infractions such as failure to signal lane changes, improper passing, or whatever.

Ask to see your CB license or transmitter identification form. If you're legal but don't have one or the other, you'll be reported to the FCC—see Chapter 16.

If this sounds like harassment, remember you could have brought it on yourself by transmissions such as these:

It's clean and green at mile marker 88. ***Put the hammer down*** *and bring it on up.*

The chicken coop is open at

Yes, m'am. A uniformed trooper is on the way to your location now. **The Ohio Highway Patrol monitors CB channel 9 around the clock from communication centers like this one in Columbus. In cooperation with REACT monitoring posts, help happens in a hurry.**

Scranton. ***Better go around it*** *via route 611 if you don't want to get on the scale.*

I'm carrying ***60 bushels.*** (60,000 pounds—each bushel is a thousand.)

I've emphasized words that may smoke out Smokey: Advising another to speed, avoiding weighing scales, advertising you are overweight. All police departments in my survey said they have general instructions to avoid harassment of CBers because most Smokey reports do more good than harm. Warning of the presence of patrols is usually not bad.

SHOULD YOU REPORT EVERY SMOKEY YOU SEE?

I don't. And I'm not a goody-goody. Here are some things I hadn't considered at first, when I had that great feeling of belonging to a national secret society of CBers protecting each other against the Smokeys. Hang your mike on the side and listen.

The Texas Department of Public Safety says, "The routine practice of many CBers is to broadcast Smokey sightings without consideration for the mission of troopers being reported. Oftentimes troopers are attempting to locate malefactors involved in serious crimes. Indiscriminate Smokey reports are capable of rendering these searches ineffective."

Robert I. Kimmel, Director of Communications for the Pennsylvania State Police, says it another way. "Many times, what appears to be a routine traffic or speed check on the highway is, in fact, a check for a known criminal. Rather than set up a road block and inhibit the normal flow of traffic, a police officer will park on the berm of the highway or off the road in such a manner as to give the impression of checking traffic and/or highway speeds. In fact he is attempting to identify the perpetrator of a crime who is reported to be traveling on the highway."

The Georgia State Patrol elaborates, "By broadcasting the officer's location, the CB operator may be providing the very information the criminal needs to escape detection and arrest. When this happens, the CB radio becomes a very real obstruction to the judicial process."

That's a big 10-4. But when do you stop broadcasting Smokey sightings?

HOW TO BE A SMART, SAFE CBer—AND SOMETIMES A HERO

Listen to your AM news radio—you don't have to turn off the CB—and remember crooks can have ears, even stolen ones. When a bank was robbed in Indiana, word was spread on CB that the robbers were at large. Smokey reports immediately stopped, not only in Indiana but in nearby Ohio. The channels remained silent about Smokey's whereabouts until the bank robbers were caught.

In some cases, you can really help. When two gunmen shot a driver in Ohio and stole his vehicle, a description of the fleeing auto was broadcast on CB. The Citizens Band went relatively silent—no chatter, no Smokey reports, just possible sightings of the wanted vehicle. Not long after the shooting the criminals were caught.

As it often turns out, not all such occurrences are as dramatic. Still, you never know how helpful you are being. In some states, the license numbers and descriptions of stolen cars are broadcast over CB channel, 10 or 19. As a result, the police have thousands of additional eyes on the lookout.

Criminals have become so bold, retaliatory and violent that few people have been willing to risk their own safety to interfere or even report them. The record of robberies, rapes, muggings and murders in full sight of onlookers who do nothing is somewhat understandable but shameful. If you have CB, you hardly have an excuse. Here's how you can help.

CB provides anonymity. Call for a break on channel 9, 19, or 10, or wherever you can raise somebody with a land line. You're safe behind the locked doors of your vehicle or base station. Raise a hue and cry. Raise hell.

In Nebraska, a state senator and a motel clerk were kidnapped by a gunman. They were rescued thanks to a trucker who spotted the car described on the local CB action channel. He simply notified the police.

You're generally safer to observe and report than to be a participant in the apprehension of a criminal.

GOOD CITIZENS WATCH AND REPORT

Despite accelerated population growth and the rise in crime, police forces continue to be undermanned. So what's new?

Back in the middle ages, one King Canute recognized the problem and proclaimed that whoever shall observe a thief in the act of flight must take up the hue and cry until the criminal is caught—or shall himself be deemed criminal.

What's new is CB, which makes it a lot safer for good citizens to report on crime and criminals from the safety of a base station or locked car. Community Radio Watch, a volunteer monitoring organization working directly with police in many areas, asks CBers to watch for and report on channel 9:

Street crimes — Assaults, robberies, riots disorder.

Fires—Buildings, vehicles, brush, trash, vacant property.

Accidents—Automobiles, people, animals.

Suspicious acts—Gang activity, loiterers, anyone seeming to maintain surveillance.

Unusual occurrences — Faulty traffic light, flooded viaduct, power lines or trees down, stalled vehicles blocking traffic.

The most prudent CBers I've yet encountered were cruising close to the legal 55 on an Interstate. Suddenly a sleek Mercedes came through with the hammer down. Knowing that Smokey had ears, one CBer grabbed his mike and broadcast the old adage: *There's never a cop around when you need one.*

The others took up the cry: *That's a big 10-4. He must have been doing a 9-oh.*

Look out for that southbound brown Mercedes passing mile marker 57!

And a bear came out of the bushes to catch him!

You don't have to be a member of a CB monitoring organization that assists police in an area. The moment you install a CB station and get a license, you are a member of the Citizens Radio Service—that's the official name, good buddy. Observe suspicious or dangerous situations and report emergencies on channel 9.

Check channel 9 for a moment or two, while you are organizing your facts, to determine if the same emergency has already been reported. For the importance of this, let me pass the mike to the Pennsylvania State Police. "Numerous reports of incidents radioed by CB have to be compared to determine if they are the same incident. Multiple reports of an incident cause police departments to respond with more resources than are necessary, or they may respond to more than one location for the same complaint."

Obviously it's important to give the exact location as a major fact in distinguishing one incident from another. It's amazing how many well-meaning CBers forget this. There were 8560 reports of stranded motorists in the first 6 months of the Missouri State Highway Patrol's CB Program. Would you believe that 1408 of them could not be located?

One CBer reported a horde of Smokeys surrounding a building and believed he had come upon a major emergency. *What's the 20 on that?* asked the emergency monitor. The building turned out to be the local police station.

Be alert when checking the channels. Conventional CBers talk about Smokeys and bears. Prowlers have broader concerns—*the man* is anybody who might catch them at work. Burglars use CB: A walkie-talkie on the inside and another portable or mobile handled by the lookout. So do bank robbers. A surveillance camera in a Montreal bank filmed a CB-equipped gang apparently in touch with a lookout. In resort areas where homes are often closed for the winter, snowmobiles with CBs are sometimes used to aid the looting of a home and transfer the stolen goods to a truck or van waiting on the nearest road where it can also serve as a lookout.

Clearly you ought to think twice about grabbing the mike to make a Smokey report every time you see a bear on the move.

Crooks with ears surely want to know where Smokey is while they're at work. That call you answered for a Smokey report may be from an inquiring crook—about to rob your house.

DON'T TALK YOURSELF INTO TROUBLE

CB is also part of the criminal's life. You're vulnerable if you have CB because crooks listen too. You're safer if you're smart about the way you use CB.

You're a stranded motorist on a lonely Interstate. You report an emergency. Your call may bring a mugger.

Or you're simply lost and in need of directions. One CBer asked for such help. He got an immediate response on channel 19 and was asked to describe his vehicle and give his location.

I'm right behind you, good buddy. Pull over to the shoulder and come back to my car. You can check my map.

That was a very big 10-4. The motorist was robbed of his wallet and watch.

These are isolated incidents, but they are typical of new opportunities for crooks due to the phenomenal growth of CB. Problems are more likely to result from communicating on the action channels than on emergency channel 9 say the monitor groups, contending they and the police are always listening on 9. If a crook hears you calling for help on the emergency channel, he is less likely to go to the scene because he can expect police or other official vehicles also to be there.

Multiple response to any emergency call brings safety in numbers. Indeed, one lady in Cleveland, harassed at her home by rowdies, gave a shout on CB even before she called the cops. Her driveway was full of CBers long before a patrol arrived.

Here's How To Play Safe—You're in trouble on the road and have asked for assistance. You expect help. Be wary of the first vehicle to arrive, unless clearly it is the police.

Don't get out of your car. Keep the doors locked, windows closed. Anybody that responded to your CB call can talk to you on CB. If you are suspicious, report the situation to the channel 9 monitor giving the description and license number of the vehicle. Ask what kind of help is on the way. If you are worried, say so on channel 10 or 19 and lots of 18-wheelers and four-wheelers will stop by. That raises the odds against a crook harming you until Smokey shows up.

CHAPTER 6

HOW TO UNDERSTAND SPECS AND EVALUATE FEATURES

There's an old saying, "If you can't stand the heat, stay out of the kitchen." In CB, a lot of the "heat" for non-technical people is the radio and engineering language that bounces back and forth in the channels, in discussions among CBers, and of course in the store where you shop for a rig.

If you choose to, you can avoid that kind of technical heat. Skip this chapter. Go to your friendly CB store and ask the person to sell you an excellent rig. Say how much you are prepared to spend and you will witness an amazing miracle. That's exactly how much an excellent rig costs! See how easy it is to avoid the heat?

Then ask the CB person to install it in your car and show you how to set the knobs and switches to talk or receive. That's it. Cool all the way. But if you're serious about what you do, if you want to do it well enough to get satisfaction and enjoyment, and particularly if you're serious enough to be poking around in this book, you will never feel right about doing that. You want to know more about choosing and using your rig.

In that case, good buddy, don't skip this chapter. If you never had any schooling in radio, you're about to get a quick shot of it—enough to solve your problem. Enough so you can understand what the controls on your set do and how to use them properly. Enough so you can read specification sheets and do an intelligent job of helping the salesman help you find that excellent rig.

HOW TRANSMITTERS WORK

If nobody is talking into the microphone, a transmitter sends out a carrier signal. This is like a single sustained pure tone from a flute or violin, except the frequency is much higher than tones we can hear. A carrier signal looks like this:

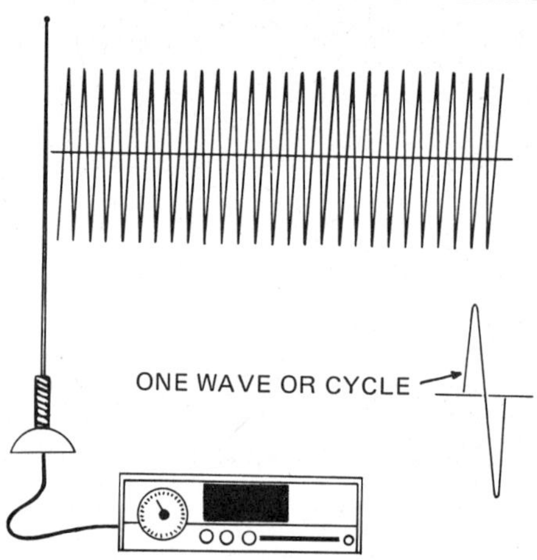

A plain carrier signal like that has no *modulation.* This unmodulated carrier has only two characteristics worth noticing: It has some amplitude, meaning how tall the radio waves are in the diagram. A more powerful signal has more amplitude, and the reverse. This signal also has a *frequency* which is how many of the individual waves happen in one second.

A single radio wave is marked on the drawing. The positive half-wave and the negative half-wave make one complete wave or cycle.

At radio frequencies, a lot of cycles happen every second. At CB frequencies, about 27 million of them leave the transmitting antenna in one second. They follow each other in one continuous chain which reaches to the horizon. A receiving antenna picks up the transmitted signal but gets only a small part of it.

When you speak into the microphone, your voice modulates the carrier and the carrier then transports the sound of your voice to any receiver which is tuned in. What happens to the carrier when it is modulated is shown like this:

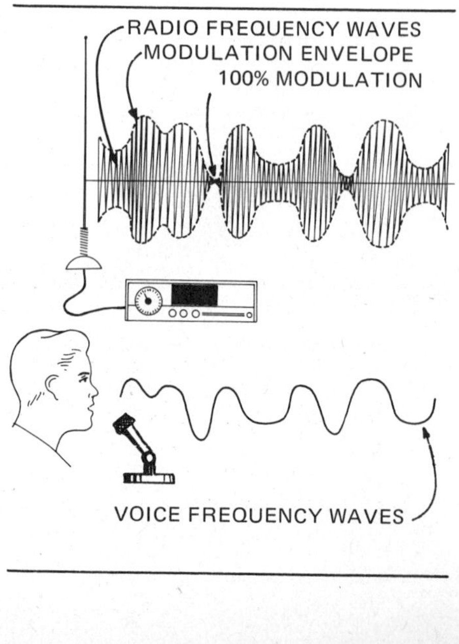

Choosing among top-of-the-line models is easy. All give you practically every feature and top performance. In other areas of the market where competition is fierce, features vary from set to set. You need to know which controls and features are important and how to read a specification sheet. This good buddy chose Browning's Golden Eagle, top-of-the-line base station with separate receiver and transmitter.

A CBer is talking into the microphone, making voice waves as shown. Some voice energy enters the mike and is used in the transmitter to modulate the radio-frequency (rf) carrier. Notice how the amplitude of the carrier is changed continuously to follow the same pattern as the voice waves. This is how the carrier "carries" your voice through the air to all those receivers listening in.

A modulated carrier has a *modulation envelope* which is the shape formed by the tips of the waves. It has exactly the same shape as the voice waves which caused it. Please notice: Both top and bottom of the carrier waves are modulated and one is the mirror image of the other. This means the carrier is carrying the voice signal in two places, once on top and again on the bottom of the waves. When the modulation is so strong that the carrier drops to zero in places, that's called 100% modulation.

This is called *amplitude modulation* because the voice signal is carried by amplitude changes of the carrier. If you could see a carrier with amplitude modulation, you would observe the amplitude changes.due to voice modulation.

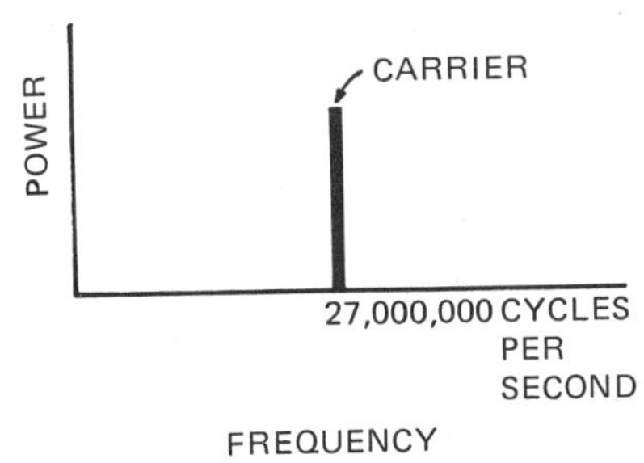

This is a graph showing what frequencies come out of the transmitter. As you can see, an unmodulated carrier has only one frequency. All of the rf power is concentrated at the broadcasting frequency, which we can say is 27 million cycles per second, usually written as 27 megahertz.

Now, when you speak into the mike, the frequency graph of the rf signal changes:

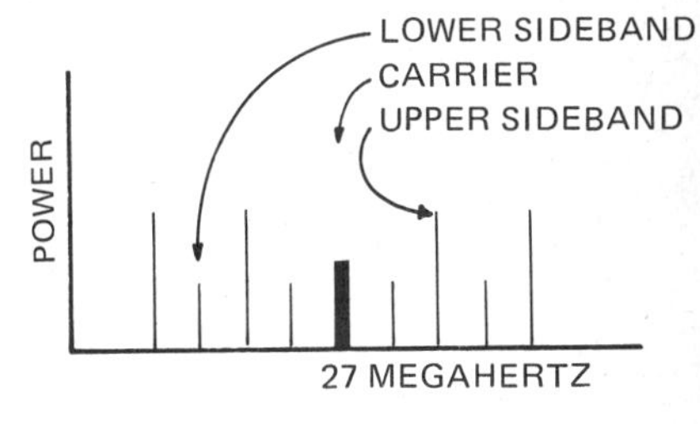

Due to voice modulation, a lot of other frequencies appear very near the carrier, some at slightly higher frequencies and some at slightly lower. These new frequencies are called sidebands, one upper and one lower. Also, the amount of power in the signal is increased. The carrier stays the same but the modulation process adds power to the sidebands.

This is not a different carrier or a different kind of modulation—it's the same one but we are looking at it differently. When you look at the frequencies, modulation of a carrier generates sidebands and the voice signal is carried in the sideband frequencies.

The upper sideband contains a complete voice signal, and the voice signal is duplicated in the lower sideband. You can extract voice from this signal by using both sidebands, or the upper only, or the lower only. When operating with only one of the two sidebands, the system is called *single sideband* (SSB) and the trend in CB today is toward SSB.

The frequency band for all frequencies in one sideband must be as wide as needed for voice. Good voice intelligibility requires voice frequencies up to about 3,000 cycles per second. If you are modulating an AM transmitter with 3,000 cycles, the total bandwidth needed for both sidebands will be 6,000 cycles—half for the upper and half for the lower sideband.

The FCC gives us a little more room than that. We get 5,000 cycles of bandwidth for each sideband. Here's an example. It's a frequency graph for three CB channels:

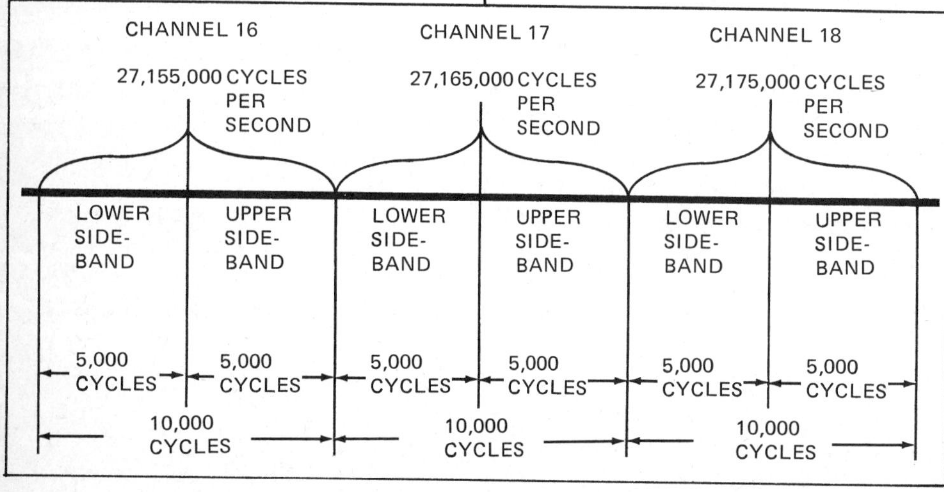

The carrier frequency for channel 17 is 10,000 cycles higher than channel 16 and 10,000 cycles lower than channel 18. The channel spacing is 10,000 cycles, so each channel has that much space in the radio spectrum. Each channel is allowed 5,000 cycles for its upper sideband and 5,000 cycles per second for its lower sideband. Because voice frequencies require only about 3,000 cycles per second, that's plenty of room for each channel to use without interference between them.

WHAT THE RECEIVER DOES

Two main things happen at a receiver. It *amplifies* the tiny weak signal it picks up at the receiving antenna. It also *converts* or changes the frequency. Amplification and frequency conversion are all mixed up together in most receivers, but all sets begin with some amplification of the rf signal because it's so weak it needs help.

Then the radio frequency is changed to a lower radio frequency by a system like this:

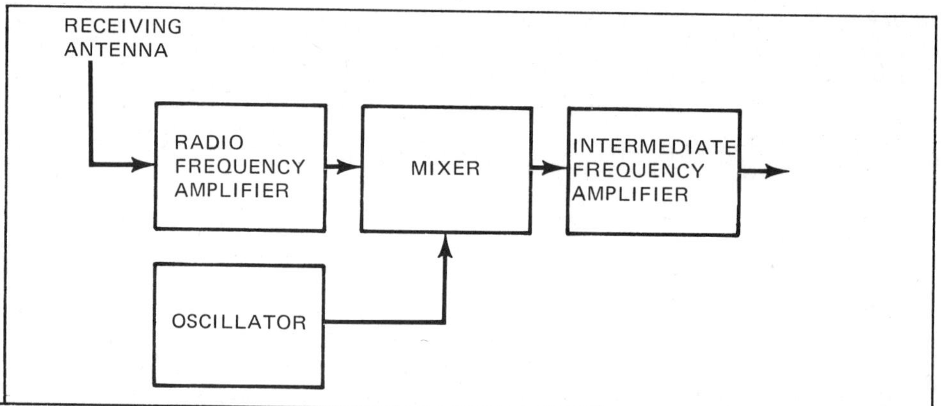

Inside the receiver is an oscillator which *generates* a radio frequency signal just like a transmitter. It would transmit if hooked up to an antenna, but we don't allow it to. We keep it inside the box and hook it up to a circuit called a mixer. Two different frequencies come into the mixer. One is the radio signal we want to tune in and listen to. The other is a different radio frequency produced by the mixer. Let's call these frequencies A and B. What comes out of the mixer is A, B, A+B, and A-B. We put in two frequencies and get out four.

The only frequency we use is the lowest of the group, which must be A-B.

Converting down to a lower frequency does good things: Additional amplification which the signal must have will not be done at the original transmitter frequency. This means the receiver is less sensitive to other transmitters at or near CB frequencies because the receiver is amplifying a lower frequency. Also, amplification is easier at lower frequencies. Incidentally, the technical name for converting with a mixer is called *heterodyning* and receivers that do it are called *superheterodyne* receivers.

The lower frequency obtained by heterodyning is in between the original transmitter frequency and the very low audio frequencies that will eventually come out of the speaker, so it is called the *intermediate frequency* (IF). The receiver is designed so the IF is

Single sideband sets look almost the same as straight AM rigs. There'll be a switch marked AM/USB/LSB indicating *Amplitude Modulation, Upper Sideband, Lower Sideband.* Be sure there is also a fine tuning control. On this Teaberry Stalker One it's called a *Clarifier,* used to clarify an incoming signal that is slightly off-frequency.

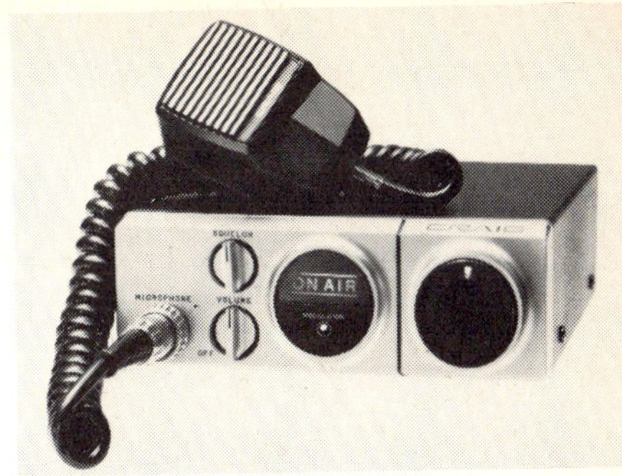

This clean-looking Craig 4101 has all the *basic* controls and indicators—there's nothing ambiguous about their functions. The channel selector lights up when the set is on, and shows which channel is in use. ON AIR **appears when you press the microphone button. When you talk the** MODULATION **light gets brighter with increasing strength of your voice.**

always the same, no matter what channel the receiver is tuned to. This is done by changing the "front end" radio frequency tuning of the set and simultaneously changing the oscillator frequency so there is always a constant difference between them. That constant difference is the IF. A common IF is 455,000 cycles per second, usually written as 455 kilohertz.

Some receivers use double-conversion, meaning there are two downward steps to intermediate frequencies. As an example, starting with a transmitter at 27.165 megahertz, the first conversion might be down to about 12 megahertz and the second conversion from 12 megahertz on down to 455 kilohertz. I'll tell you the advantage of this a little later.

Finally the inaudible modulated carrier is *demodulated* or *detected* which means the audio signal is extracted from the rf waves. Following that, the audio is amplified enough to drive a loudspeaker in your car.

Besides these basics, there are circuits to reduce noise in receivers, meters to read signal strength and other things, meters or blinking lights to indicate modulation, channel selectors and a front panel full of controls, buttons and switches.

CONTROLS

I am going to tell you what each control does and evaluate its importance. I have some strong opinions; so do the professional mobile-communication engineers who ran our performance tests in Chapter 8. We spent hundreds of hours operating various sets to determine how each could be made to perform best under identical conditions. Our dial-twiddling, knob-twisting and switch-jiggling experience adds up to this comment on the control panels: On most sets it's too hard to tell where the controls are set.

What bugged us also might bug you, so let me tell you about some bad features before I go on to evaluate control functions.

Concentric knobs always make you wonder whether to turn the inner or the outer. Sure they're marked, but you have to take your eyes off the road and try to read the tiny print.

Dual-function push buttons also make you check markings on a depressed button to find out what you've turned on or off, unless there is an indicator light for each function.

Ambiguous markings don't make it clear what a control does. There's a lot of confusing nomenclature.

You rightly expect that the more controls you have, the more performance you can wring out of a CB rig. Some controls you simply set and forget; others you have to operate.

THE BASIC CONTROLS

The *power switch* may be separate or part of the volume control. An indicator or panel meter should light up to tell you the set is on.

The *volume control* adjusts audio volume from the loudspeaker; it does not control your transmitted signal.

Squelch quiets the audio output of the receiver so you don't have to listen to background noise. When a signal stronger than noise appears, the squelch circuit releases so you can hear the message. This control can be set so only strong signals will be heard. I use it on the highway as a range control when I'd rather listen to music but don't want to miss transmissions from CBers within 3 miles or so. Conversely, when you want to hear weaker stations, even with bothersome noise, turn the squelch control off.

A *channel selector* switch seems obvious. There's more to it than meets the eye. If you can't look at the channel indicator because you're driving in tight traffic, you can always count clicks so you should be able to feel them. Channel indicators should have numbers large enough and bright enough so you can read them even in bright daylight.

Some channel selectors, like the one shown on the cover of this book, are stepping controls. You can step upward or downward, one channel per flick of the switch; or you can operate the switch and hold it until automatic stepping circuits reach your channel. Stepping selectors become important as new channels are authorized.

You should be able to see the channel number from a reasonable distance. This Royce rig uses light-emitting diodes (LED's) to form an illuminated channel-number display. It has a switch to make the channel number brighter when day-light is coming through the windows of your vehicle.

This Digicom 100 was designed for channel expansion in the Citizens Band because the two selectors can handle up to 99 channels. You set both dials—here they're set for channel 11. Newly-manufactured Digicoms will have additional circuitry for all 40 channels and require a new FCC type-approval.

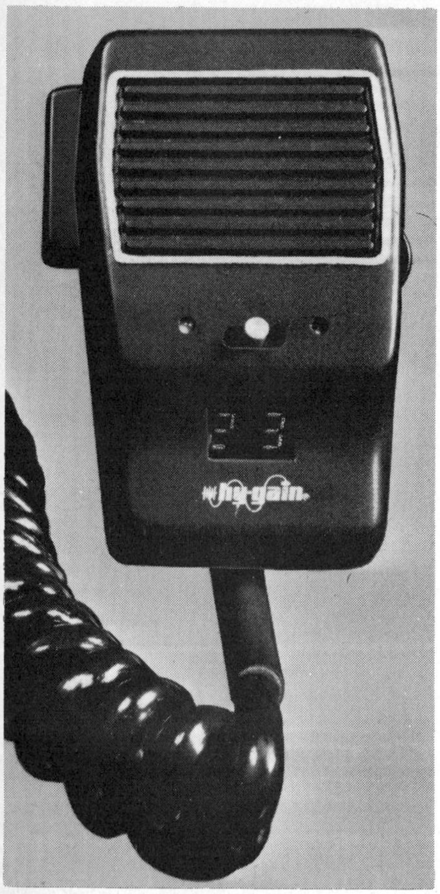

For a remote-control transceiver like this Hy-Gain 9, the microphone contains the channel indicator, selector, speaker and all other controls. You step higher or lower channels using the control just above the indicator. Push it to the left and it automatically steps down to lower channels while the LED display shows you what's happening. Move control to the right to step up. Release control when you reach desired channel.

It's hard to count clicks when you have 40 channels. What's more, no single rotary dial can show a lot of channels with numbers large enough to read. One set we tested, the Digicom 100, has two dials—one for the first digit, another for the second. It's legible and even expandable—up to 99 channels if there ever are that many.

Some stepping selectors may step when you turn the set on and will not be at the channel you used last. It's not important if you remember to glance at the setting when you turn on the set.

The controls listed above are probably the minimum you'll get on any set. In addition, you need some sort of *modulation indicator.* Inexpensive sets provide either a small meter or a transmitter indicator light that glows dim or bright according to the level of your voice as you talk into the microphone.

CONTROLS THAT IMPROVE PERFORMANCE

Noise is the CBer's enemy. You need to be able to control whatever noise you can't eliminate—see Chapter 18. More expensive transceivers, have the following two controls. Sometimes they are not selectable as on/off functions but as either/or, or one plus the other.

Automatic noise limiter is usually marked ANL or NL. In some sets it simply may be always on. There are many types of noise-limiting circuits; the spec sheets don't tell you exactly what you are getting, so I won't try to categorize them here. The noise they deal with consists of extremely short pulses—sharp spikes of unwanted radio emissions from lightning, spark plugs, electric motor arcs and static. When such pulses are louder than the average voice signal you are listening to, they mask the message your ears are straining to hear. A good noise-limiting circuit clips (reduces the amplitude of) the noise spikes so they are approximately the same strength as the average desired signal. You'll still hear them if you listen for them, but your ears tend to ignore the noise in favor of the message. When noise pulses become so rapid that they seem almost continuous, the noise limiter won't be very effective.

A *noise blanker* or *noise silencer* is labeled NB or NS on control panels. It can be more effective than a noise limiter because it operates in a different part of the transceiver circuit to chop out noise rather than just reduce its level. Essentially a blanking circuit detects noise pulses and momentarily silences the receiver for the duration of each pulse. Your ears won't recognize any difference in the voice signal with holes chopped in it. A good noise blanker really improves CB performance in mobile stations.

Fine tuning controls, sometimes labeled *delta tune* or *clarifier,* help you improve reception of stations that are broadcasting slightly off the correct channel frequency. You know what your home broadcast AM radio sounds like when you tune away from a station. In this case, your CB receiver may be set right but the other guy's transmitter is detuned. This control allows you to change your receiver tuning a little bit in either direction, to match up with the off-frequency transmitter.

Fine tuning is often accomplished with a continuous control

so you can choose any frequency within certain limits. Sometimes it's done with a three-position switch which gives you the correct channel frequency in the middle position. Flip it one way and you change tuning a certain number of cycles downward. Flip it the other way and it changes the tuned frequency upward by a set amount. Incidentally, in engineering language, the word *delta* means a small change. That's what delta-tune controls do.

Continously variable fine tuning is best. The control range should be at least 1,000 Hertz (cycles) above and below the nominal channel frequency on AM.

Fine tuning is a lot more critical on SSB sets. With these you must tune very carefully to make voices intelligible—otherwise everybody sounds like Donald Duck. Which is fine for the guy who uses "Rubber Duck" for a handle, but how about "The Quarterback?"

RF gain controls are useful. For receiving strong signals under noisy conditions or when the band is crowded, you can reduce the sensitivity of the receiver to signals so strong that they overload and distort—a problem that may occur when you are very near the station you are receiving. You can also reduce the sensitivity of the set to noise and splatter. I find this control especially helpful on single-sideband. For weak signals, of course you'll want to adjust rf gain for maximum sensitivity.

A *distance/local switch* usually reduces your transmitter power so you don't overload nearby stations you talk to, particularly if they don't have rf gain controls. If you are going to use CB on a boat, this control is a must when operating in a harbor.

There are some automotive in-dash sets with distance/local switches that are simply rf gain controls for the receiver and don't have any effect on transmitter power. Read the instruction manual or spec sheet carefully to be sure what the switch does.

Microphone gain is a front panel control on some sets. If there is one, it controls a voice preamplifier which makes the microphone output stronger before the voice signal is used to modulate the transmitter. Turn up the mike gain and you get more "talk power" on the air because you get more modulation on the rf carrier.

The FCC says it's your responsibility not to overmodulate the carrier—which means not to exceed 100% modulation. A microphone gain control invites overmodulation unless the CB set also has a built-in modulation limiting circuit which will prevent overmodulation. Check the specs. If it has a modulation limiter, it's a better set because you won't have to keep an eye on the modulation indicator.

You won't miss emergency calls on channel 9 with the second-channel monitoring feature on Lafayette's Micro 923. The flashing CH-9 indicates an emergency call even while you're communicating on another channel. Push the button above the flashing signal and you're immediately set to monitor or transmit on channel 9.

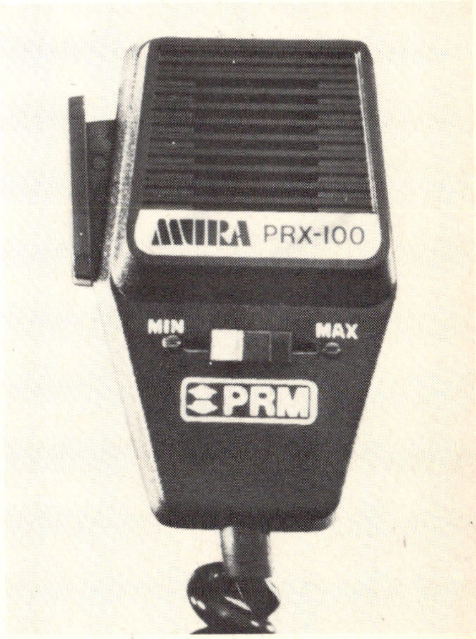

This variable-gain power microphone can upgrade a set without a microphone gain control. PRM identifies Mura's patented Peak-Redistribution Modulation circuit inside the mike which produces higher overall modulation—more talk-power—without clipping.

If you often get so busy you need more hands, try *VOX*—voice-operated transmitter control. This Tram D-201 which our engineers used for the performance tests in Chapter 8 features vox controls you adjust for your style of speaking.

If you normally speak softly or at some distance from the microphone, you need a mike gain control more than someone with a booming voice or one who likes to speak right into the mike. These people normally get plenty of modulation without electronic help. In noisy environments, it's best to turn down the mike gain control and talk louder or closer to the microphone. That's because a microphone preamplifier makes background noise louder just as it makes your voice louder.

If you buy a set without a front panel microphone gain control and later wish you had it, you can buy an accessory *power microphone* with a built-in voice preamplifier and gain control on the microphone. It does the same thing and will overmodulate the carrier unless there is built-in protection against overmodulation.

A *voice-sensitive* control or *vox* makes a deluxe base station. Instead of pushing a button to transmit, your voice keys (turns on) the transmitter so you can operate hands-off. I don't recommend the vox control unless the front panel also has a *vox-sensitivity* control and a *vox-delay* control. The sensitivity control lets you set the voice level at which the transmitter will be turned on. Set it for the distance from the mike you find comfortable, or for the volume of your speaking voice, but not so high that background noises in the room may falsely key the transmitter. The delay control is set to keep the transmitter on during normal pauses in speech.

Tone controls aren't a necessity but do make listening more comfortable and affect the apparent audio level to some extent. Some transceivers also have a transmitter tone control in the mike circuit to tailor your voice for a more audible signal.

Calibration controls are important on sets with a built-in test meter. Used to adjust the test meter so it reads accurately.

Antenna load matching controls are rare but useful and offered primarily on deluxe base stations with special FCC type approval. This control is used to get maximum radiated power from your transmitter.

Second-channel monitoring is a system that gives priority to one channel while you listen and talk on another. If a call comes in on the monitored priority channel, you are interrupted automatically and switched to the priority channel to hear the message. Usually transceivers with this feature are set to monitor channel 9 and give it priority, so you will hear emergency calls. Some transceivers have a control which allows you to assign priority to any channel while you work another one.

Selective calling by code alert gets you off the CB party line. You equip one of your channels with a device which recognizes only a certain code. Your friends or business associates have transmitters which send that code. Only when the code is sent can you hear messages on that channel. Some operate by flashing a light or operating a buzzer which tells you to turn on the speaker. Others do the whole thing automatically in various ways—such as turning off the squelch which you have previously set to squelch everything.

When you are fed up with CB chatter, selective calling brings a lot of peace and quiet. There can be different tones for different stations so you need not hear calls addressed to other stations in the same network, or you can set a control to monitor any messages within the net. This feature is attractive for business users of CB equipment.

Caution. Code-alert operation is restricted by FCC rules to tone signals which activate the receiver circuit. You can't simply broadcast a recognition tone or whistle to attract the attention of another station.

CB interrupt is a feature of some combination AM/FM/CB radios made for in-dash mobile use. You can listen to the AM or FM standard broadcast band programs while the CB circuits listen for calls on whatever channel you choose. This is *not* a selective call system; *any* signals received on CB will interrupt the regular program to which you are listening. This can become annoying if you're tuned to an action channel when the band is busy, but the solution is to set your squelch so you'll only be interrupted by strong signals from nearby stations. Presumably they will be reporting something of interest within a few minutes driving time.

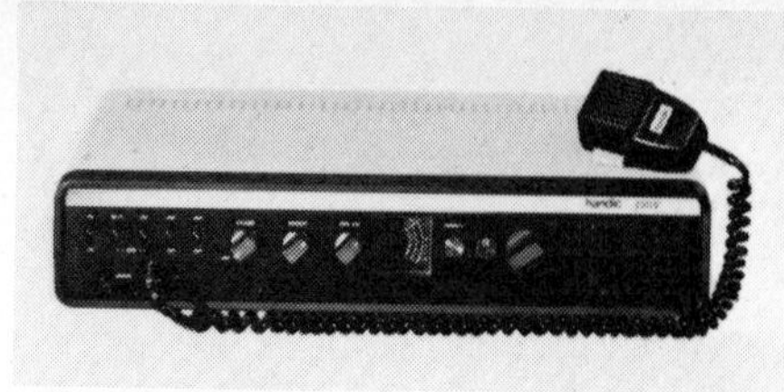

Handic's 2305 base station has second-channel monitoring. You pick any channel to be priority monitored by ordering a crystal for that channel. The priority light flashes and you hear the selected channel in the loudspeaker no matter what other channel you are working.

Weather channels are available on some marine CB transceivers. The National Weather Service stations provide 24-hr. local weather coverage in many parts of the United States. You'll hear them identified as NOAA Radio, short for National Oceanic and Atmospheric Administration. There are two VHF channels well above the CB band where you won't encounter much noise or interference. I get reliable reception 40 miles away from most NOAA radio transmitters and listen to the weather announcers reading their radar screens or giving conditions reported far beyond my CB operating range. Storm alarms with a priority-interrupt tone system are broadcast by the weather service. The necessary receiving equipment is in newer CB transceivers offering weather channels. Check the specs carefully; some sets give you both weather channels, others require you to change a crystal.

CHECK THE SPECS

If you are shopping for a CB, the front panel controls will tell you a lot about what's inside, but not everything. Back up your careful inspection of various brands by collecting the spec sheets. Take the time to look them over carefully and make comparisons.

We believe most specifications are honest and not intended to mislead you. However, specs are normally measured in the laboratory and they are taken for individual parts of a CB set. Doing that is like checking the lifetime statistics on each player of a baseball team. The statistics may say what each is capable of individually, but they sure don't tell you if *the team* is gonna win the series.

Checking the *overall operation* of a CB rig must be done in the real world if it means very much. That's what we've done very carefully and reported in Chapter 8.

Don't throw away the spec sheets though, because they have a lot of good information you need. Usually a manufacturer's specification sheet covers three main topics. General information covers weight and size, power requirements, circuits and control features.

A transmitter section tells about the signal, power output and antenna matching. The FCC does not permit a CB set to be sold unless it has approved the transmitter. This part of the spec sheet will never give numbers that exceed FCC limits, but it may be vague about performance within those limits.

A receiver section describes characteristics of signal reception and sound reproduction.

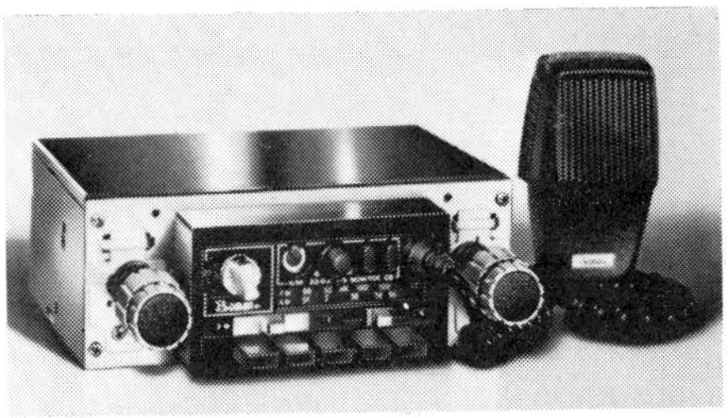

If you select an in-dash rig, the overall dimensions of the set are important because you want it to fit. Dashboards are padded, but they're not elastic. The control shafts should be adjustable in position and projection.

GENERAL INFORMATION

Circuit features are usually described in terms meant for the technically sophisticated, or in numbers of parts–like 17 transistors, 12 diodes. Forget the number of parts; clever engineers can make a mere handful do wonders, or use hundreds in pre-fabricated integrated circuits.

Dual-conversion Superheterodyne circuitry–Dual conversion is used by most manufacturers to make tuning more selective. Such sets cost slightly more to make than single-conversion rigs but that doesn't mean you'll find dual conversion only at the top-of-the-line and single conversion only in the least expensive sets. Some receivers at the top of a manufacturer's line use single conversion because noise blanking is easier.

Personally I'd rather read a signal through noise than through interference; it seems easier to make out the message. But I wouldn't decide on dual or single conversion from specs until I listened to the set on a busy channel in a noisy electrical environment, like in a vehicle in heavy traffic.

Synthesized Circuitry or Crystals for Every Channel?–Not long ago, CB sets required two crystals for every channel, one to transmit and one to receive. You switch among crystals by turning the channel selector. There are still CB sets around that require crystals in matched pairs for all channels, and lots of sets are still on the market that provide 3 or 6 or 12 crystal-controlled channels. Such sets are attractively-priced–until you discover you pay up to $5 extra for each crystal, *if* you can get crystals for the channels you want. Forget it. There's a crystal crunch.

To get around the shortage of crystals, manufacturers use *frequency-synthesizer circuitry.* This lets them use fewer crystals and all are installed in the set. Clever circuitry makes them work on send or receive for all channels.

Phase-lock Loop (PLL)–This system drops the crystal requirement to 1 and brings extraordinary frequency stability to CB. When there are enough rigs around with PLL, you won't need fine-tuning controls. The trend toward phase-lock-loop is in your favor. PLL began to show up in sets manufactured in late 1975.

Size–Ought to be simple, but isn't. Too many spec sheets toss off a statement like 3x7-1/2x9 but don't say which is which. For CB mobile use this hardly helps if you're going to hide a transceiver in the glove box or shoehorn it under the dash of a subcompact car. Some CB sets are narrow across the front but rather deep; they'll take up little leg-room under the dash but won't fit in a shallow glove box. In-dash units are critical to fit.

Be sure you know which numbers mean height, depth and width –and whether these dimensions include the mounting bracket.

Weight–May not seem important unless you're going to be carrying a portable. In that case, read the spec carefully to determine if the number means with batteries–they're heavy. I think weight is important for mobile sets, too. If you're going to mount one on quick-disconnect brackets so you can remove it to stow in the trunk, or carry it around with you, then lightweight is vital. One day you'll be just too tired to disconnect a heavy one, and that's the day a CB snatcher is waiting for.

Power Supply—Specs are often too skimpy. Transmitters tend to give more power output with more volts input from the power source—and less power with reduced supply voltage—which is usually the problem.

Part of a well-designed CB rig is an internal voltage regulator which protects the transmitter from supply-voltage changes. The voltage supplied by the generator in a moving automobile should be 13.8 volts even though we normally call it a 12-volt system. Within 10% of that voltage, a CB transceiver which meets Electronic Industry Association (EIA) standards should put out within 2 dB of full power—4 watts into the antenna.

In many cars, the generator voltage drops when the engine is idling. It may drop to 13.5 volts or less. The rf power you get from your CB may or may not drop, depending on how good the built-in voltage regulator is.

With engine off, the voltage is even less. It may be around 12 volts. You have the same situation. Some sets will still pump out rated power. Some will still work but at greatly reduced transmitter power.

In addition to power output of the transmitter, other kinds of CB performance are affected by supply voltage fluctuations. At reduced voltage, receiver sensitivity may drop and you will not be able to "pull in" distant stations. The audio amplifier section which drives the loudspeaker will make less audio power so it will be harder to hear the faint signals the receiver does pull in.

Most important, fluctuations of supply voltage may change the operating frequency of the transmitter. The FCC says *you are responsible* even though you didn't build the transmitter.

All specs say something about electrical power and the tolerable range of variation. What they say *exactly* is sometimes hard to decipher. You may get the feeling that this is done on purpose.

What you would like to know is this: Over a specified range of input voltage variation, the set will operate and meet every specification. That means it works the same at any voltage within the specified range.

Sometimes the brochure seems to say that but actually means this: The set will operate *satisfactorily* over a specified range of input voltage variation. That does not mean with 4 watts output; it does not mean with full receiver sensitivity or audio output; but it must mean at the correct transmitter frequency because no other operating frequency is satisfactory.

It's up to you to interpret specs you see and ask questions of the manufacturer or salesman until you are sure you have the facts and understand them.

The input voltage specification may be stated in more than one way:

13.8 V DC (11.6 – 15.6 V)—or some other voltage range

13.8 V DC ± 10%—or some other percentage

In both cases, the nominal input voltage is stated, followed by the range over which operation will be at specs or maybe just satisfactory.

Manufacturers adhere to EIA standards voluntarily, but most consider it important to do so. According to EIA standards, the transmitter should be stable and remain on frequency over a voltage input drift of plus or minus 15%. Some set makers may offer you more stability than that.

Base-Station Power Supply Range—Theoretically the electric company sells you 110 to 120-volt alternating current, which is why 117 V AC is the nominal voltage. In these days of power brown-outs, the power company may reduce the voltage by 5% to reduce heavy loads such as widespread use of air conditioning on a hot day. If you are some distance from a power company distribution transformer, you may be getting only 112 V on an ordinary day. When this voltage is cut 5%, your base station gets only 106 V.

That bulge in the case of Midland's 13-861 mobile/portable is the external battery pack with a built-in charger for use with Nicad batteries.

Look for a notation in the spec sheet that the base station transceiver has a *regulated* power supply. That means it will compensate automatically for line voltage changes. A good range for power supply operation is: 105 V to 125 V AC.

Positive Or Negative Ground—A lot of specs for mobile equipment say 12 volts DC but don't say if the transceiver can be hooked up to a positive or negative ground system. This refers to which side of the battery is connected to the metal frame of the vehicle.

If you're going to transfer CB equipment from one vehicle to another, look for a set that operates on *either* positive or negative ground. Also look for *reverse-polarity protection,* a way of preventing damage to the set if you connect it backwards. In some designs, a fuse will blow, which is simple but inconvenient protection. Others, just won't work unless you hook up correctly, but the set won't be damaged.

Today, most vehicles have a 12-volt system. You may have an oldie with a 6-volt battery. Let me save you the trouble of poring over spec sheets. You'll need an accessory called a *DC converter* because manufacturers don't make 6 V mobiles.

Marine Power Supply—CB sets aren't designed for the variety of power sources aboard various boats and ships. But that's no obstacle with an appropriate DC converter that can handle 4 amperes of current or more. The spec sheets will say 12 V or 13.8 V DC and 117 V AC for base station or dual-purpose models. Many vessels have both power sources, so you're almost home free. Some use a higher-than-normal voltage to charge the onboard batteries. So you may need a *voltage stabilizer* to protect your CB rig. This does the same thing as a voltage regulator built into a CB rig, except it does it separately and reduces vehicle voltage fluctuations before they reach your rig. Marine equipment catalogs and supply stores call them voltage stabilizers. To buy one, pick the stabilizer that gives the correct output voltage for your CB and can use the input voltage that the vessel supplies, which may not be 12 volts nominal. Other marine power sources you may encounter include 12, 14, 24, 32 and 36 V DC and 230 V AC. You'll need a step-down transformer to convert 230 V AC into 117 V AC.

If you want to operate a 117 V AC-only base station transceiver on your 12 V-only vessel, get a *power inverter.* This changes DC into AC current.

Portable Power Supply—Portables have their own batteries—either dry-cells or rechargeable. Read the spec carefully to determine if the charger is in the set or if you have to remove the batteries and put them in a separate charger which may be an extra-cost option. You'll find both 9 V and 12 V DC sets available. I advise getting a 12 V unit with an external power jack. Then you can use a nominal 12 V source when on a boat or vehicle thus saving your batteries. I have yet to see a spec that tells you the number of hours you can operate on batteries in receive or transmit modes.

Caution: There are *mobile* units that will take an externally attached battery pack so they can be operated as portables. Be sure the pack furnishes close to 13.8 V—such as 9 cells rated at 1.5 volts each—or performance will suffer. Some packs have only enough room for batteries that add up to 12 volts.

Power Supply Adapters—At one time these were accessories found in catalogs. Now you'll find mobile-base station adapters on transceiver spec sheets. These are stands or cabinets into which you pop a mobile rig so it can be used as a base station on 117 V AC. At least one transceiver on the market operates on batteries as a portable, fits in mount for mobile use, and in a stand for base station operation and recharging.

Temperature Range—Not many specs give this information. Your CB set doesn't have to be in the desert or in the Arctic to encounter extreme temperatures, especially when you leave it in a closed, parked vehicle. Even in temperate climes, summer sun can run the interior temperature past 122°F, or 50°C. Extremely high or low temperatures can affect frequency stability. Most will tolerate a temperature range of −22°F to 122°F, or −30°C to 50°C; this is the minimum temperature range required by the Electronics Industry Association.

Channels—This can confuse the unwary. Some sets come equipped to operate on all currently authorized channels of the Class D Citizens Band. Some operate on fewer channels such as 23 which was formerly the limit. Some use even less, such as 9 or 6 or 3. A few are built

Most mobile sets simply aren't made for 117-volt alternating current so you'll need an adapter to provide the 12-13.8 volts DC when you want to use the rig indoors. Radio Shack makes this Realistic TRC 24-C and the Micronta 12-volt power supply.

strictly for emergency use on channel 9.

Besides the basic number of channels available, use of single-sideband seems to magically multiply the channels into three times as many. That's because SSB sets can operate as AM (double sideband) on a channel and *also* operate USB (Upper Sideband) and LSB (Lower Sideband) on the same channel.

Indicators and Meters—There are several types of *channel indicators.* Most common is an illuminated dial on the rotary selector switch. When black numbers are silhouetted against the light, they are easy to read unless they are tiny. When the numbers are clear in a black dial—sometimes called a black-out dial—they are difficult to read in bright daylight unless the indicator is in shadow under the dashboard. Electronic digital indicators have larger numbers. Most are readable in bright shade but can disappear in sunlight. Numbers on an electronic display may require a second look because they are made up of straight-line segments.

Transmit indicators are of many types. Some sets light up a panel that says "on-the-air." Others turn on a red light when the transmit button is pressed. There are sets that change a meter scale from white to red. You need a transmit indicator, but the type isn't important.

Modulation indicators tell you something about "talk power" when you transmit. The simplest indicator is a light that changes brightness when you talk but it's hard to know how much brightness is enough modulation. You get a better indication from a row of LED's—light-emitting diodes. The more that are lit, the higher the modulation. While a row of LED's is easy to see at a distance, or from an angle when the set is not in front of you, LED's are hard to see in very bright light.

Modulation meters are usually one scale of a multi-purpose meter on the front panel. The needle will flicker as you talk, showing the percentage of modulation from 0 to 100%. I think this is best.

Multi-purpose meters also show some or all of the following information: *S-units* tell you the strength of received signals. In most sets, the reading doesn't mean much—it only tells you one signal is stronger than another. The best sets are calibrated carefully so the S-unit reading is a measure of actual signal strength. *Power-output* meter scales usually show relative readings rather than actual watts. If the scale has numbers only, it's not calibrated in watts. *SWR*—Standing Wave Ratio—is a valuable indication of antenna matching. You're more likely to find an SWR meter on an expensive mobile or base station.

Specs tell you about the meter on a CB set but you still won't know how convenient it is to read. Some meters are recessed in the front panel, making it hard to read the top scale. Others are tiny and can't be read from a distance, which makes for difficulty when you're driving and somebody asks for a radio check. Curved-face meters mounted horizontally are difficult to read from the driver's seat in a vehicle. Vertical meters of the same type are better.

When the surrounding light level is a mite too bright, channel numbers made of illuminated line segments can fool you. Mount your mobile under-dash for maximum shading.

Microphone—Surprisingly, a lot of specs won't tell you the kind of microphone you get with a transceiver. Of the two kinds, a *ceramic mike* gives crisp highs, making your speech more understandable when reproduced in somebody else's loudspeaker. But a ceramic mike is fragile and in mobile use it's sure to be dropped more than once. A *dynamic mike* is more rugged and its voice reproduction is just a shade less crisp; not enough to worry about. Dynamic is better.

Power mikes rarely come as original equipment. If the CB set maker figures on amplifying the voice signal as it comes out of the microphone, he normally does it with a microphone preamplifier built into the CB rather than into the microphone. As discussed earlier, if you need help to get enough voice power to modulate the transmitter, amplification will do it. Accessory power mikes normally have the amplifier in the mike itself, if hand held, or in the base if a table-top type. They work on self-contained transistor-radio-type batteries.

Noise-cancelling mikes are sometimes part of the original equipment. They're good for noisy mobile environments like the cab of an 18-wheeler. But they're not an absolute requirement; you can talk close to a good ordinary mike at a 45° angle across the face of it and get great results with not too much background noise.

Speech-compressor mikes will provide a higher average modulation and therefore an apparently louder signal. The output of the mike is held relatively constant despite the uneven strength of your voice or differences in distance between mouth and mike. I don't advise such mikes for noisy environments because they raise the level of background noise, too. Some base stations have built-in voice compressor circuits. You may find mobile transceivers equipped with speech-compressor mikes. Usually it's optional equipment in a manufacturer's line or available from accessory makers.

Nothing in a specification will tell you whether you are getting a good *microphone cord* or a cheapie. Almost all cords today are of the coiled type. They should return to the original shape and length after being stretched. Cheap cords won't recoil tightly. Look for strain-relief fittings at *both* ends of the cord; tension at these points eventually breaks the wires inside.

Loudspeakers—Almost all transceivers have small built-in speakers. The exception is the combination radio/CB in-dash transceiver which connects to external speakers intended for music. External speakers may not be furnished with the set. The restricted range of small CB speakers is good for hearing speech clearly, though some speakers are harsh. Check the spec for an external speaker jack—the majority of transceivers have one so you can plug in a better speaker. You'll need one for glove-box mounting a set.

The spec won't tell you where the built-in speaker is mounted so look at the CB unit. A few mobile CB rigs have the speaker on the left so the driver will hear best. Most sets point the speaker downward which reduces clarity somewhat because the sound bounces around your feet and is partially absorbed by floor mats. Some have top-mounted loudspeakers which in under-dash mounting directs the sound behind the dashboard. Top-mounting is fine if your CB set sits on top of the transmission hump.

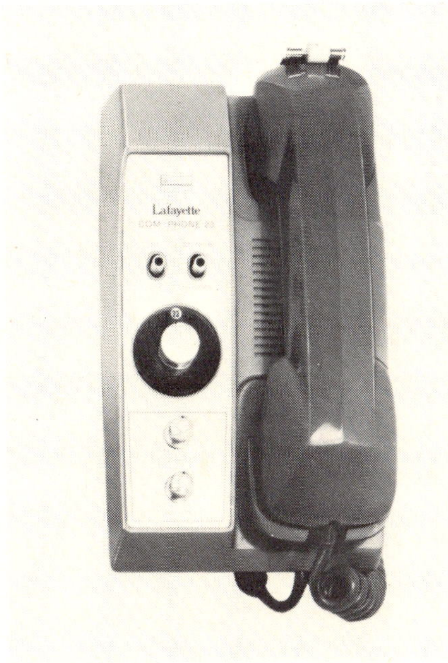

Mobile radio-telephone handsets are good for privacy and eliminating background noise. Lafayette's Com-Phone is compact. The Realistic TRC-56 is shaped for mounting on the dash or on a shelf behind the rear seat. I think these are best as a passenger's CB.

Telephone Handsets—For base stations, these are becoming popular for privacy. You listen for calls on the built-in loudspeaker, then switch to the handset for conversation with another station. Handsets are fine for base stations and the passenger seats of automobiles. I think they are clumsy for drivers.

Ear-phone/lip-mike Combination—Standard only on CB rigs made for motorcycles. Otherwise they are accessory items.

Audio Output—Runs from 2 to 4 watts in most sets. Less than 2 watts may not be enough in noisy vehicles. You'll need at least 3 watts if you intend to use the public address feature on some CB sets to drive external speakers. Mobile public address systems, even in a CB, are regulated by law in some places. Check before you buy.

Audio-output figures will usually be lower for sets that feed earphones and handsets to be kind to your ear drums. Maximum acceptable audio distortion in any set is 10%; a lower figure is also kind to your ears.

Clocks—Useful in a control panel for more than just answering a 10-36. You'll find base stations equipped with clock-operated controls to turn the set on and off so you can keep a contact schedule with other stations.

Twelve-hour mechanical clocks with the on-off feature have some way to switch the clock controls from AM to PM. If you use the control right, you can be sure you have set it to turn on at 3 PM rather than 3 AM. I have done it backwards a few times and missed some scheduled communications—and some sleep.

I prefer clocks with electronic digital displays which can be set to work on a 12 or 24 hour time base. Flick the switch and the clock timer will count off all 24 hours in each day so there is no ambiguity about AM and PM. All you have to remember is, add 12 to each hour after noon to get 24-hour time. For example, 3 PM is 15:00 hours.

TRANSMITTER SPECIFICATIONS

Power output is too often stated in terms intended to impress rather than to inform. FCC rules are rigid: Power available at the output terminals of a CB transmitter may not exceed 4 watts in the AM mode.

If you see a claim of 5 or 6 watts, that means power input to the transmitter, not carrier power out of it.

In our performance tests, we carefully measured the rf power output of all transceivers in Chapter 8. They produce 3 to 4 watts, but rarely precisely 4 watts as given in the specifications.

Single-sideband power may be given as 25 watts PEP input or 12 watts PEP output which is the legal maximum. PEP (Peak Envelope Power) is the maximum power developed on SSB at peaks of the voice signal. 25 watts input PEP usually yields 12 watts output PEP.

Modulation percentage is mentioned in some specs. There is no meaningful difference in performance between 85% and 100%. Transceivers type-accepted after May 24, 1974, by the FCC must contain a device limiting modulation to 100%.

Harmonic suppression is governed by FCC requirements. Harmonics are unwanted signals related to the actual transmitted frequency.

A few transceivers have television interference (TVI) traps to suppress any harmonics that might leak into nearby TV sets.

Frequency tolerance of your transmitter is important. You want to be right in the center of the channel as much as possible. The FCC wants you there too, and has a rule requiring it. They insist that your transmitter frequency be maintained within ± 0.005% of the assigned channel frequency.

Because 0.005% is required by law, all manufacturers claim at least that much tolerance. Some offer more, such as 0.003% plus or minus. When shopping, buy the highest tolerance—signified by the smallest tolerance percentage number you can find in a set that meets your other requirements.

SSB transceivers are more intolerant of frequency drift than AM sets so be especially watchful for low numbers in the tolerance specs.

RECEIVER SPECIFICATIONS

Because FCC type acceptance applies primarily to the transmitter and not to the receiver, you'll find spec sheets take a lot more latitude in what they say about receivers.

Sensitivity is the ability to detect a signal and produce a specified audio output. Because a signal overcomes and quiets noise in the receiver, sensitivity is measured in a laboratory shielded against outside rf noise so the engineers can measure the weakest signal that produces a given amount of quieting.

Sensitivity figures involve two numbers—the signal input, expressed in microvolts (mv or μV) and the quieting ratio given in dB. You'll see it written in a variety of ways. Just watch the combination of numbers. The lower the input signal number, expressed in microvolts fractions of microvolts, and the higher the quieting in dB, the more sensitive the receiver. You need at least 10 dB quieting to extract 80% of the intelligence from a voice signal.

Good numbers for "hot" sets are 1 microvolt or lower for 10 dB quieting.

Selectivity is the ability of a receiver to tune to one channel and reject adjacent channels so you don't hear them. The specification of selectivity is technical, but here's the idea. Two numbers are given which should *both* be large or both be small. For example, 6 dB at 5 KHz is a typical spec. It means another radio frequency only 5 KHz away from the channel you are tuned to will be reduced in volume by 6 dB. 50 dB at 10 KHz in also a typical spec. It says if the other frequency is 10 KHz away, its volume will be reduced by 50 dB. Because channel spacing is 10 KHz, this is more meaningful, however 6 dB at 5 KHz could very well become 50 dB if the frequency were 10 KHz.

What you want is a large number of dB for whatever frequency is used in the spec. 10 dB at 6 KHz is better than 5 dB at 6 KHz. Look over the specs in Chapter 8 and you'll see a pattern. Look hard at the specs for mention of the type of filter used to let signals through the rf end of the set. I prefer the sharper selectivity of a crystal-lattice filter to the mechanical or ceramic filters.

Image rejection is rarely a cause for concern. It's expressed in dB and the larger the number, the better. Expect 50 to 60 dB in most double conversion sets. Lower numbers OK in single conversion sets.

Automatic gain control (AGC) is in all transceivers. It automatically adjusts the strength of received signals so you don't have to keep changing the volume control.

The spec should tell you how much the speaker volume will change for a specified amount of received signal change at the antenna. Example: When the rf signal changes from 10 mv to 100,000 mv, the speaker volume will change only 10 dB. This is acceptable. An AGC spec will probably say that backwards—10 dB for 10 to 100,000 mv—but it means the same thing.

The *lower* the number in dB, the less likely you'll miss a signal or be blasted out of your seat. You won't always find AGC in specifications and when you do, it may simply be given in dB without specifying the input range. In that case you are reading an AGC *figure of merit*; the AGC action gets better with a *higher* number.

HOW TO CHOOSE YOUR CB

You'd like to have everything—top performance and all the useful features. But buying a CB set is not like acquiring Detroit iron. Unlike an auto, you can't pick a basic set and order performance options and convenience or luxury features installed at the factory. You have to choose among models that have the performance capabilities you need and most of the features you want, then try to purchase accessories that add other features to your system. There are few dream sets, except for certain deluxe base stations.

You have to accept the fact that ordinary strategies don't always work. Picking the top-of-the-line gets you more controls, finer features, but our performance tests show that you don't always get maximum capability. Reading the specs is helpful up to a point. As you can see it's hard to compare numbers unless they are written the same way and result from similar measurements.

You have to decide where you will be using the CB set most. In some operating environments, you won't get better performance out of a more sensitive set because the band is too crowded.

Mobile Use—For *city* use look for selectivity; all channels are usually busy. Sensitivity won't matter much because weak signals will be overridden by the abundance of strong signals. Noise blankers are vital, not only for suppressing ignition noise from your own vehicle but also noise from others in tight traffic. Single-sideband will pump out a stronger signal to begin with; lots of it will be reflected or absorbed among all those vehicles and buildings. Be sure there are other sidebanders to talk to, or your base has SSB. Look for ease of fine tuning in SSB and good frequency stability. You'll be too busy avoiding vehicles and pedestrians to fiddle with controls.

In *suburb and country* use, sensitivity becomes as important as selectivity. You want range so you can receive weaker signals. Good AGC action is important because there will be a mixture of weak and strong signals. Some may be so strong that an rf gain control will help. Noise limiting and noise blanking are important. Single-sideband will get you 50% more range but good fine tuning is a must or you'll miss most of a message while trying to clarify the signal. A noise-cancelling mike will help 18-wheelers and four-wheelers with open windows. Out in the open country it's brighter inside your vehicle so look for a channel indicator you can see easily.

Base Stations—Receiver sensitivity and transmitter power output are a little less important in base stations because you can make up for small deficiencies by using a more efficient antenna. Still, if you want maximum performance, don't compromise on these two items. Noise controls aren't quite as important as for mobile, but you really ought to have them in densely populated areas. Selectivity counts heavily because a superior antenna system will pull in signals from all over. For this reason you'll want an rf gain control to reduce interference by controlling sensitivity; it's helpful for SSB and some mobile-to-base communication.

Price and performance should be compared among sets with the capabilities you want. Not all base stations have SSB. Most base stations have about the same controls. Additional controls and features justify higher prices, like autos loaded with accessories, but they don't always guarantee better performance. Not all manufacturers' top-of-the-line base stations are equal. CB has its Rolls-Royces among base stations—and waiting lists to buy them.

Portables—Sensitivity, power output and selectivity count in about that order. You won't have a very efficient antenna. Normally you operate from batteries so you have a power source with gradually deteriorating performance. Your best strategy is to concentrate on the three items listed above because they are critical to range. Then look over the few features portables have. Six channels and selective call may be more valuable to you than more channels.

Marine—Over-water communication by radio covers greater distance than on land so you might think sensitivity and power output are less important. Don't kid yourself. In an emergency you'll want all the range you can get if CB is your only on-board two-way radio. Opt for SSB. Moreover, you need good selectivity because shore-based CB traffic will carry right out over the water and clutter whatever channel you're working. Noise controls are important. Features you might especially want include public address for loud hailing, *both* NOAA weather channels if you cruise well beyond home port, and a distance/local switch for harbor transmissions.

Price and performance should be checked among *marine* CB sets and not among just any mobiles or portables you can put on a boat. Marine CB's are rugged with watertight construction of corrosion-resistant materials. You can upgrade performance of a marine CB with a more efficient antenna. You can't upgrade a mobile that swallowed salt water.

"WHAT'S A DECIBEL?"

Abbreviated dB, it is an engineering unit of measurement related mainly to the way the human ear hears things. It is also involved with logarithms which is heavy math for most of us.

If you deliver **2** times as much power to your loudspeaker, you increase the amount of power by **3** dB. When you listen to the result, it won't sound **2** times as loud. It won't sound **3** times as loud either. It will be noticeably louder.

When comparing specs using dB, just consider dB as a curious way to measure something—like a pound of water. When things are going up, 50 dB is *more* than 40 dB. When things are less than zero, –50 dB is *smaller* than –40 dB.

CHAPTER 7

HOW TO SELECT AND INSTALL MOBILE TRANSCEIVERS

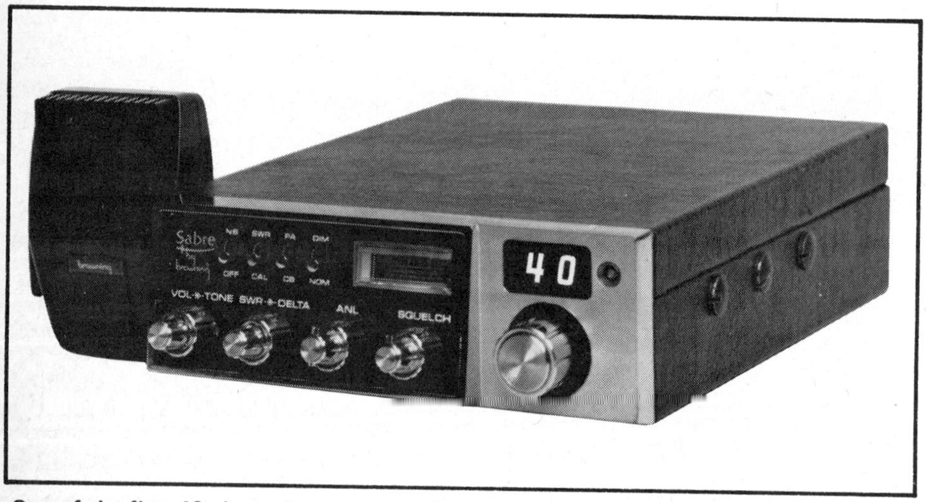

One of the first 40-channel type-accepted mobile sets from Browning will be the Sabre. This set, at press time, is not type accepted and is not offered for sale.

Choosing a set is a compromise. You'll have to make trade-offs among performance and features, but those aren't the only factors. You want to mount the set conveniently in your vehicle and secure it against theft. The hardest sets to steal are also among the hardest to install, and they may lack features you want. You can't do much to change the shiny new transceiver you buy, so take a little time to make your decision.

CB lines are priced to offer something for everybody. The least expensive sets in each line have the fewest features, but don't assume they always have the poorest performance. As the price goes up in steps of ten, twenty or thirty dollars, you get more features. Some additional features—like noise blankers—improve performance significantly. Meters simply improve operating convenience. Carefully compare what you get in improved performance—in addition to more features—as you move up the price ladder.

You get what you pay for—usually—but sometimes you get more in one set than another at the *same* price. Your best strategy is to compare performance first, then compare controls and features.

How do you compare performance and observe the effect of circuit features which may not be mentioned on the spec sheet? The best way is to make comparisons of actual sets in actual operation in a real-world environment—and that isn't easy to do.

Because CB sets are for communicating, the basic test for performance is to measure how well you can communicate under various operating conditions. The next chapter is filled with our test data on a large number of sets from many different manufacturers.

Don't buy a mobile rig until you have decided where you will mount it and how you will mount it.

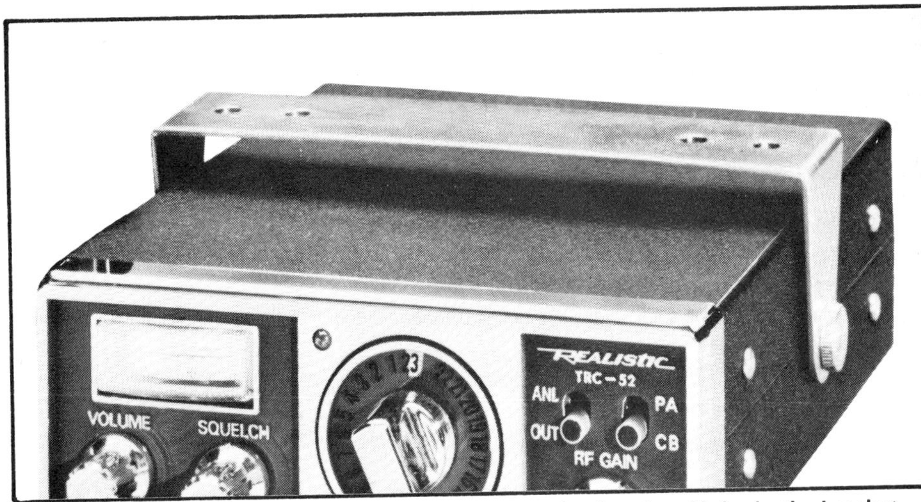

Most mobile CB rigs come with a U-shaped mounting bracket. Holes in the bracket are for screws to mount it.

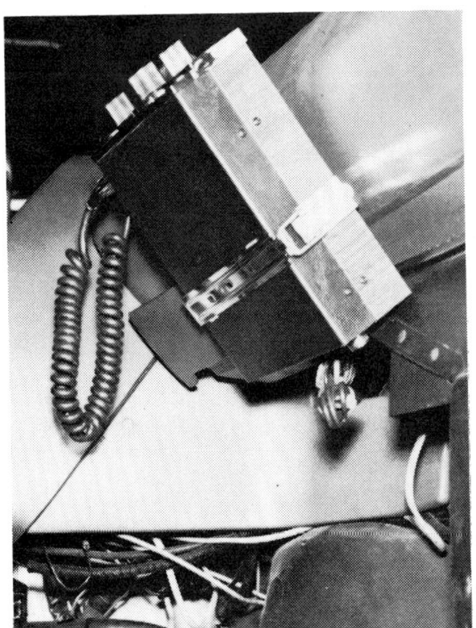
Some U-shaped brackets hold the transceiver firmly with a metal strap. The buckle holding the strap of this Pace under-dash mount opens for speedy removal of the set once wires to the rig are disconnected. The buckle can be secured with a padlock.

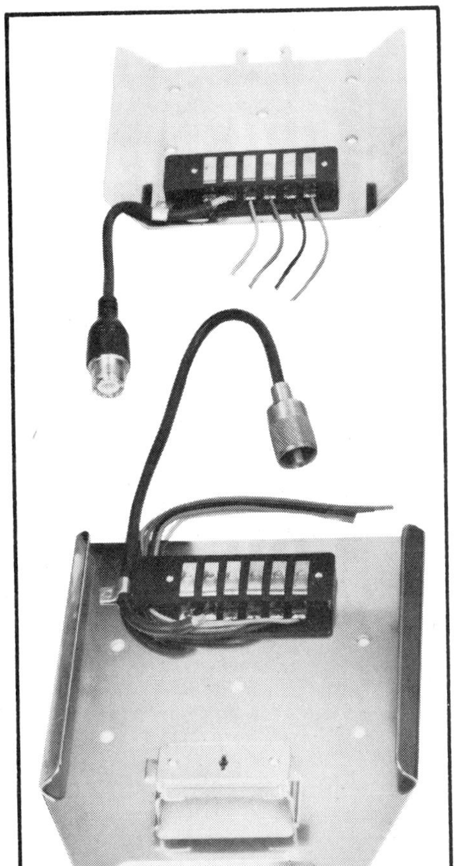
Slide brackets with wiping contacts for all connections except the antenna fit between the U-bracket on the CB set and the vehicle mounting surface. This one from Magnadyne has a lock. You can purchase extra slides so one CB may be moved among several vehicles.

WHERE TO MOUNT A MOBILE CB

Mounting means attaching the set to the dash, floor, transmission hump, roof-beam, glove box, floor, or interior of a trunk. I do not advise simply *placing* the rig on the seat beside you, on top of the dash or on the floor. It's hard to use the controls of a set that slides around. Worse—if you hit the brakes hard your CB set will continue to do the double-nickle until it is stopped by you, your passengers or maybe the windshield. Secure the set in a mount so it can't move around.

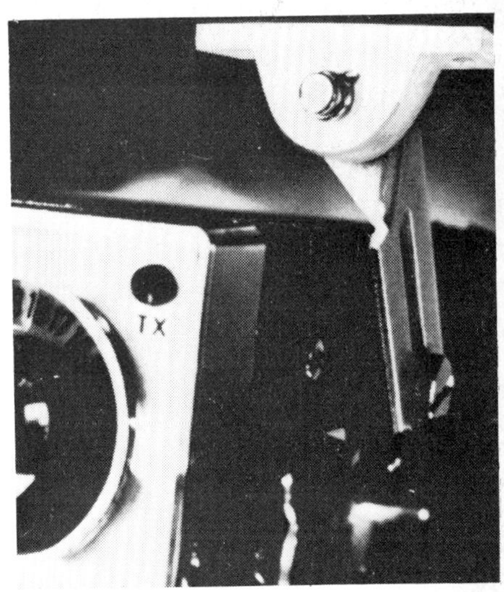

Spring-loaded pins that snap into under-dash hanger blocks make a sturdy mount allowing easy rig removal if you put quick-disconnects on the power leads. This BSI mount is a good idea for rough-riding vehicles.

IN-DASH MOUNTING

You don't use scarce dashboard real estate with an in-dash mount. This AM/FM/Tape/CB unit by J.I.L. blends with the vehicle's interior trim.

Choose among sets made specifically for in-dash mounting. Usually these are combination AM/FM/CB units designed to replace your present auto radio. Most have minimum control features for CB. You can expect a channel selector, on-air indicator, squelch control and audio volume control.

There's really not much room on the front-panel of an in-dash set but some designs make indicators and controls do double-duty. In the CB mode, a local/distance switch for FM becomes an rf gain switch to prevent overloading by nearby CB stations. A stereo indicator light serves as the on-air indicator. With luck you'll get a tiny meter which you might even be able to see if it's mounted vertically and not too recessed. Speakers will be separate, wide-range units intended for music. Under noisy conditions they won't be quite as intelligible as speakers restricted to voice frequencies.

Safety—With an in-dash set, you won't have to worry about banging your knees or catching your clothes on a protruding rig. But if you have to take your eyes off the road and bend over to check the channel setting or other controls, you're asking for trouble. When you shop, stand to the left of the set and look down on it from about the distance your head would be when you are in the driver's seat. If you can't read the controls, keep on shopping. Some in-dash combinations have remote controls and indicators in the microphone.

Convenience—Any in-dash CB is almost a permanent part of the vehicle, ready to use when your antenna is in place. The mounting uses no additional dashboard space—an increasingly important consideration as autos get smaller. Auto radios have gotten smaller for the same reason, so the CB combination designed to go into the same space has tiny knobs that may be hard to see and manipulate. The type with remote controls in the mike may prove easier to use.

Theft-proofing—Mounting a set in the dash discourages all but the most determined CB snatchers. Some will wreck your dash to get the set out in a hurry. Most insurance companies recognize in-dash mounting as "permanently affixed to the vehicle" or similar fine-print legalese which gets coverage under an auto-theft policy clause. But some insurers won't cover a CB under any circumstances unless you pay for a specific rider on your policy. Read your insurance policy and its endorsements carefully. If you are puzzled—and you will be—ask your agent. If he sounds puzzled, ask him to put it in writing. That way he'll be sure to check with the company he represents before committing to you.

Installation and Servicing Ease—It's trickier to put a set in the dash than any other location—even if it's an exact fit. The problem is working in the crowded space behind the dash unless there is a section that pulls out. If a rig is hard to put in, it's hard to take out for service.

UNDER-DASH MOUNTING

The most popular location for add-on CB installations is under the dash. You *should* be able to reach it without leaning over or taking your eyes from the road for more than a second or two.

If you have room under the dash you can mount any set you like. Fortunately, CB transceivers come in all shapes and sizes. Some have their narrowest dimension across the front which helps fit them into today's compact cars. Built-in speakers are most often on the bottom of the set and squirt the sound at floor mats and feet. *10-9, good buddy, you were stepped on.* Speakers on the left side of sets aim somewhat crisper sound right at the driver. A few sets have the speaker on top, which is like shouting into a cave under the dash. All these problems can be solved with an external speaker mounted so you can aim the sound where you want it.

Safety—Protruding knobs and sharp edges are a hazard in any panic stop or crash. Any mounting position requiring the driver to reach, bend or twist while driving reduces safety. If a quick-disconnect mount is used, be sure power contacts are not exposed if the vehicle is operated with the CB out of its mount.

Convenience—Mount the set so it's close to the driver and angled upward with controls easily seen and reached. Finding such space in compact cars is more difficult because you must allow room for the driver's feet to move without obstruction to operate accelerator, brake and clutch pedals.

Theft-proofing – An under-dash mount is a cinch for any hard-working thief. Locking mounts just slow up the process. Insurance companies do not regard CB sets in an under-dash mount as a permanent part of the vehicle no matter how securely the set is fastened. It probably won't be covered under auto theft provisions.

Installation and Servicing Ease—If you can find existing screws in the right places under the dash for a convenient mounting bracket, installing and removing a CB rig is a snap. If not, it helps to be a contortionist. Caution: The trend toward plastic parts in the dash brings two problems. First, you're better off mounting the bracket to sturdier metal. Second, you can't depend on a proper ground to the vehicle frame if power comes through a single wire. Most sets come with two power leads, so you can ground one directly to the car frame or at the battery, as I'll explain shortly. Drilling holes for under-dash mounts requires care so you don't penetrate ducts or cables.

DASH-TOP MOUNTING

Dash-top mounting won't obstruct vision from the driver's throne in this motor home. But don't leave a goody like this Kris XL-23 where CB snatchers can window shop.

In advertisements and publicity photos, you see CB rigs on top of dashboards because it's nice to put the product where it will show. Mounting holes in the dash-top may blow your chances of getting a good price for the vehicle when you sell it or trade it in.

Safety—Unless you have a very thin set, it may obstruct vision. You also defeat the purpose of a padded dash by placing a hard metal box with sharp edges where your forehead can hit in a panic stop or crash.

Convenience—Mounted just a bit below eye level, you can easily see all the controls, dials and meters, especially if you angle the set toward the driver. Daytime light levels may be so high at dash-top that you won't be able to see illuminated black-out dials or digital indicators.

Theft-proofing—Your rig will catch a crook's eye for sure. Dash-top is no place for a *permanent* installation unless there is a compelling reason for it.

Installation and Servicing Ease—Very easy to remove for servicing. Power leads and antenna cable will come through holes in the dash. A quick-disconnect mounting bracket makes a neater installation without wire-ends flopping around when the set is out of the vehicle. Be sure not to leave live electrical contacts uncovered or they may be accidentally short-circuited. In sunshine, top of the dash can reach the temperature limit for survival of electronic gear, so you may be taking in a well-cooked set for frequent service. A top- or side-mount speaker in the set is preferable. If you use a bracket to raise the set a couple of inches above the dash so you can hear a bottom-mounted speaker, your vision is further obstructed. Use an extension speaker instead.

TRANSMISSION-HUMP MOUNTING

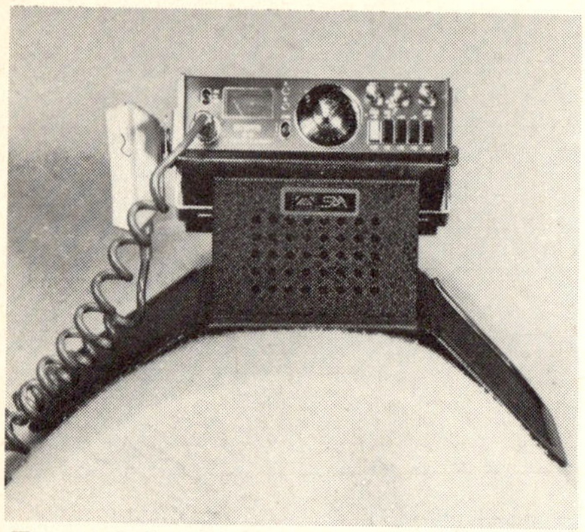

Transmission-hump mounts run from plain to fancy. This SMA mount sticks to the rug with Velcro which makes it easy to install without risk of drilling into the transmission. For some types of mounting, the CB set is grounded to the car through the mount. Here, it is not and a separate ground wire may be needed.

The transmission hump between the front seats is popular for lock-mounts and quick-disconnects. Stick shifts and sports-car consoles may prevent using this method.

Safety—This mounting position is safe if both you *and* the set are securely fastened. Wear your seat belt. Some mounts merely rest on the hump like a saddle—bolt them in place.

Convenience—Unless the floor-level set is tilted almost vertical, it will be hard to see dials and indicators. Even then you'll have to crane your neck a bit. With floor space used by a CB rig, say goodbye to the possibility of a third passenger in front unless he fits a kiddie seat.

Theft-proofing – A hump-mount won't safeguard your set for long, even if it locks. There's lots of room to pry with a crowbar.

Installation and Servicing Ease—Drop-on saddle-mounts without bolts are easy as pie to put on the hump—but they are a safety hazard. A theft-resistant mount bolts through the floor on both sides of the transmission hump. Installation is not for amateurs who don't know autos. One became famous by drilling 16 holes into the transmission casing. In favor of this location, you do have easy access to the rear of the set. Obviously a top or left-side-mounted speaker is best here.

ROOF-BEAM MOUNTING

There's that Kris XL-23 again, hanging by its thumbscrews from a U-bracket attached to the roof of an 18-wheeler. This location is convenient for driver and helper.

Vans and trucks have lots of headroom for attaching a CB to the roof-beam above the windshield. If you put really fancy equipment up there, you get the joyful feeling you're piloting a 747 jumbo jet. Howboutcha, Boeing?

Safety—The front edge of a CB is no substitute for a padded sun visor or upholstered roof-beam. Fasten seat and shoulder belts if you favor this mounting location. No driver deserves the distraction of a swaying mike cord dangling from above. Hang the mike cord out of the way or use a very short coiled cord.

Convenience—Reaching up to the controls is easy for the driver—and passenger, too—if the set is mounted in the center. The front panel of the set may have to be angled downward to see the indicators and meters.

Theft-proofing—A quick look is all it takes to spot a CB on the roof-beam. Smart snatchers know a roof-beam enclosure is strictly cosmetic and they'll have your set in a jiffy. Better to mount the set on a quick-disconnect bracket and take it with you.

Installation and Servicing Ease—You'll have a lot more fuss with power and antenna leads to the roof-beam location. Behind the covering of the windshield posts is a handy duct for wiring on some vehicles. It's accessible by removing a few screws. Bottom-mounted speakers work fine. Any enclosure to provide a built-in look needs a speaker grille.

GLOVE BOX

Would you believe my Mazda RX-2 has room for an SBE Brute, cassette player, tape library, weather radio, tire gauge, maps—and one cough drop?

As vehicles get smaller, so do glove boxes—with a few exceptions. Obviously you need a small transceiver to begin with. Losing the storage space is the price you pay for a concealed CB—when the glove-box door is closed. But when was the last time you saw a pair of gloves in a glove box?

Safety—Leaning over from the driver's seat to see indicators or adjust controls is just plain hazardous when driving. If you see a vehicle weaving down the highway it may not be a drunk but a fellow-CBer fiddling with a glove-box rig.

Convenience—You'll have to operate with the glove box lid open, an inconvenience to your front-seat passenger. An external speaker is necessary unless you want to cut a hole in the bottom of the compartment so you can hear.

Theft-proofing—Keeping your CB out of sight is good protection—unless you leave a tell-tale mike hanger on the dash or your antenna is a tip-off. Even so, if the bad guys can't see your set they may think you took it with you and they won't bother to break in. Locking your glove box door is some protection against theft, but not much.

Installation and Servicing Ease—Glove boxes are often shallow and rarely rectangular. Make measurements to select a set that fits—and watch those slanting surfaces! Allow 1-3/4 inches extra depth for the antenna connector. The compartment is usually plastic or fiberboard. Sometimes you can get needed clearance by cutting away part of the liner. You may need to use a reinforcing strap or bar on the outside of the box—top or bottom—to attach a mounting bracket. This requires removing the compartment.

One alternative to lots of dismantling and drilling is to use double-faced adhesive mounting strips. Hardware and art supply stores sell them for picture mounting. I find Scotch Brand or Magic Mounts have the best adhesive properties. Press the strip on the bracket, remove the protective paper and press the bracket in place. Mount the bracket on the glove-compartment floor so the weight of the CB rig will help keep it in place.

The bracket may take up more room than you have available. If you can fit the set on the compartment floor without a bracket, you can secure it with adhesive-backed Velcro® tape. Attach hook-tape Velcro to the bottom of your transceiver and attach the loop-tape Velcro to mating positions on the glove-box floor.

Removal of a bracket-mounted set for servicing may require turning knurled attachment screws with your fingers—frustratingly difficult in a confined space. The Velcro fastening used without a bracket is simply peeled apart by lifting up on one side or one corner of the set.

If you are not familiar with Velcro tape, more information is given a little later.

UNDER-SEAT MOUNTING

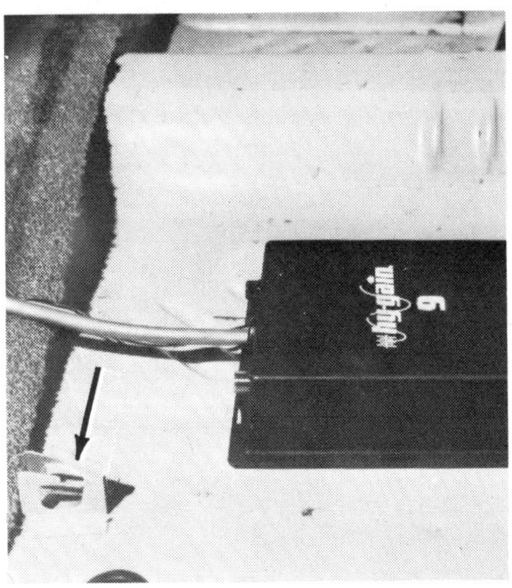

Under the front or back seat is a good location for a remote-control set such as this HY-Gain 9. Notice the seat clip (arrow). To remove seat, you have to defeat these little clips.

Even a thin set may not fit under the seat of a small car, but most will fit in standard cars. Measure before you buy. Built-in speakers are muffled in this location. *10-9, good buddy, you were sat on.* Better to use an external speaker permanently mounted under the dash near the firewall where it won't be seen easily.

Check the location of the exhaust system to avoid putting the CB where it will be cooked from below. Any exhaust system and muffler gets hot. Late models with catalytic converters under the floor get much hotter. In some cases there is insulation under the floor mat to keep the mat from catching fire. Use the other side of the vehicle if possible, or find an alternate location, or use a different type of CB unit.

Safety—Any CBer who drives with his head between his legs deserves what he gets. *10-96.*

Convenience—You won't be able to get at the controls of standard sets. With a remote-control unit you won't have any problem.

Theft-proofing—Hide the power and antenna leads under the carpet so they don't leave a trail to your set. Shrewd parking attendants will know where to look anyway—they read the CB ads too. A microphone, mike hanger, or visible CB antenna is a tip-off, but casual thieves usually won't go after a CB they can't see so you are better off with a concealed set.

Installation and Servicing Ease—It's simple to slide a set under the seat, but hard to keep it from sliding around when the vehicle is in motion. You'll have difficulty installing a mounting bracket without removing the seat, which is hardly worth it to some people. Here's another way:

With the seat in the normal driving position, place the CB set underneath and mark the location on floor or carpet using a grease pencil. If the floor is carpeted, you can usually move the seat to one end of its travel and pull the carpet out from under. Sew, pop-rivet or staple the Velcro to the carpet along the front and one side of the CB location. If there is no carpet under the seat, roughen the surface and use self-adhesive Velcro or apply it with epoxy glue.

Then fasten adhesive-backed Velcro hook tape to the bottom of the transceiver so it will mate with the tape on the floor. Press the CB into place. It should not slide around and is easy to remove for service.

TRUNK-MOUNTING

A vehicle trunk is really out of sight for a remote control rig. This is the preferred mounting location—where it won't get clobbered by objects rolling around the trunk or be rained on when the lid is up.

Only a remote-control rig will do. This location is especially worthwhile in subcompact cars and sports cars where passenger space is at a premium.

Safety—A remote-control mike puts every CB function at your fingertips. You can hang the mike where it can be seen and reached easily while driving. I prefer it near the top of the dash, just right of the instrument cluster. There's no hazard to knees and shins from a set in the trunk.

Convenience—You or your passengers can use the transceiver with equal ease. No interior space is taken by the CB set, so you have more room to carry other things or stretch out and relax.

Theft-proofing—Out of sight, out of crook's mind—unless you've left other clues. Use a disguise antenna or a removable one. Hide or disguise the otherwise prominent mike hanger; hang a small clip board in place of the control head. Put the control-head socket where it won't be noticed. Take the remote-control head with you instead of

TIPS ON USING VELCRO® FASTENERS

I've suggested Velcro as a handy way of fastening CB rigs, brackets and mikes. Velcro consists of two woven tapes—one covered with tiny hooks and the other covered with tiny loops. When pressed together, they cling very strongly because the hooks interlock with the loops. However, they can be separated conveniently. Velcro can be pressed together and separated again many thousands of times without deterioration, so don't hesitate to use it for a mike hanger if you want to.

You'll find Velcro strips and tabs at sewing supply counters, RV supply stores and marine supply stores. The strongest tapes are sold in marine and RV stores and I recommend them instead of the type used for sewing clothing.

Properly mated, marine Velcro holds about 15 lbs. per square inch. Here are five ways to mate it in order of decreasing holding power:

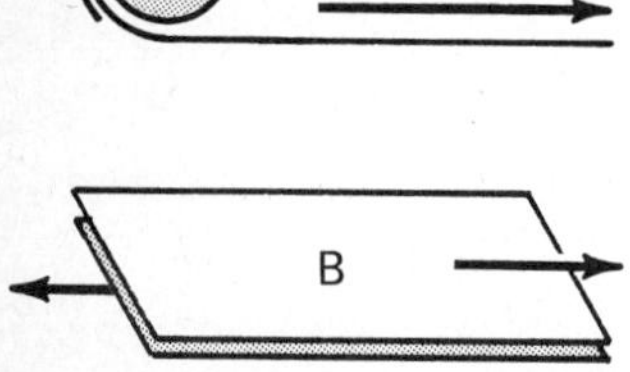

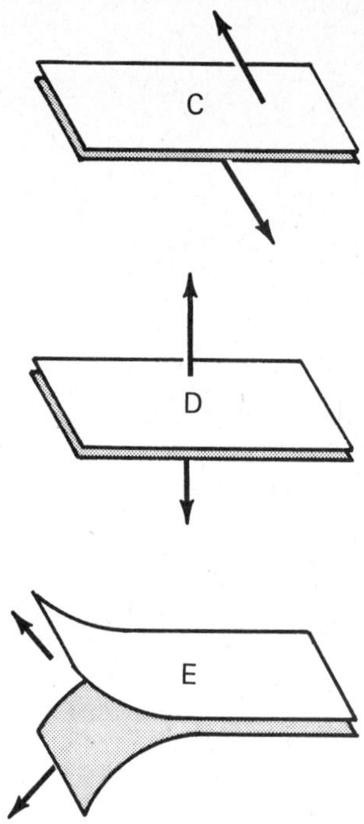

When sewing Velcro, for example, to a carpet for under-seat mounting, use Nylon or polyester thread and stitch along *all* edges to form a box.

Velcro is available with precoated adhesive backing, activated by methyl ethyl ketone (MEK) or an adhesive with an MEK base such as Velcro V40 adhesive or Goodyear Pliobond. Be sure the surface for attachment is free of dirt, oil or detergent, and is dry. Wet the back of pre-coated Velcro with any of the listed solvents or cements. When it is tacky, press in place for about 1 minute.

stowing it. Of course, your trunk can be pried open, but there's no point in stealing half a CB rig.

Installation and Servicing Ease—The cable for a remote-control rig is thick and the connecting plug for the control head is thicker. You'll find it hard to feed into the passenger compartment and difficult to run the cable underneath the sills at the edge of the carpet. You may have to drill a hole from the trunk into the passenger compartment to push the fat plug through. If you have a long cable run to the trunk or if you want to move a trunk-mounted remote-control rig from vehicle to vehicle, get extension cables for the "umbilical" cord from your CB dealer. Leave an extension cable permanently in place in each vehicle with the rear socket accessible in the trunk. A Velcro 36-inch tie-down strap, usually sold for RV's, will keep the set from sliding around.

For a more permanent installation fasten the transceiver to a strut with through-bolts or to the floor with sheet metal screws and lockwashers. Make certain you don't drill holes in your gas tank. Some trunks are lined with fiberboard, tempting you to mount the transceiver vertically. Don't do it. A CB is too heavy and eventually will tear loose.

HOW TO INSTALL IN-DASH CB

1. Get your tools together. You'll probably need a flat-bladed screwdriver, perhaps a cross-head screwdriver, pliers with wire cutter, adjustable wrench, wire stripper.

2. Disconnect the battery ground lead, the one leading to a bolt on the engine block. If you have trouble removing it, or you're a bit leery of working with electrical wires, read the section about how to make power connections farther along in this chapter.

3. Reach up behind the dash and disconnect the wiring of the auto radio you're about to replace with an AM/FM/CB unit. Scrunch your head beneath the dash and use a flashlight, or a trouble light on an extension cord. The antenna is the thick wire plugged into the set. Disconnect the speaker leads; they're usually on a little round plug. Trace the power lead from the set to the nearest connector; it may be to another wire or to the fuse block. There may be another lead that dims the radio dial light as the dash lights are dimmed; disconnect it.

4. Pull the knobs off the radio. Remove the hex nuts on the control shafts. Remove the trim plate; it may be in front of or behind the hex nuts and any washers on those shafts. Put your head under the dash again and remove any bolts or braces holding the radio. Remove the radio—you may have to wriggle it out from the front or from the back. The parts you removed should be put in a plastic bag and taped to your old radio. They make handy spares.

5. If the new set is to be installed from the back, you may find a plate in the dash with hole-spacing that doesn't quite fit the spacing between control shafts on your rig. Or you have an extra-large opening for front-dash installation which you fill with the back-up plate supplied with the set. In the first case, re-set the shaft-spacing on your new set. Loosen the *first* nut on each shaft and move the positioning washer along the horizontal slots. Correct spacing will allow the shafts to pass through the slotted holes in the instrument panel or back-up plate and be centered in the shaft holes of the trim plate matching your car. The set will come packed with trim plates for major automaker's vehicles and with a universal trim plate. Re-tighten the first nut on each shaft.

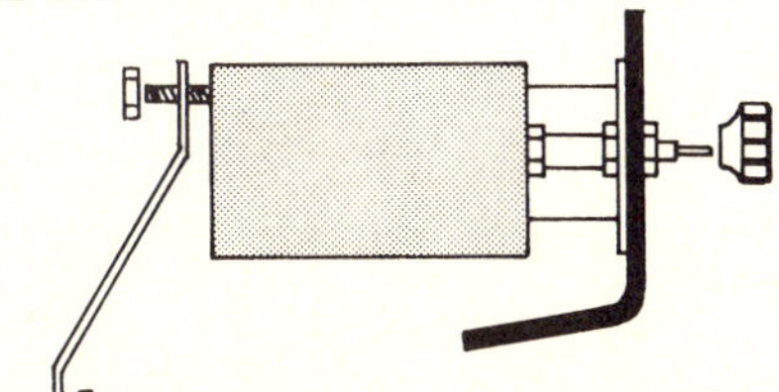
BEHIND DASH INSTALLATION

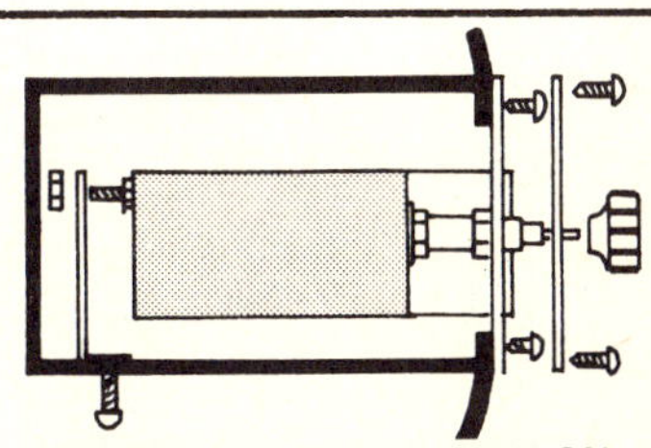
FRONT DASH INSTALLATION

In-dash mounting Step 4/There are only two ways to get a set in or out of the dash—from behind or from the front. Look behind the control knobs for shaft nuts.

6. Now check the shaft length. Test by loosely assembling all the plates and trim hardware that fits over one shaft. Do the knobs fit back on? You can usually adjust shaft length by loosening the *second* nut on each control shaft and gently pushing or pulling the control. Tighten the second nut again. Remove plates and hardware.

7. Work the set into the dashboard opening, from front or back as required. Don't use brute force. Be careful of that maze of wires in the area behind the opening; you don't want to yank any of them loose or parts of your car won't work.

8. On the back of the set you'll find a sturdy threaded shaft with nuts and washers on it. Remove the hardware from this shaft and slip one of the lower holes of a support strap over it; tighten by replacing the washer and nuts. As the drawing for Step 5 shows, a support strap is a piece of metal with lots of closely spaced holes. Bolt the other end of this strap to an *un-insulated* part of the dashboard. If you can't find an un-insulated spot that's convenient, fasten the support strap anywhere so it mechanically secures the rig. Then run a separate ground wire from the threaded shaft on the set to base metal on the car.

9. Connect the red power lead from the set to the same place the old radio got power. You may have to use one of the connectors packed with the set to match the vehicle connectors, or get one to fit from an auto or electronic parts store.

10. Connect the CB antenna and the AM/FM antenna lead you removed from the previous set. Connect loudspeaker leads according to the set manufacturer's instructions. The wires are color-coded and also marked on some sets with plus and minus. Don't mix up polarity; most in-dash combination rigs are also stereo. Re-connect the battery.

11. Put all trim and hardware on the front of the set. Follow the precise order given in the set manufacturer's instructions. On some sets, adjustment of shaft length may cause an erroneous setting of the CB channel indicator. Check the manufacturer's instruction book for this possiblility; fix as instructed by manufacturer.

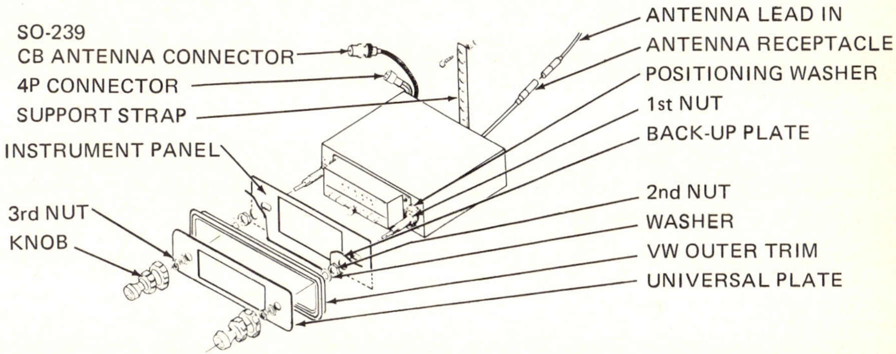

In-dash mounting Step 5/This exploded view of a J.I.L. JN852CB is typical of most mounting arrangements. J.I.L. pioneered in-dash CB. We field-tested this one in Chapter 8.

HOW TO INSTALL CB UNDER THE DASH

1. Get all your tools together before starting: Electric drill, center punch, 5/32" and 1/8" drill bits, tape measure, pencil, hammer, wire stripper, soldering iron, *rosin-core* solder.

2. Look over the vehicle and plan the *complete* layout. See Chapter 12 for details of antennas and their installation.

3. Temporarily attach the mounting U-bracket to the top of the set. Hold set in position under the dash at the mounting location you plan to use. Is the rig back far enough from the seat so it won't gash shins or tear hosiery? With a pencil, mark the U-bracket location.

4. Remove the U-bracket from the set and hold it in position under the dash, as marked, and mark the location of the bracket mounting holes on the dash.

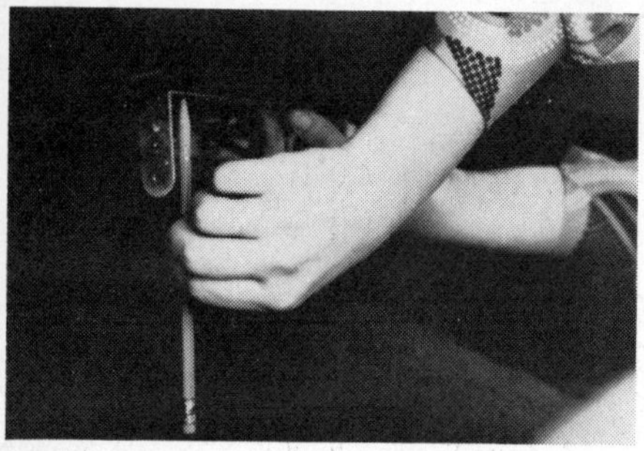

Under-dash mounting Step 4/Mark the holes for drilling by using the bracket as a template. Hold it firmly and mark carefully.

5. Feel behind the dash in the vicinity of the mounting location to see if the drill bit may chew into ducts or wiring. If you're the least bit unsure, scrunch your head under and take a look. If there is interference, relocate the set. If not, centerpunch the holes to be drilled.

Photos courtesy of E.F. Johnson and Hy-Gain

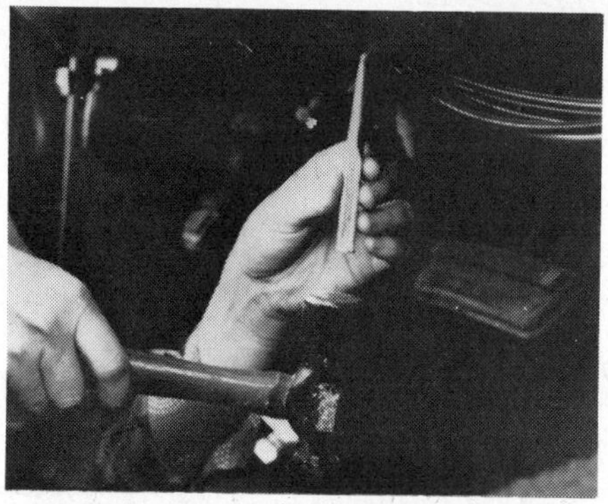

Under-dash mounting Step 5/Centerpunch before drilling so the drill doesn't slip and mar your dash.

6. Drill each hole using the bit size suggested by the set manufacturer for the mounting screws packed in the transceiver carton.

Under-dash mounting Step 6/Drill each hole carefully after checking to be sure your drill bit won't penetrate wiring or ductwork once it gets through the metal dash.

7. Fasten the U-bracket firmly to the dash, using screws furnished by the set manufacturer. If they are not Phillips-head type, consider substituting Phillips-head screws of the same size and type. This will reduce the chance of your screwdriver slipping.

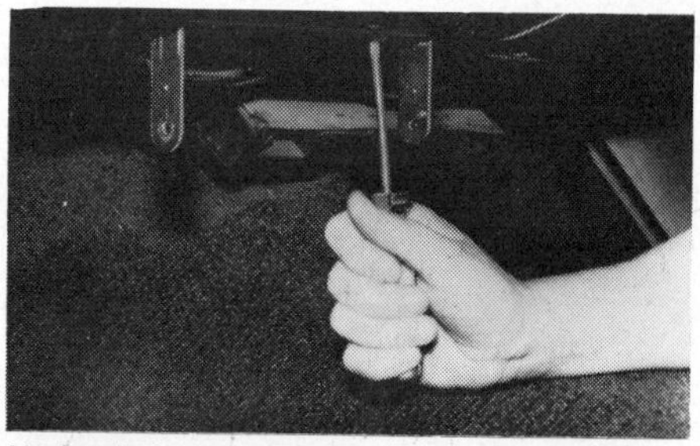

Under-dash mounting Step 7/Fasten the U-bracket to the dash. Most manufacturers supply sheet-metal screws. Four screws are better than two. Don't skimp.

8. Route the coax cable from the antenna behind the dash to the transceiver mounting location, allowing about a foot of slack. Use plastic ties at several points behind the dash to prevent the coax from dangling.

9. Disconnect the battery ground cable.

10. Connect the red power lead from the set to the vehicle electrical system as described later in this chapter. Connect the black lead to vehicle ground. Read the set manufacturer's instructions carefully to be sure you don't hook up with reversed polarity. Use plastic ties to keep the power leads from dangling.

11. Connect the antenna lead by firmly seating the coax connector in the socket of the set and tightening the locking collar. Mount the set in the U-bracket.

12. Reconnect the battery ground. Turn on your rig and adjust for lowest SWR with the set's meter or an accessory SWR meter as described in Chapters 11 and 12. It should read better than 2.5:1.

HOW TO PUT A CB RIG ON YOUR MOTORCYCLE

1. Gathering tools won't take long. Just one small, flat-bladed screwdriver.

2. The Beltek CB set-up is simple. You wear an earphone inside your helmet. The mike is on a boom attached to the helmet. The CB set is strapped on the gas tank. The push-to-talk switch is clamped to a handlebar. The antenna is fastened to the vehicle frame toward the rear.

3. Strap the mounting bracket loosely over the gas tank taking care not to cover the filler cap with the belt fastener. To fill the tank later you'll have to lift the transceiver off the mount. Position precisely and then push down on the mount with even pressure all around to make the four suction cups adhere to the top of the tank. Tighten the strap around the tank.

4. Attach the remote control switch to the handlebar and tighten its clamp.

5. Attach mike-earphone assembly to helmet. Plug cord into transceiver.

6. Clamp antenna to frame and connect to transceiver. Be sure antenna base is grounded to frame.

7. Connect power leads to battery terminals, observing polarity. A CB rig like the Beltek Enduro shown here has lots of noise-control circuitry because it's right on top of all that ignition wiring. Connecting the set directly to the battery helps suppress noise.

8. Plug in battery/remote-switch cord to transceiver.

9. Adjust for lowest SWR as described in Chapters 11 and 12.

Motorcycle mounting Step 6/With this Beltek Enduro 23, 2-wheelers have the same CB benefits that 4- and 18-wheelers enjoy. After mounting the antenna, secure cable under or along the lower edge of seat.

HOW TO INSTALL REMOTE-CONTROL CB IN YOUR TRUNK

1. Gather the same tools required for under-dash installation.

2. Look over your vehicle to plan the system layout for trunk-mounted CB. See Chapter 12 for antenna details and installation. These instructions are for the Hy-Gain **9** as shown in the photos, but apply to most rigs with full remote control.

3. Mount the transceiver in the trunk where it won't be harmed by luggage, loose objects or a leaky lid. Just back of the rear seat partition is good. Use the transceiver case as a template to mark the mounting holes. Cut holes in the carpet where you'll drill, otherwise carpet threads wind up on the drill bit.

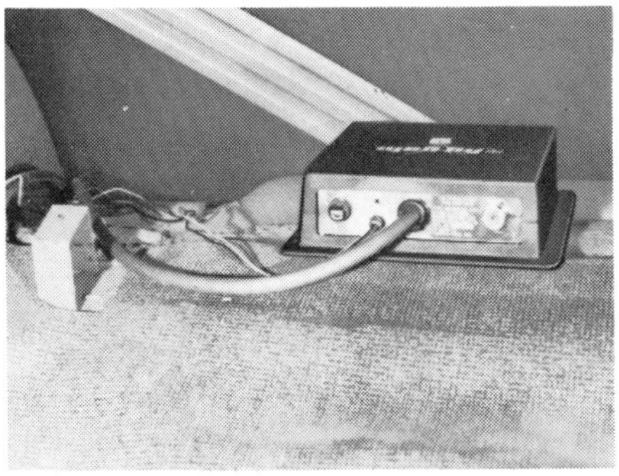

Trunk-mounting Step 3/The closer you mount the transceiver to the rear-seat back, the safer it is from damage.

4. Centerpunch the marked area to prevent drill from wandering or slipping. Use drill bit size recommended by manufacturer for the mounting screws supplied. Drill carefully and hold the drill straight.

5. Fasten the transceiver firmly with at least 4 screws.

6. Remove the back seat. It may be held by bolts or catches. Put your head down at floor level to see. Some catches require pressing down on the top front edge of the seat with one hand while pushing the bottom toward the trunk and then lifting. Do this as close as possible to each catch.

7. Route the cable from the transceiver through any available opening revealed by removal of the rear seat. See Chapter 12 for a blow-by-blow description of this procedure.

8. Remove door sills and peel back carpet edges along cable route through vehicle. Avoid passing over exhaust system or catalytic converter—too much heat. You'll need an extension umbilical cable for long runs; put the connection of the two cables under a seat where it won't be damaged. Bring cable end to dash where you'll plug in remote-control head. I advise using the bottom edge of dash. The connector is too visible to thieves when mounted on the console. The forward lower edge of the front seat seems convenient but you'll flex and wear the cable every time the seat is moved.

9. Install the connector on the end of the umbilical cable. Control head plugs into this connector. Secure any dangling loop of umbilical cord to a point under the dash using plastic ties.

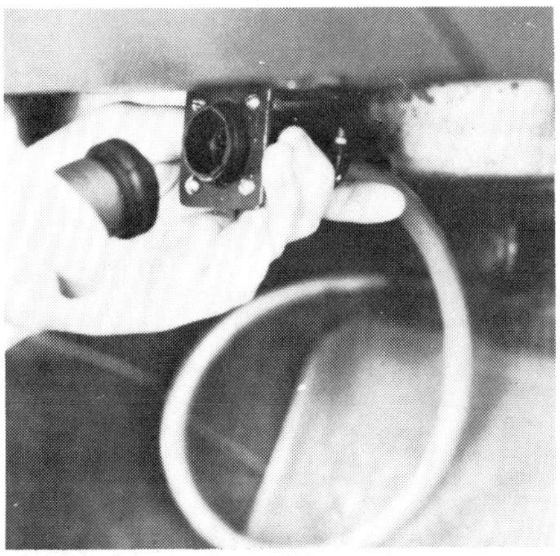

Trunk-mounting Step 9/Mount the connector for the control head so it's out of sight but accessible. To plug the remote-control cable into the receptacle, line up the pins and slots.

10. Disconnect battery ground cable.

11. Connect power leads from set following manufacturer's instructions. For trunk mount you may have to extend the leads that come with the set. Use black and red # 16 or # 18 stranded primary wire. Route along sills to fuse block under dash or through firewall.

12. Reconnect battery.

13. Install hanger for remote control head on dash close to driver. Choose a location high enough to keep cord from dangling near driver's feet. When drilling for screw-holes check to see what the drill bit could bite into behind the dash. You can use Velcro tabs instead of the hanger furnished with the set.

Trunk-mounting Step 13/Locate the remote-control head high on the dash where you can see the indicator and hear the built-in speaker. This dash is padded, so I'd glue on a pair of Velcro tabs for a mike hanger.

14. Connect antenna. If it's on the forward cowl or fender, the coax cable can be brought back to the trunk along the umbilical cable route.

15. Put back carpeting, sills, rear seat.

16. Connect control head. Turn on rig and adjust for lowest SWR as described in Chapters 11 and 12. Because meter will be connected to rig in trunk, you'll need a helper to key the transmitter while you take a reading. Don't step on other stations when testing.

BRIEFCASE UNITS

Some owners put CB sets in carrying cases so they can be quickly moved from vehicle to vehicle or used in rented autos. They can also be purchased in cases, sometimes with other electronic gear included, such as a tape player. There isn't any handy way to hold these in place on the seat or floor although if the passenger's seat is unoccupied you can sometimes use that seat belt to tie down your briefcase CB.

If you use this type of unit, place it so you can reach the controls without leaning over and so you can read the dials without taking your eyes from the road for a long time. The best set for use in a briefcase is the type with controls in a remote head, along with the microphone/speaker unit. Hung on the dash, the remote-control unit is as convenient as any other set.

Normally these units get power by plugging into the cigarette-lighter socket on the dash. Don't use a cheap cigarette-lighter plug. Good ones cost only a little more and make a reliable connection.

WHERE TO CONNECT FOR ELECTRIC POWER

Common places to connect the power wires from a CB in an automobile are: The car battery, the ignition switch, the fuse block, and a plug fitting the cigarette-lighter socket.

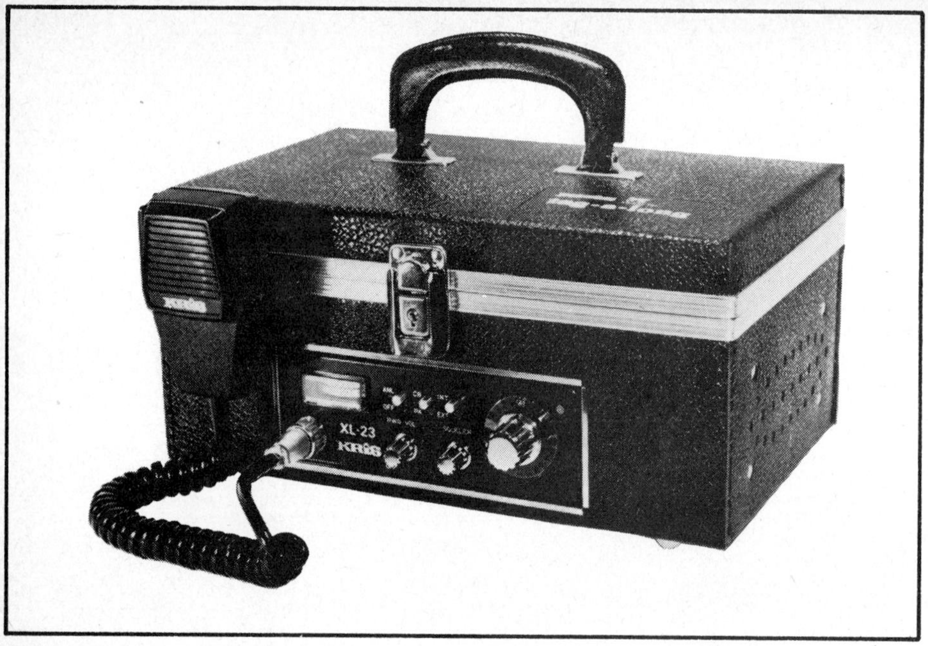

Kris packages some mobile rigs in Tag-a-long cases. There's room inside for a speaker, antenna, microphone, and a peanut-butter sandwich.

Permanent mobile installations are best wired so the set is connected directly to the vehicle battery. As you'll see in Chapter 18 on coping with noise and interference, this method greatly reduces the amount of noise you hear in your CB speaker—caused by electrical noises in your own vehicle—such as ignition.

Connecting direct to the battery also makes your CB useable in an emergency without your having to turn the ignition key to get power. A disadvantage of this method is battery drain when you accidentally leave the CB set turned on. This won't hurt the battery for short periods—even overnight if your battery is in good condition. Over a weekend, you'll be late to work Monday morning.

Normally CB sets have their own fuses, usually in a holder in one of the power wires for the set. This fuse should furnish adequate protection so a direct connection to the car battery is satisfactory.

Connecting to the ignition switch is difficult and is done only to wire the system so your CB is inoperative when the switch is turned off—which I think is the wrong way to do it.

Connecting to the fuse block in the car is usually easy but your set will be quieter if you take the time to run the power wires all the way to the car battery.

NEGATIVE GROUND OR POSITIVE GROUND

Check the specs or instruction booklet. Some sets are designed to work with negative-ground car systems, some with positive-ground systems, and some will work either way. Some sets will be damaged if you connect them the wrong way; some will merely blow a fuse; some won't work but are not damaged.

How To Tell Which—If you still have the owner's manual for your vehicle, the electrical specifications may say whether it is positive or negative ground. Most autos today have negative-ground systems. To be sure, look in the engine compartment for the plus and minus signs on the battery. They may be stamped into the battery terminals or marked adjacent to them. Trace the cable from the negative terminal. If it goes to a bolt on the engine block, you have a negative-ground system. If not, the positive cable will be fastened to the engine block and you have a positive-ground system. Don't bother to trace the other lead. It may disappear into a maze of wires.

HOW TO MAKE POWER CONNECTIONS

At one time or another, you have probably seen somebody make a mistake when working around the electrical circuits or battery of a car. Sparks fly, along with bits of molten metal, and sometimes wires get so hot they burn up. If you worry about this happening to you, that's good. It probably won't if you worry a little. Here's a way to prevent it:

Finding the fuse block isn't always easy. With luck, yours may be on the firewall rather than up under the dash. In my Mazda, it's on the front of the dash under an easy-to-remove panel.

Find the heavy wire that goes from one battery post to a bolt on the engine block or car frame. This is the ground wire for the battery. Loosen the clamp bolt and remove the wire terminal from the battery post.

Gently wiggle the terminal back and forth while lifting. If the terminal remains clamped to the battery post, spread the connector slightly with a screwdriver blade inserted in the open part of the U through which the bolt passes. Corroded battery terminals may have to be pried up *gently* from the bottom with a broad-bladed tool. If the connection remains stubborn, get your local service station mechanic to use a terminal puller or battery terminal pliers so you don't damage the battery or its tapered posts. Some vehicles have spring-clamp battery connectors which can be released by careful squeezing with pliers; if you have difficulty, get help at the service station.

Once the battery connector is loosened by your friendly service-station attendant, have it replaced so you can drive home to finish installing your CB. It won't be difficult to remove a second time, at home.

Remove the battery ground terminal, at the battery, wrap it in a cloth and tuck it somewhere so it can't possibly move back and contact the battery post again. Now you can safely work on wiring without any chance of sparks until you reconnect the battery wire. The only thing you must be sure not to do is connect a wire between battery posts.

Are The CB Power Leads Long Enough?—If you haven't already, now's the time to decide where you will make the power connection for the CB. Look around the car and engine compartment and start thinking how you will run the wires.

If the wires supplied with the transceiver are not long enough to reach, add wire of *at least* the same gage. For short runs it's usually #16 or #18 stranded. Not just any wire will do, particularly if you're going to route it through the engine compartment. I prefer the larger 14-gage plastic-insulated "primary" wire—available in auto stores. Twist and solder the connection, wrap with tape and then tape or clamp the joint to something solid so it can't vibrate. Buy a length of red and a length of black wire so you don't get polarity mixed up. Be sure you keep the CB set's line fuse in the circuit when you add or change wiring.

HOW TO MAKE BATTERY CONNECTIONS

Run the power leads from the CB set through an existing hole in the firewall if you can find one. You usually can but, if necessary, drill a new hole being careful what you drill into. Usually, you can push wires through a grommet which is sealing another cable or control housing.

Double check to be sure the CB set has its own fuse so you can safely make a direct connection to the car battery.

You can run both power wires all the way to the battery and connect with proper polarity. Or, you can connect the grounded wire to metal on the firewall, engine or frame and run only the "hot" wire to the ungrounded terminal of the battery. The CB fuse should be in the "hot" wire whether the system is negative ground or positive ground.

Wherever you route wires, be sure they are securely tied down with clamps or tape so they don't vibrate and perhaps cut through the insulation by rubbing against metal parts of the car.

Get The Polarity Right—Check the installation instructions with your CB set and follow them exactly. You should already know which battery terminal is positive and which grounding system your car uses.

If your CB is built using the standard color codes for electronic equipment, the red power wire connects to the positive (+) side of the battery and the black lead goes to the negative (−) side.

With a negative-ground automobile system, the negative terminal of the battery is connected—"grounded"—to the engine and frame of the car. The black wire which is supposed to connect to the negative battery terminal can be connected anywhere on the metal frame or body of the car (not doors) and that is equivalent to connecting at the negative battery terminal.

With a positive-ground system, the same idea applies, but the colors change. If the positive battery terminal is grounded to the engine and frame of the car, connect the red wire from the CB set to the car frame.

Either way, the wire not connected to the car frame should run all the way to the battery and connect to the *ungrounded* battery terminal.

Prepare The Connection—Remove the insulation from the end of the power wire, unless it has already been done. Insulation can be removed with a pocket knife if you are very careful not to nick the wires while cutting through the insulation. If you do nick them they are likely to break at this point after you have driven the car a while.

Electronic supply stores sell small adjustable wire strippers that are inexpensive and handy. If you buy one, you will probably use it many times and be glad you did.

Solder a terminal on the end of the CB power lead. Use an O-terminal with a round hole in it, rather than a U-terminal or any other kind. Select a wire terminal with a hole in it large enough to pass over the bolt which clamps the battery cable onto the battery post.

Remove the nut on the battery clamp bolt. If there is corrosion with some white or green chalky material on the battery post and connector, make a solution of ordinary baking soda in warm water and wash off the corrosion. Use about 1 teaspoon of baking soda to

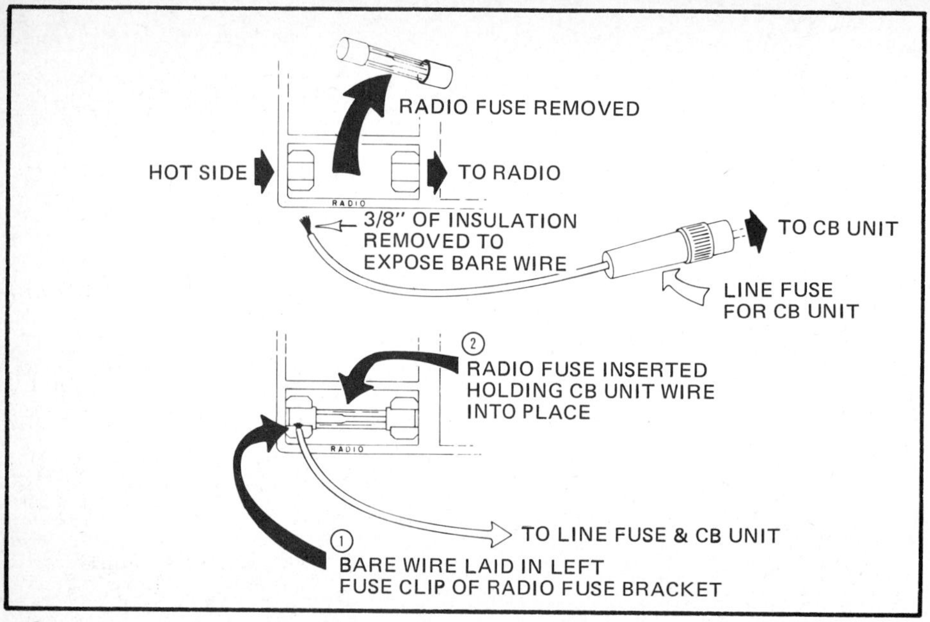

Connecting to the fuse block is easy. Disconnect the battery first so you don't draw sparks. Be sure the wire end is firmly held when you snap the fuse back in its clip. To be sure it won't be pulled out accidentally, tape the power lead from the CB set to the side of the fuse block or some other convenient point.

a glass of water. Wear old clothes and be careful not to get material from the battery onto your skin or clothes. It eats holes in clothing which appear later, after laundering. It may damage your skin, so wash it off promptly with soap and running water. If you get it in your eyes, get medical attention as fast as possible.

When most of the corrosion is removed from the battery connector, finish the job with a wire brush or sandpaper, giving a final rinse with the soda water.

Slip the O-terminal of the CB power wire onto the bolt, thread the nut back on and tighten.

Clean the other battery terminal the same way if it is badly corroded. Protect the CB power connection and the freshly-cleaned battery terminals from corrosion by liberally smearing with grease, petroleum jelly, or spray them with a commercial product such as CB Climate Barrier which I recommend for use on antenna connections also.

CONNECTING TO THE VEHICLE FUSE BLOCK

This is easy but you should be sure you connect on the correct side of the existing fuses in the car. Current from the battery flows to the fuse block and then branches and flows through several fuses leading to individual circuits in the car, such as the instrument lights. Each fuse is chosen so it will blow if the current goes too high in the circuit it feeds.

Each fuse in the fuse block therefore has a *battery side* where current comes into the fuse and a *load side* where current flows out to the electrical devices it protects. If you connect a CB on the battery side of the fuse, the current for the CB does not flow through the fuse—but it should flow through the fuse built into the CB power wire. If you connect a CB on the load side of a fuse, the current drawn by the CB must flow through the fuse in addition to normal current to supply the circuit in the car. The additional CB current may blow the fuse. CB sets draw about 0.5 amp when turned on but not receiving any signal and about 2.5 amp when transmitting with full modulation.

You'll find the fuse block on the firewall inside the driver's compartment, or on the underside of the dash, or in a compartment on the dash. If you can find an unused screw terminal on the battery side of the fuse block, you can make the most reliable kind of connection. Solder an O-terminal on the end of the CB power wire and attach it to the screw terminal on the fuse block. Finding such a handy screw terminal is unusual but check for it first.

Another way is to trap the bare end of the CB power wire between the metal end of a fuse and the spring clip which holds the fuse. First identify the battery side of the fuse. This is usually impossible by wire tracing and you may need a test light or voltmeter to figure it out. Tip out that end of the fuse; place the bare wire-end into the fuse clip; snap the fuse back in; hope the wire stays there a long time. Tie down the wire with clamps or tape so it isn't likely to get pulled out or jiggled loose.

The other lead from the CB can be connected under a metal screw on the firewall or into metal behind the dash.

DON'T TURN IT ON WITHOUT AN ANTENNA!

Resist the urge to turn the CB on to see if it works. You have one more important thing to do—connect the antenna. The set won't work, even on receive, without an antenna. If you accidentally switch it to transmit and don't have an antenna connected, you may cause expensive damage to the transmitter section of your brand new toy!

If you are shopping for a CB, the next chapter gives you our field-test results on a lot of different sets. This chapter is intended to help you decide on the kind of set and installation method so you will have that in mind while reviewing the performance data in Chapter 8.

If you already own a set and are using this book as an installation guide, read Chapters 11 and 12 to learn about mobile antennas and how to install them.

Our engineers begin a day of field-testing. At all test base stations, the standard reference transceiver was a Tram D 201 feeding a 5/8 wave Hy-Gain Super Penetrator.

CONSUMERS PERFORMANCE GUIDE – EXCLUSIVE FIELD-TEST REPORTS

You want to know what to expect of a CB set before you buy it, or whether the rig you have is up to specifications. You're awash in data about transceivers—manufacturer's specs, FCC type acceptance criteria, product reviews and discussions in magazines, and commercials on TV. Put all that together and you should be able to pick the best CB for what you want to do. But you can't.

Printed specifications and laboratory measurements won't tell you how a CB will operate in the real world. The CB world is cluttered with signals from other transceivers, man-made and atmospheric noise, much of it bounced around by reflections from vehicles, buildings and hills.

How well do CB rigs operate? The best way to find out is try out the sets you're considering. But few dealers will let you take one, and certainly not 3 or 4 rigs out for a comparison check in your vehicle.

Performance reports in this chapter make that comparison for you—probably considerably more precisely than you could without help. I worked with professional mobile-radio engineers from a well-known communications laboratory to make actual field tests of leading brands of CB sets.

The goal was to measure and rate CB sets for intelligible two-way communication with another station at various distances—which is what really counts. We tested in typical CB operating environments found in the city, suburbs and country. Here's how we made the tests.

WHO PICKED THE TRANSCEIVER MODELS?

Many CB manufacturers have a product line spanning the entire range of price and performance. We asked each manufacturer represented here to choose two mobile models: The best performance for your dollar—which we call *best bang for the buck,* and the best-performing unit regardless of price.

We asked some manufacturers to pick base stations, too, because base stations suitable for mobile use are sometimes used in large RVs and in deluxe cabin cruisers.

OUR PERFORMANCE FIGURES

We set up fixed base stations at standard locations for all communications to the sets in the field. Each base station, with its superior antenna could always communicate with all mobile sets even when some of those sets were on the fringes of their own transmitting or receiving range to the base station. Sets were evaluated at checkpoints near the fringe of communication range, so no set tested perfect.

The base station evaluated the intelligibility of communications received from mobile sets.

The mobile sets were evaluated on ability to receive intelligible communications from the same base station. This tested the two-way performance of transceivers.

We used figure-of-merit ratings

like professional communication engineers use to rate the quality of speech transmission. The word, *readable,* means understandable in the rating descriptions shown in the accompanying table.

The checkpoints had to be within the range where *two-way* communication was still possible even with the poorest performing transceivers. A checkpoint had to yield *at least* a merit number of 2 to be useful for testing. At the field checkpoints we used, the lowest overall rating tended to be somewhat higher than 2. So the checkpoints are not just locations where a signal *could* theoretically penetrate, but places where all sets could transmit and receive communications with the test base—for better or for worse. As you will see, the FCC type-acceptance requirements and the standards of the Electronic Industries Association voluntarily observed by most manufacturers bring about fairly consistent performance among most brands.

TRANSCEIVER TEST CRITERIA

Merit number	Listener's judgment of speech and noise
5	Perfectly readable. Noise interference absent or negligible.
4	Readable without repetitions. Noise noticeable but not seriously annoying.
3	Readable but requires occasional repetitions. Noise quite annoying.
2	Readable only with difficulty, requiring frequent repetition.
1	Unreadable, though presence of speech is discernable.

HOW WE TESTED

Our mobile test unit was a station wagon equipped to test groups of sets by switching among them at each checkpoint. A precision, low-loss coaxial switch connected each set in turn to the same antenna—the Hy-Gain Gypsy 1, a top-loaded fiberglass whip of a type found on 18-wheelers and four-wheelers.

It was mounted at the center of the test vehicle roof. SWR was measured for each set to be sure the match was good. During the tests our engineers turned on power to only one set at a time to avoid unpredictable interaction from stray rf.

The station wagon engine was not given special noise-suppression treatment and was representative of vehicles used by most CBers. The engine was given a major tune-up before testing began, and voltage regulation was carefully set to the standard 13.8 V DC reference used by the electronic industry.

In testing each set we used the microphone furnished with the transceiver, speaking into it from the same distance each time. If a set had a microphone gain control, it was adjusted according to the manufacturer's instructions.

Loudspeaker mounting position can make a difference in speech intelligibility. We tape-recorded received communications, using a microphone directly in front of each set's speaker. The tapes were reviewed to resolve differences in perceived intelligibility. You'll find the actual mounting position of each rig's speaker noted in the test reports so you can allow for loudspeaker aiming and its effects on intelligibility where you plan to mount the set in your vehicle.

Tests were made first with switchable noise controls off, then on, to note their effectiveness. Some sets do not have switchable noise controls but the circuitry is built-in and always operative. All figures of merit in the test reports show performance with switchable noise controls *on.* At no time at the distant checkpoints did any set perform better with noise controls off.

Our base test stations all used the same transceiver, a Tram D 201, known among CBers for excellent performance. Its signal has a quality all its own and we had lots of breakers asking about the rig. We had a clean, well-modulated signal reaching out to the transceivers under test. Each base station used a 5/8-wavelength omnidirectional antenna—a Hy-Gain Super Penetrator. These were so easily tuned to an SWR of 1.1:1 we never had to use the handy load matching adjustment on the Tram D 201.

WHERE WE TESTED

We picked operating environments likely to be encountered by CBers almost anywhere. Here's a rundown on topography, building density and CB traffic density:

Urban testing was downtown in a large city. Streets were narrow to medium width with a high density of buildings. Steel-frame buildings up to 20 stories were interspersed with smaller 5 to 12-story buildings of steel or wood framing and covered with wood, brick, glass, stone or concrete. Our base station antenna tip was 77-1/2 ft. above the street because it was on the roof of a building. The urban checkpoint for mobile sets was 2.7 miles from the base station.

Heavy vehicular traffic didn't bother us—but heavy CB traffic did. Strong signals predominate in a city and you move in and out of radio-shadow areas created by buildings.

The limit on performance in a city appears to be interference from other stations—not man-made or atmospheric noise. Even the least-sensitive sets, or those equipped with the least-effective noise controls, worked well when not stepped on by other stations on the same channel.

Suburban testing was conducted in a flat to gently rolling area of medium to wide streets and low to medium density of buildings. Structure heights varied from 1 to 4 stories. Most buildings were wood, brick or cinderblock. The tip of our base antenna was 41 ft. above the street, and 64 ft. above sea level. Vehicular traffic was medium. CB traffic typically was medium to heavy.

Final evaluations were made at a checkpoint 6.5 miles from the test base where the mobile test station's antenna was 48 ft. above sea level. This is well within the theoretical line-of-sight to the base station antenna. Between the base and mobile test sites lay 4 towns of varying density–typical suburbs of 1 to 3-story homes, scattered shopping centers, some golf courses, creeks and marshes. The limitations on two-way communications tended to be interference and noise.

Country testing was conducted at two sites. The farthest was too far for reliable AM but within range for good single sideband, depending on the set. Our base test station was atop a 22-ft. building on a small hill. The antenna tip was 198 ft. above sea level compared to 178 ft. for the mobile antenna at our 8.5-mile checkpoint for AM and 176 ft. at our 12.2-mile checkpoint for SSB. The intervening area was flat to gently rolling open country with wooded patches, sparsely populated with two-story houses, farms, and occasional light industry. A loose network of two-lane roads threads through the area. There is also a state highway with light to medium traffic. CB traffic was medium.

AM transceivers evaluated at the 8.5-mile checkpoint displayed significant performance differences because of interference and noise. Sets with SSB were hardly strained, so their ratings at this checkpoint are not reported. In general, SSB provided 50% more range for reliable two-way communication than did AM at all checkpoints. In country testing, our second site at 12.2 miles turned out to be strictly SSB territory. Theoretically we had line-of-sight communication, though the intervening area had a few water towers and one 200-ft. hill near the base station. The limitations on open-country communication tended to be interference and man-made or atmospheric noise.

HOW TO USE THE PERFORMANCE RATINGS

Although these field tests reflect typical CB operating environments, they are not absolute figures on which to judge transceivers. Consider where you'll be operating and whether the electrical noise and CB traffic will be better or worse there.

In sparsely populated wide-open country, you may want the most sensitive set you can get. In the city and suburbs, pay more attention to selectivity. Operating features and controls are a matter of personal preference.

We've provided the most significant data from the specification sheets of the sets we've tested. These are manufacturers' specs, not ours. Guidance on how to read a spec is in Chapter 6. In practically all cases we report specifications given in the owner's manuals packed with the sets. Sometimes we found these figures more conservative than advertising literature for the same models.

The differences between our measurements of power output and power output specified by the manufacturers may be attributed to differences among transistors. Transceivers from the same production run can easily vary in performance because of differences among components, and we tested only production models. As you can see from the reports, some sets that measured less than the full 4 legal watts turned out to be excellent performers.

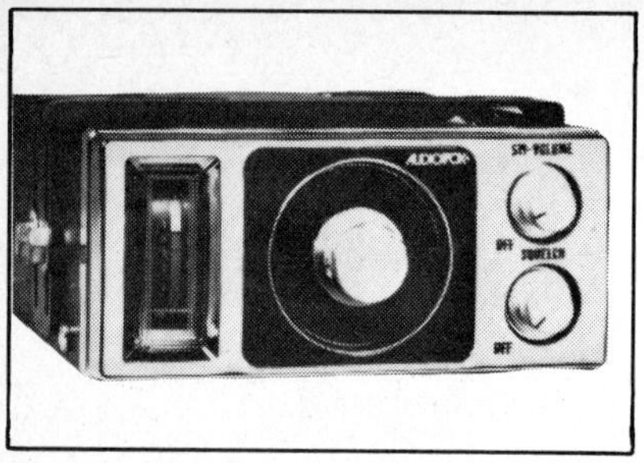

Audiovox MCB-500

Manufacturer Says: Best bang for the buck.
1976 Price: $149.95, suggested retail
Features: Very compact AM mobile transceiver. Built-in automatic noise limiter. S/RF meter. Accepts battery pack.

MANUFACTURER'S SPECS
Dimensions: 4-7/16"W x 2"H x 5-15/16"D
Weight: 3 lbs.
Channels: 23
Power supply: 13 V DC, negative or positive ground. Operating range 12.6 DC to 13.8 V DC
Circuitry: Frequency synthesizer, single conversion superheterodyne
Operating environment: NA
Transmitter
Power output: 4 watts maximum
Output impedance: 50 ohms
Power consumption:
Transmit: NA
Receive: NA
Standby: NA
Harmonic suppression: Meets FCC requirements at time of type acceptance
Frequency tolerance: Within 0.005%
Modulation capability: 100%
Receiver
Sensitivity: 1 μV for 10dB (S+N)/N ratio
Selectivity: 30 dB at 10 KHz
Image rejection: 10 dB
Audio output: 3 watts
Speaker location: Bottom

FIELD TEST RESULTS

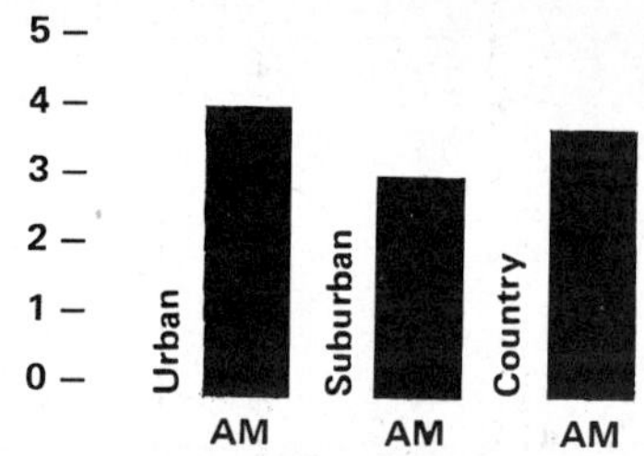

Measured AM Transmitter Power: 3.60 watts
Test Engineer's Comments: Narrow width makes for easy mounting in subcompact cars. Small meter easy to read because of vertical mounting.

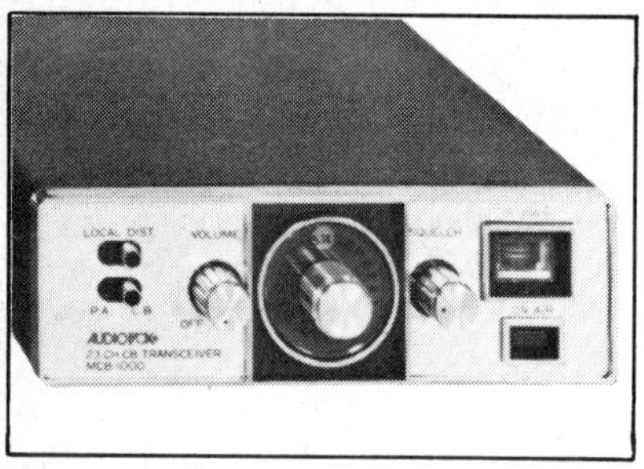

Audiovox MCB-1000

Manufacturer Says: Best performer.
1976 Price: $159.95, suggested retail
Features: S/RF meter, distance/local switch, built-in automatic noise limiter.

MANUFACTURER'S SPECS
Dimensions: 6-3/4"W x 2-1/4"H x 8-1/4"D
Weight: 3-3/4 lbs.
Channels: 23
Power supply: 12 V DC, negative or positive ground; operating range 1 to 15 V DC
Circuitry: Frequency synthesizer, single conversion superheterodyne
Operating environment: 14°F to 122°F (–10°C to +50°C)
Transmitter
Power output: 4 watts
Output impedance: 50 ohms
Power consumption:
Transmit: 1.5 amp, full modulation
Receive:
Standby: 0.2 amp, squelched
Harmonic suppression: 50 dB or more below carrier level
Frequency tolerance: Within 0.005% or less
Modulation capability: 95%
Receiver
Sensitivity: 0.5 μV for 10 dB (S+N)/N (30% 1 KHz mod.)
Selectivity: 6 dB at 5.5 KHz minimum
Image rejection: 50 dB minimum
Audio output: 3 watts
Speaker location: Left side

FIELD TEST RESULTS

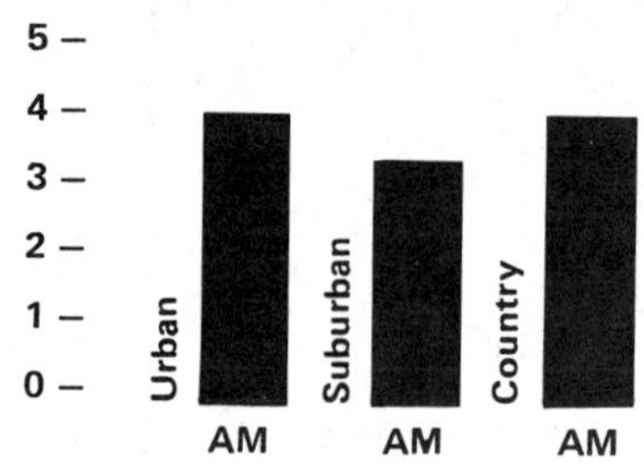

Measured AM Transmitter Power: 3.40 watts
Test Engineer's Comments: Meter judged too small for easy reading. Distance/local switch cuts RF gain by fixed amount.

Beltek W 5398 ENDURO 23

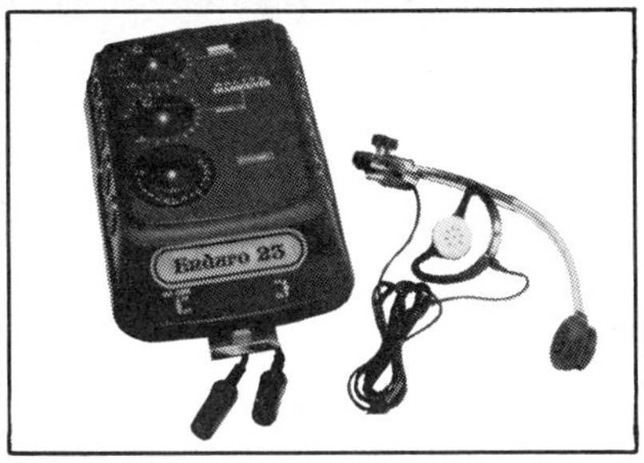

Manufacturer Says: Best bang for the biker's buck.
1976 Price: $229.95, suggested retail
Features: Designed for rough-riding service on motorcycles, snowmobiles, farm and construction machinery. Wind-cancelling microphone on helmet boom, dynamic earphone.

MANUFACTURER'S SPECS
Dimensions: 9-1/8"W x 2-1/2"H x 5-7/8"D
Weight: 3.3 lbs.
Channels: 23
Power supply: 13.8 V DC nominal 12-16 V DC range
Circuitry: Frequency synthesized
Operating environment: Transceiver sealed, knobs guttered to drain water
Transmitter
Power output: 3.8 watts
Output impedance: 50 ohms
Power consumption:
Transmit: 17 watts
Receive: 11 watts
Standby: 0.7 watts
Harmonic suppression: NA
Frequency tolerance: Within 0.003%
Modulation capability: to 100%
Receiver
Sensitivity: 1 μV for 10 dB (S+N)/N
Selectivity: 6 dB at 5 KHz, 40 dB at 20 KHz, 60 dB at 40 KHz
Image rejection: NA
Audio output: 300 mW with dynamic earphone
Speaker location: Inside helmet

FIELD TEST RESULTS

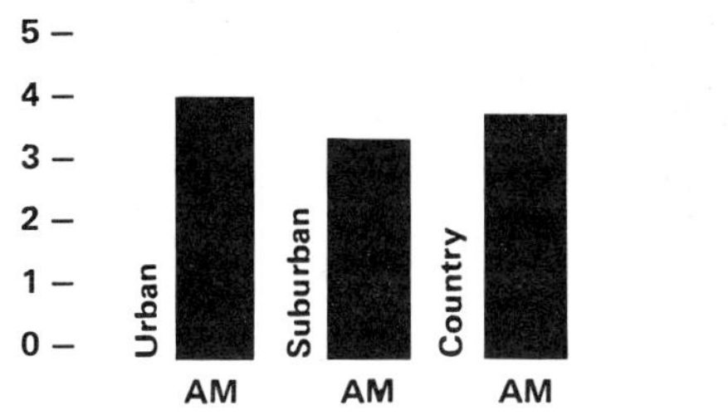

Measured AM Transmitter Power: 3.2 watts
Test Engineer's Comments: Evaluated in mobile service using top-loaded fiberglass 48-inch whip. Microphone and earphone judged especially effective in noisy environments. Remote push-to-talk switch for motorcycle handlebars.

Browning Laboratories, Inc. BROWNIE

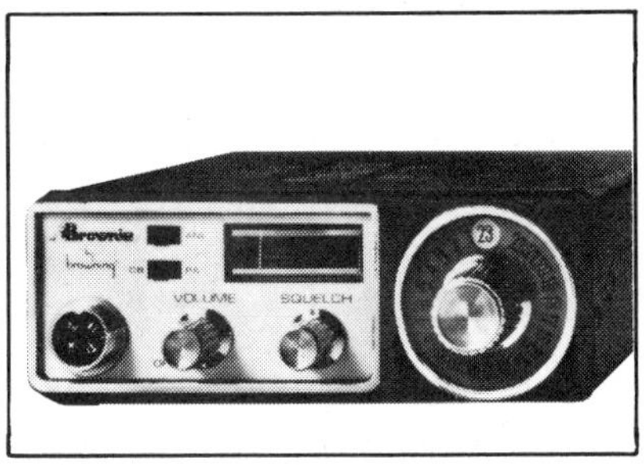

Manufacturer Says: Best bang for the buck.
1976 Price: $139.95, suggested retail
Features: Switchable automatic noise limiter, S/RF meter, very compact, theft-resistant mounting bolts.

MANUFACTURER'S SPECS
Dimensions: 5.49"W x 1.97"H x 6.50"D
Weight: 5 lbs.
Channels: 23
Power supply: 13.8 V DC nominal, negative or positive ground, operating range 12.4 to 15.2 V DC
Circuitry: Frequency synthesizer, dual conversion superheterodyne
Operating environment: –25°F to +125°F (–30°C to +54°C)
Transmitter
Power output: 3.5 watts minimum
Output impedance: 50 ohms
Power consumption:
Transmit: NA
Receive: NA
Standby: NA
Harmonic suppression: Better than 55 dB
Frequency tolerance: Within 0.005%
Modulation capability: 100%
Receiver
Sensitivity: 1 μV for 10 dB (S+N)/N
Selectivity: NA
Image rejection: NA
Audio output: 6 watts
Speaker Location: Top

FIELD TEST RESULTS

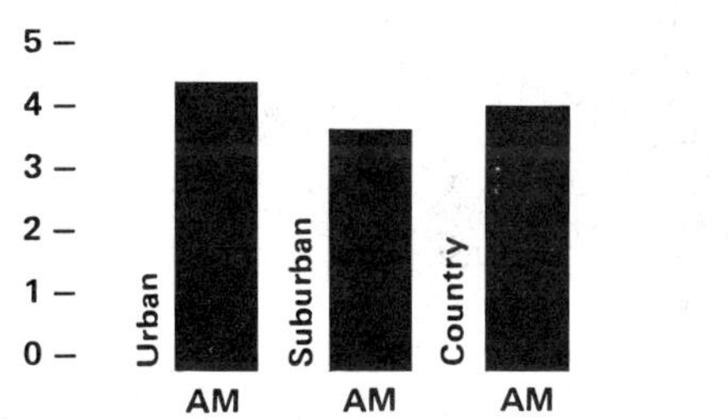

Measured AM Transmitter Power: 3.80 watts
Test Engineer's Comments: Meter judged somewhat small for reading at a glance. Good noise limiting action.

Browning Laboratories, Inc. LTD

Manufacturer Says: Best performer.
1976 Price: $359.95, suggested retail
Features: AM/SSB mobile transceiver. RF gain control, delta tuning, switchable noise blanker, S/RF meter. Theft resistant mounting bolts.

MANUFACTURER'S SPECS
Dimensions: 6.49"W x 2.36"H x 9.87"D
Weight: 8 lbs.
Channels: 23
Power supply: 13.8 V DC, negative or positive ground, operation range 11.5 to 16 V DC
Circuitry: Frequency synthesizer, dual conversion superheterodyne
Operating environment: –22°F to +131°F (–30°C to +55°C)
Transmitter
Power output: 3.8 watts AM, 12 watts PEP
Output impedance: 50 ohms
Power consumption:
Transmit: NA
Receive: NA
Standby: NA
Harmonic suppression: Better than 56 dB
Frequency tolerance: Better than ±0.005%
Modulation capability: 100% AM
Receiver
Sensitivity: AM: 0.5 μV for 12 dB (S+N)/N
SSB: 0.5 μV for 10 dB (S+N)/N
Selectivity: 55 dB minimum adjacent channel rejection
Image rejection: NA
Audio output: 6 watts
Speaker location: Bottom

FIELD TEST RESULTS

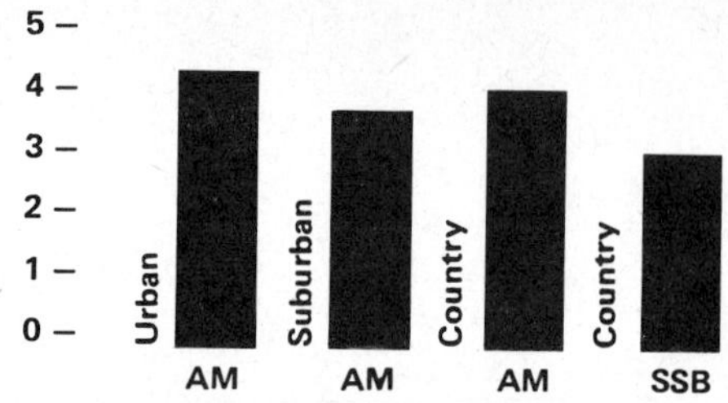

Measured AM Transmitter Power: 3.60 watts
Test Engineer's Comments: Delta tune varies over 3 KHz on AM, over 1200 Hz on SSB for faster adjustment of signal. Meter judged somewhat small for reading at a glance.

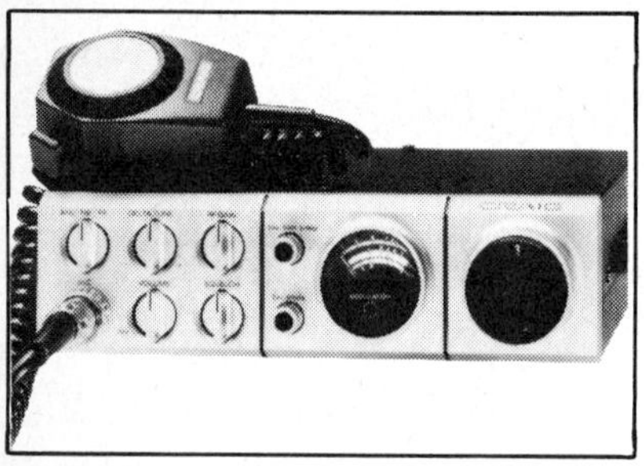

Craig Corporation 4103

Manufacturer Says: Best performer.
1976 Price: $199.95, suggested retail
Features: Switchable automatic noise limiter, switchable noise blanker, large S/RF/SWR meter, RF gain control, meter calibration control, delta tuning, LED modulation indicator, speech compressor, quick-disconnect mount.

MANUFACTURER'S SPECS
Dimensions: 8-1/4"W x 2-3/8"H x 8-7/8"D
Weight: 4.5 lbs.
Channels: 23
Power supply: 13.8 V DC negative or positive ground
Circuitry: Frequency synthesizer, dual conversion superheterodyne
Operating environment: –22°F to +122°F (–30°C to +50°C)
Transmitter
Power output: 4 watts
Output impedance: 50 ohms
Power consumption:
Transmit: NA
Receive: NA
Standby: NA
Harmonic suppression: 59 dB minimum
Frequency tolerance: ±0.005% from –30°C to +50°C
Modulation capability: 100%
Receiver
Sensitivity: Better than 0.5 μV for 10 dB (S+N)/N
Selectivity: 6 dB at ±3 KHz
Image rejection: NA
Audio output: 4 watts
Speaker location: Bottom

FIELD TEST RESULTS

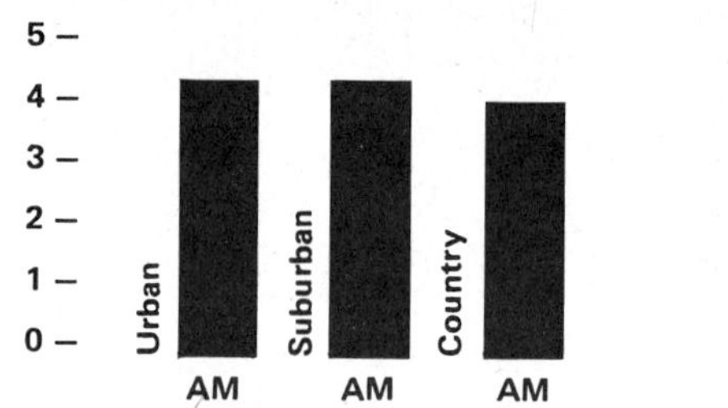

Measured AM Transmitter Power: 3.50 watts
Test Engineer's Comments: Transceiver furnished with well designed quick-disconnect mount; optional extra brackets would make easy transfer to other vehicles or boat. Automatic protection against transistor overload by open or shorted antenna. Noise controls very effective.

Dynascan COBRA 19

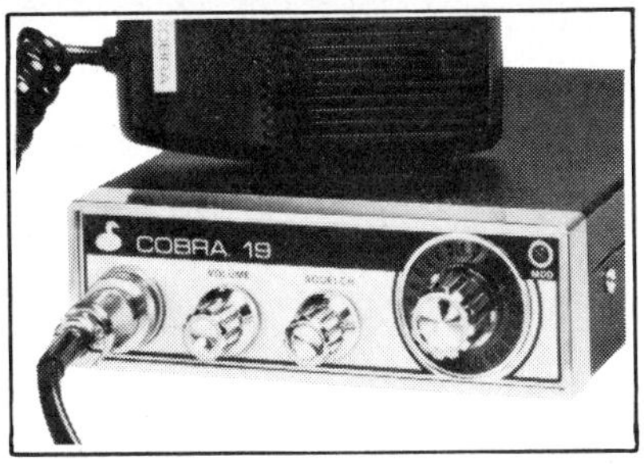

Manufacturer Says: Best bang for the buck.
1976 Price: $129.95, suggested retail
Features: Compact AM mobile. Automatic noise limiter built-in. Speech compressor on microphone circuit for 100% modulation.

MANUFACTURER'S SPECS
Dimensions: 7-7/8"W x 1-9/16"H x 5-1/8"D
Weight: 2.2 lbs.
Channels: 23
Power supply: 13.8 V DC negative or positive ground
Circuitry: Frequency synthesizer, dual conversion superheterodyne
Operating environment: –22°F to +122°F (–30°C to +50°C)
Transmitter
Power output: 4 watts
Output impedance: 50 ohms unbalanced
Power consumption:
Transmit: 1.5 amp fully modulated
Receive: 1.1 amp full audio output
Standby: 0.34 amp squelched
Harmonic suppression: NA
Frequency tolerance: Within 0.005%
Modulation capability: 100% Automatic Modulation Control feature
Receiver
Sensitivity: Less than 1 μV for 10 dB (S+N)/N
Selectivity: 6 dB at 4 KHz, 40 dB at 20 KHz
Image rejection: 30 dB
Audio output: 2.5 watts into 8 ohms
Speaker location: Bottom

FIELD TEST RESULTS

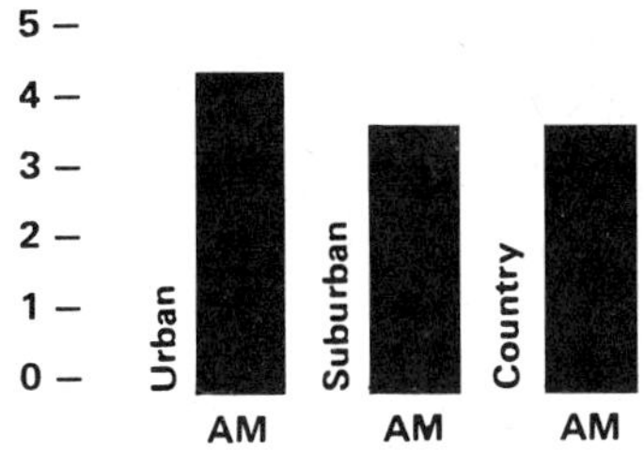

Measured AM Transmitter Power: 3.40 watts
Test Engineer's Comments: Lack of meter relatively unimportant in transmit mode because of automatic modulation control.

Dynascan COBRA 29

Manufacturer Says: Best performer.
1976 Price: $199.95, suggested retail
Features: AM mobile transceiver. Switchable automatic noise limiter and noise blanker. RF gain control, microphone gain control, meter shows relative RF output power, fine tuning control.

MANUFACTURER'S SPECS
Dimensions: 8-1/2"W x 2-1/4"H x 6-7/8"D
Weight: 5-1/2 lbs.
Channels: 23
Power supply: 13.8 V DC positive or negative ground
Circuitry: Frequency synthesizer, dual conversion superheterodyne
Operating environment: –5°F to +122°F (–20°C to +50°C)
Transmitter
Power output: 4 watts
Output impedance: 50 ohms unbalanced
Power consumption:
Transmit: 1.5 amp full modulation
Receive: 1.2 amp full audio output
Standby: 0.3 amp squelched
Harmonic suppression: NA
Frequency tolerance: Within 0.005%
Modulation capability: 100%
Receiver
Sensitivity: Less than 1 μV for 10 dB (S+N)/N
Selectivity: 6 dB at 4 KHz, 40 dB at 20 KHz
Image rejection: 30 dB
Audio output: 2.5 watts into 8 ohms
Speaker location: Bottom

FIELD TEST RESULTS

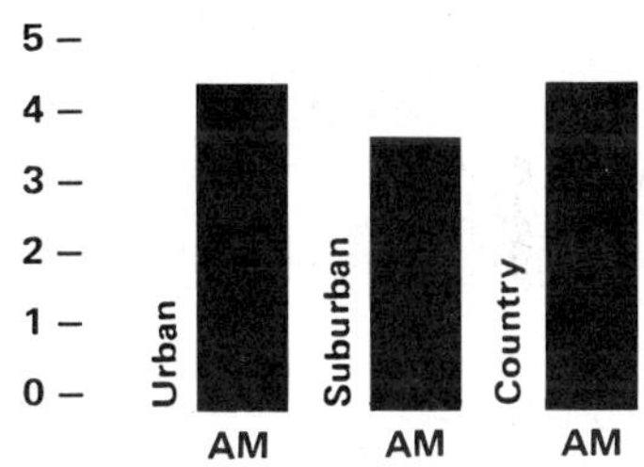

Measured AM Transmitter Power: 3.45 watts
Test Engineer's Comments: Meter judged small but readable. RF gain control useful for preventing overload from nearby stations. Reception generally good—noise controls make moderate improvement.

Dynascan COBRA 135

Manufacturer Says: Base station suitable for mobile operations.
1976 Price: $479.95, suggested retail
Features: AM/SSB base/mobile transceiver. Automatic noise limiter built-in, switchable noise-blanker. Large meter shows relative RF power output, received signals strength. Digital leaf-type clock can turn on set automatically. Clock operates only on 117 V AC. RF gain control.

MANUFACTURER'S SPECS
Dimensions: 13-3/8"W x 5-3/8"H x 12"D
Weight: 14.6 lbs.
Channels: 23
Power supply: Mobile: 13.8 V DC nominal 15.9 V DC maximum
Base: 120 V AC nominal, operating range 108 V AC to 132 V AC
Circuitry: Frequency synthesized, dual conversion superheterodyne
Operating environment: −22°F to +122°F (−30°C to +50°C)
Transmitter
Power output: AM: 4 watts SSB: 12 watts PEP
Output impedance: 50 ohms unbalanced
Power consumption: Base Station: / Mobile:
Transmit: AM: Full mod, 60 watts / AM: Full mod, 2.0 amp
SSB: 7 watts PEP, 70 watts / SSB: 7 watts PEP, 2.4 amp
Receive: 25 watts squelched, 45 watts full audio / 0.6A squelched, 1.4 amp full audio
Frequency tolerance: Within 0.005%
Modulation capability: AM: 100%
Receiver
Sensitivity: AM: Less than 0.5 μV for 10 dB (S+N)/N at greater than 1 watt of audio
SSB: Less than 0.25 μV for 10 dB (S+N)/N at greater than 1 watt of audio
Selectivity: AM: 6 dB at 5 KHz, 50 dB at 20 KHz SSB: 6 dB at 2.2 KHz, 60 dB at 5.0 KHz
Image rejection: More than 50 dB
Audio output: 3 watts into 8 ohms
Speaker location: Bottom

FIELD TEST RESULTS

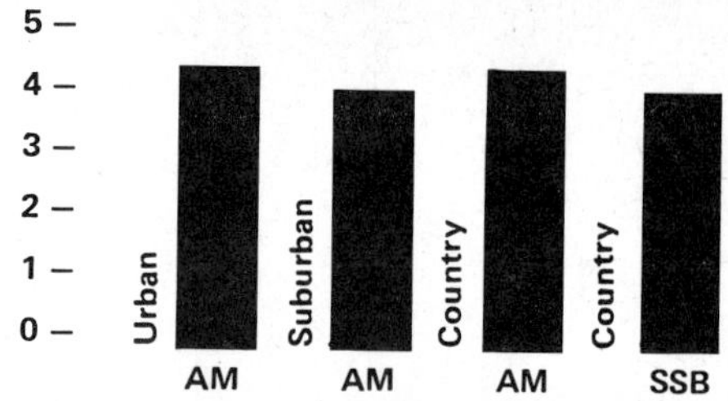

Measured AM Transmitter Power:
On AC as base station: 3.8 watts
On DC as mobile station: 3.35 watts
Test Engineer's Comments: Noise controls moderately improve generally good reception. Crystal lattice filter provided good selectivity.

Gemtronics GTX-3323 APACHE

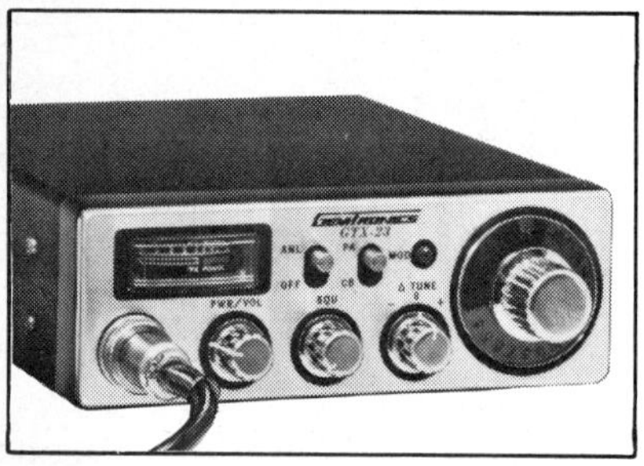

Manufacturer Says: Best performer.
1976 Price: $169.95, suggested retail
Features: Switchable automatic noise limiter, delta tune switch, S/RF meter, modulation light.

MANUFACTURER'S SPECS
Dimensions: 6"W x 2"H x 8-2/3"D
Weight: 3-1/2 lbs.
Channels: 23
Power supply: 13.8 V DC negative or positive ground
Circuitry: Frequency synthesizer, dual conversion superheterodyne
Operating environment: −22°F to +122°F (−30°C to +50°C)
Transmitter
Power output: 4 watts
Output impedance: 50 ohms
Power consumption:
Transmit: NA
Receive: NA
Standby: NA
Harmonic suppression: 50 dB
Frequency tolerance: Within 0.005%
Modulation capability: 100%
Receiver
Sensitivity: 0.7 μV for 10 dB (S+N)/N
Selectivity: 6 dB at 6 KHz, 60 dB at 20 KHz
Image rejection: At least 60 dB
Audio output: 3 watts
Speaker location: Bottom

FIELD TEST RESULTS

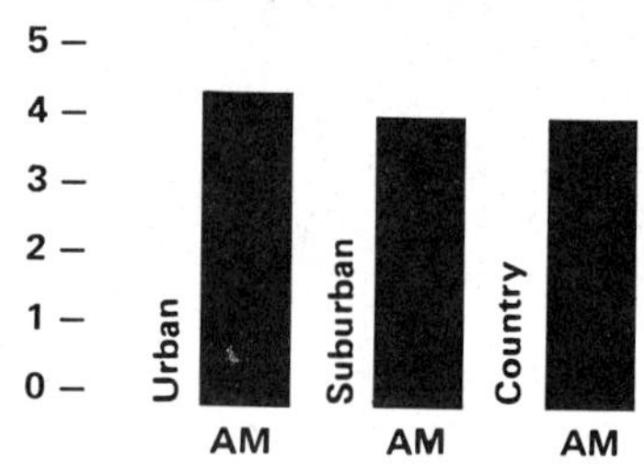

Measured AM Transmitter Power: 3.40 watts
Test Engineer's Comments: Meter judged too small for driver viewing despite good positioning.

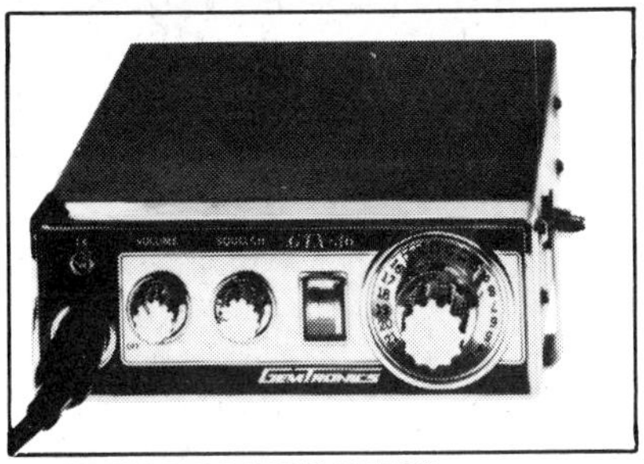

Gemtronics GTX-36

Manufacturer Says: Best bang for the buck.
1976 Price: $139.95, suggested retail
Features: S/RF meter, detachable microphone, transmit indicator.

MANUFACTURER'S SPECS
Dimensions: 6″W x 2″H-2/3″D
Weight: 3-1/2 lbs.
Channels: 23
Power supply: 13.8 V DC negative or positive ground
Circuitry: Frequency synthesizer, dual conversion superheterodyne
Operating environment: –22°F to +122°F (–30°C to +50°C)
Transmitter
Power output: 4 watts
Output impedance: 50 ohms
Power consumption:
 Transmit: NA
 Receive: NA
 Standby: NA
Harmonic suppression: 50 dB
Frequency tolerance: Within 0.005%
Modulation capability: 100%
Receiver
Sensitivity: 0.7 μV for 10 dB (S+N)/N
Selectivity: 6 dB at 6 KHz, 60 dB at 20 KHz
Image rejection: 60 dB or better
Audio output: 3 watts
Speaker location: Bottom

FIELD TEST RESULTS

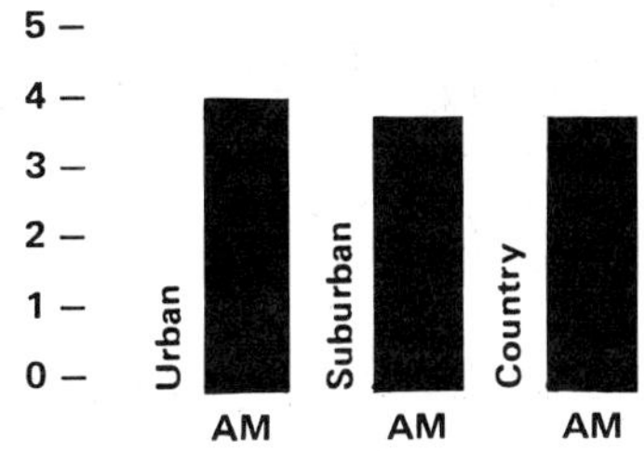

Measured AM Transmitter Power: 4.3 watts
Test Enginner's Comments: S/RF meter small but judged readable because of vertical mounting.

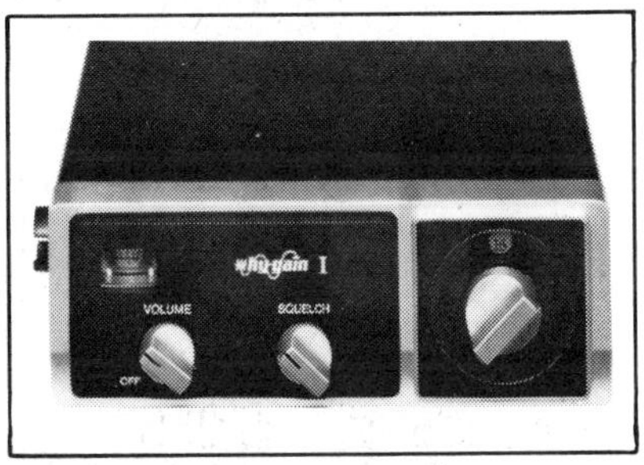

Hy-gain Electronics Corporation HY-RANGE Ia

Manufacturer Says: Budget-priced set.
1976 Price: $139.95, suggested retail
Features: Compact mobile AM transceiver. S/RF meter, squelch, PA. Optional telephone handset.

MANUFACTURER'S SPECS
Dimensions: 6-3/8″W x 2-1/4″H x 7-5/8″D
Weight: 3-1/2 lbs.
Channels: 23
Power supply: 13.8 V DC negative or positive ground, operating range 11.5 to 14.5 V DC
Circuitry: Frequency synthesizer, dual conversion superheterodyne
Operating environment: NA
Transmitter
Power output: 4 watts
Output impedance: 50 ohms nominal
Power consumption:
 Transmit: Less than 1 amp at 12 V
 Receive:
 Standby: 0.1 amp squelched
Harmonic suppression: NA
Frequency tolerance: Within 0.005% or better with PLL
Modulation capability: 90% typical
Receiver
Sensitivity: 0.7 μV for 10 dB (S+N)/N
Selectivity: 50 dB at ±10 KHz
Image rejection: NA
Audio output: 3 watts maximum
Speaker location: Bottom

FIELD TEST RESULTS

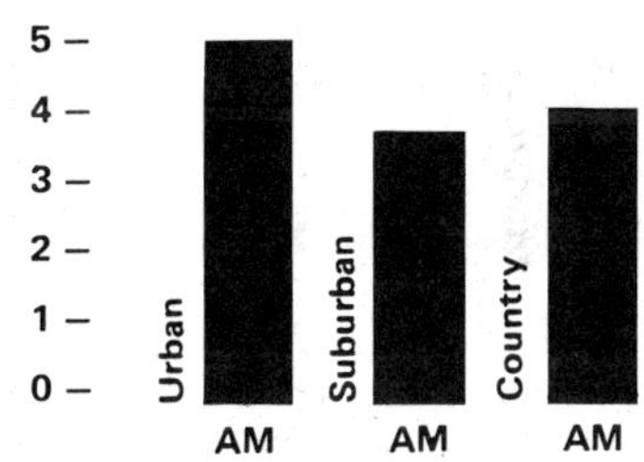

Measured AM Transmitter Power: 3.40 watts
Test Engineer's Comments: Models designated Ia found to have PLL circuitry; superior frequency tolerance to be expected. Meter judged too small.

Hy-gain Electronics Corporation
HY-RANGE IIa—682 PR

Manufacturer Says: Best bang for the buck.
1976 Price: $179.95, suggested retail
Features: Switchable automatic noise limiter, continuous delta tuning, automatic modulation control, S/RF meter, RF gain control.

MANUFACTURER'S SPECS
Dimensions: 6-3/8"W x 2-1/4"H x 5-5/8"D
Weight: 3-1/2 lbs.
Channels: 23
Power supply: 13.8 V DC negative or positive ground, oprating range 11.6 to 14.5 V DC
Circuitry: Frequency synthesizer, dual conversion superheterodyne
Operating environment: NA
Transmitter
Power output: 4 watts
Output impedance: 50 ohms
Power consumption:
  Transmit: Less than 1.0 amp at 12 V DC
  Receive:
  Standby: 0.1 amp squelched
Harmonic suppression: NA
Frequency tolerance: Within 0.005% or better with PLL
Modulation capability: 90% typical
Receiver
Sensitivity: 0.7 μV for 10 dB (S+N)/N
Selectivity: 50 dB at ±10 KHz
Image rejection: NA
Audio output: 3 watts maximum
Speaker location: Bottom

FIELD TEST RESULTS

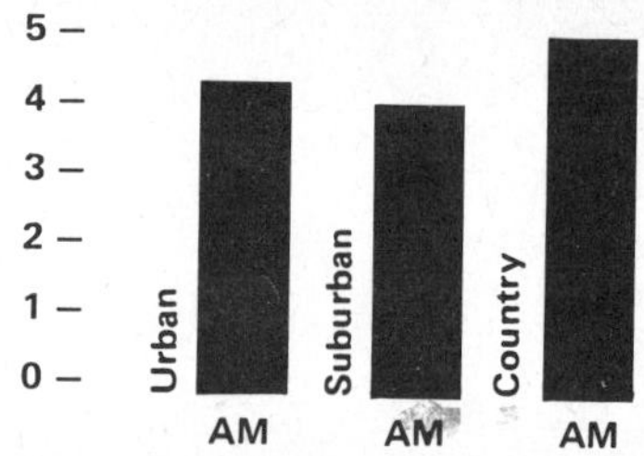

Measured AM Transmitter Power: 3.6 watts
Test Engineer's Comments: Automatic noise limiter judged especially effective. Panel meter too small for reading at a glance. Model IIa has PLL for superior frequency tolerance.

Hy-gain Electronics Corporation
HY-RANGE V 674B

Manufacturer Says: Best performer.
1976 Price: $349.95, suggested retail
Features: AM/SSB mobile transceiver. Switchable automatic noise limiter and noise blanker, automatic modulation control, RF gain control, fine tuning, optional telephone handset, large S/RF meter.

MANUFACTURER'S SPECS
Dimensions: 8-1/2"W x 2-3/4"H x 11-1/2"D
Weight: 8 lbs.
Channels: 23
Power supply: 13.8 V DC negative or positive ground, operating range 11.5 to 14.5 V DC
Circuitry: Frequency synthesizer, single conversion superheterodyne
Operating environment: NA
Transmitter
Power output: AM: 4 watts; SSB: 12 watts PEP
Output impedance: 50 ohms
Power consumption:
  Transmit: NA
  Receive: NA
  Standby: NA
Harmonic suppression: 50 dB down
Frequency tolerance: Within 0.005%
Modulation capability: 90% typical
Receiver
Sensitivity: AM: 1 μV for 10 dB (S+N)/N; SSB: 0.25 μV for 10 dB (S+N)/N
Selectivity: AM: 6 dB at 7 KHz; SSB: 6 dB at 2.4 KHz
Image rejection: NA
Audio output: 3.0 watts
Speaker location: Bottom

FIELD TEST RESULTS

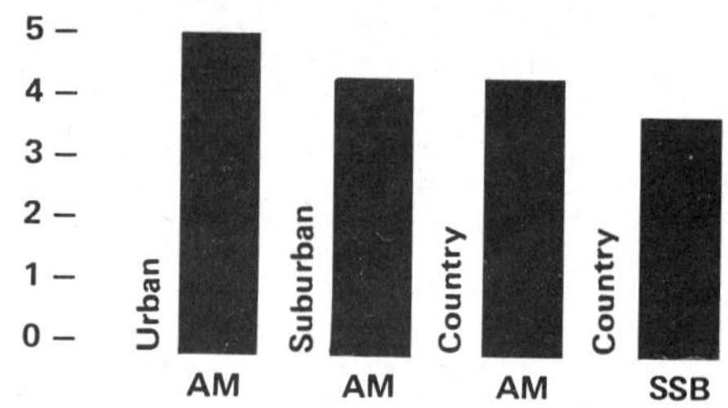

Measured AM Transmitter Power: 3.55 watts
Test Engineer's Comments: Noise limiter and noise blanker especially effective but switch positions confusing. Meter easy to read from driver's position. RF gain control particularly useful on SSB.

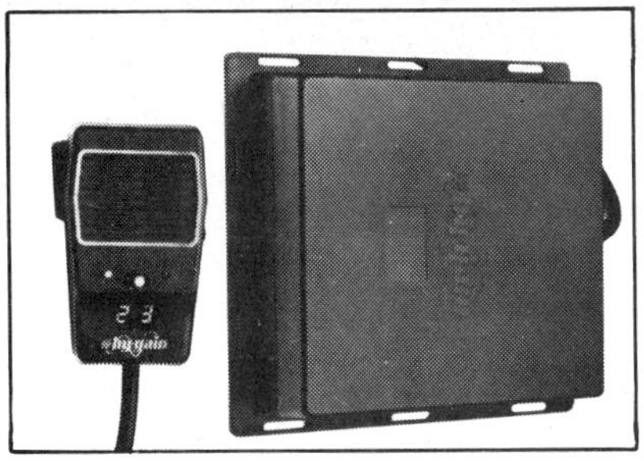

Hy-Gain Electronics Corporation
HY-GAIN 9 MODEL 2679

Manufacturer Says: Fully remote mobile transceiver
1976 Price: $219.95, suggested retail
Features: All controls in single remote-control head including speaker/microphone, channel selection and indication.

MANUFACTURER'S SPECS
Dimensions: 8"W x 2-3/8"H x 9"D
Weight: 3 lbs. remote transceiver, 1/2 lb. control head with cable
Channels: 23
Power supply: 13.8 V DC, negative or positive ground, operating range 11.5 to 14.5 V DC
Circuitry: Frequency synthesizer, dual conversion superheterodyne, phase lock loop.
Operating environment: –22°F to +122°F (–30°C to +50°C)
Transmitter
Power output: 4 watts
Output impendance: 50 ohms
Power consumption:
  Transmit: 1 amp at 13.8 V DC
  Receive: NA
  Standby: NA
Harmonic suppression 50 dB
Frequency tolerance: Within 0.005%
Modulation capability: 80 to 95%
Receiver
Sensitivity: 0.7μV for 10 dB (S+N)/N
Selectivity: 40 dB adjacent channel
Image rejection: NA
Audio output: 3 watts
Speaker location: In remote-control head

FIELD TEST RESULTS

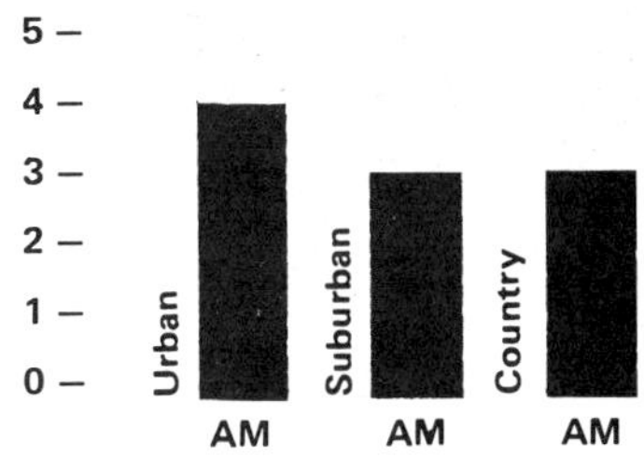

Measured AM Transmitter Power: 3.7 watts
Test Engineer's Comments: Large LED's very readable. Electrical noise in fringe areas affects intelligibility because of speaker size in remote-control head. Suggest using external speaker.

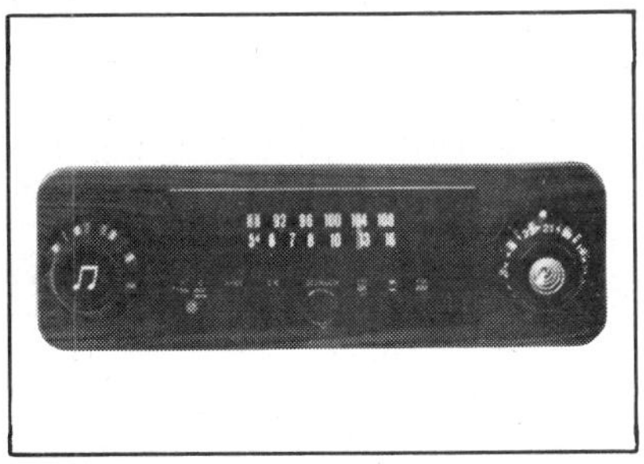

J.I.L. Corporation of America
852 CB and 606 CB

Manufacturer Says: In-dash AM/FM/CB units with tape players.
1976 Price:
  Model 852 CB: $399.95, suggested retail
  Model 6060 CB: $419.95, suggested retail
Features: CB circuitry identical in both units. Model 852 contains 8 track stereo cartridge player; Model 606 contains stereo cassette player. Stand-by switch, squelch control. Requires external speakers (not included).

MANUFACTURER'S SPECS
Dimensions: 7-1/2"W x 2"H x 7"D
Weight: NA
Channels: 23
Power supply: 11 to 16 V DC negative ground
Circuitry: Frequency synthesizer, dual conversion superheterodyne
Operating environment: –22°F to +122°F (–30°C to +50°C)
Transmitter
Power output: 3.5 watts
Output impedance: 50 ohms
Power consumption:
  Transmit: NA
  Receive: NA
  Standby: NA
Harmonic suppression: More than 50 dB
Frequency tolerance: Within 0.005%
Modulation capability: More than 80%
Receiver
Sensitivity: 1 μV for 10 dB (S+N)/N
Selectivity: 6 dB at 5 KHz minimum
Image rejection: NA
Audio output: 3 watts
Speaker location: External

FIELD TEST RESULTS

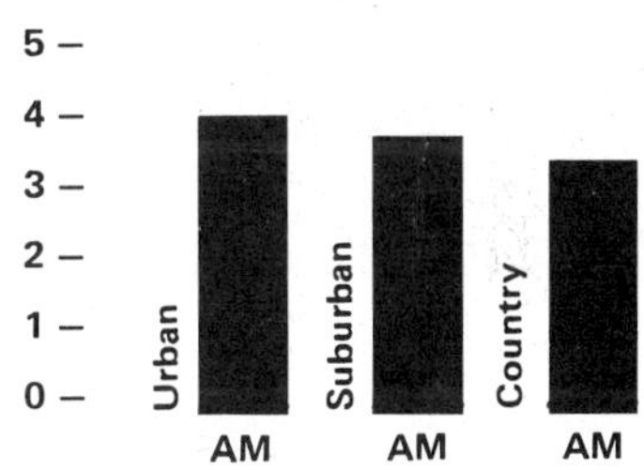

Measured AM Transmitter Power: 3.20 watts
Test Engineer's Comments: Somewhat difficult to judge modulation level.

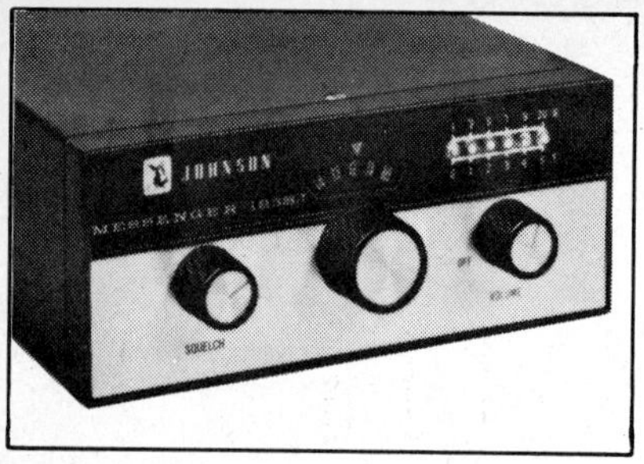

E.F. Johnson
MESSENGER 123A/123SJ

Manufacturer Says: Best bang for the buck.
1976 Price: 123A – $159.95, suggested retail
123SJ – $169.95, suggested retail
Features: Models identical except: 123A has panel-meter for S/RF. 123SJ has LED display for same information. Both: Speech compressor for higher average modulation on transmit, audio clipping to cut splatter, mechanical selectivity filter, battery polarity reversing switch.

MANUFACTURER'S SPECS
Dimensions: 6.3"W x 2.5"H x 9.1"D
Weight: Approximately 3.6 lbs.
Channels: 23
Power supply: 13.8 V DC negative or positive ground, optional 117 V AC and portable power packs available from manufacturer
Circuitry: Frequency synthesizer
Operating environment: –22°F to +140°F (–30°C to +60°C)
Transmitter
Power output: 4 watts maximum at 13.8 V DC
Output impedance: 50 ohms
Power consumption:
Transmit: 1.3 amp
Receive: 0.35 amp squelched
Harmonic suppression: Meets FCC requirements at time of type acceptance
Frequency tolerance: Within 0.005%
Modulation capability: 100%
Receiver
Sensitivity: 0.5 µV for 10 dB (S+N)/N (30% modulation at 1 KHz)
Selectivity: 6 dB at 6 KHz, 60 dB at 30 KHz
Image rejection: 10 dB
Audio output: 3 watts
Speaker location: Bottom

FIELD TEST RESULTS

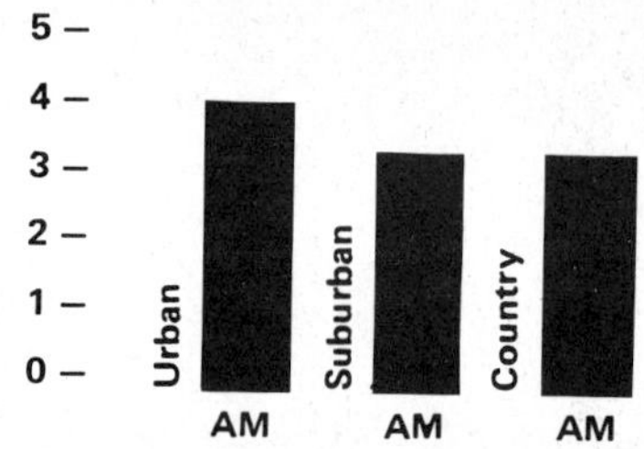

Measured AM Transmitter Power: 3.25 watts
Test Engineer's Comments: LED display judged more visible than equivalent size panel meter viewed at same distance. However, LED's were less meaningful at a glance. Only "hot" DC power lead provided. Be sure to ground chassis when connecting set.

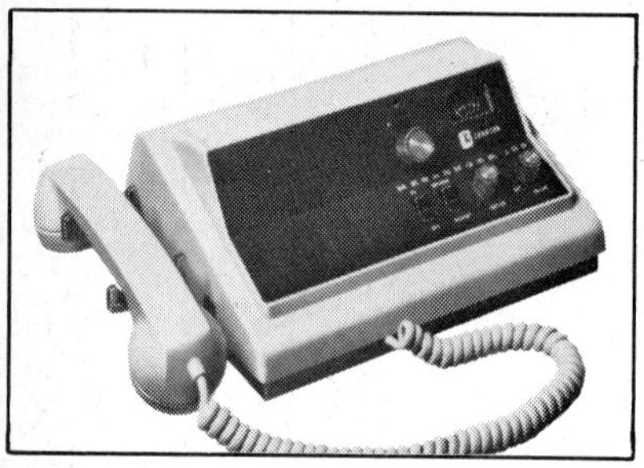

E.F. Johnson
MESSENGER 132

Manufacturer Says: Base station transceiver suitable for mobile operation.
1976 Price: $259.95, suggested retail
Features: Radiotelephone handset. Simultaneous handset/loudspeaker operation for conference calls. Built-in automatic noise limiter. Speech compression.

MANUFACTURER'S SPECS
Dimensions: 13-3/4"W x 5-1/4"H x 9" D
Weight: 8.5 lbs.
Channels: 23
Power supply: 117 V AC, 13.8 V DC negative ground
Circuitry: Single conversion superheterodyne
Operating environment: –22°F to +140°F (–30°C to +60°C)
Transmitter
Power output: 4 watts
Output impedance: 50 ohms
Power consumption:
Transmit: NA
Receive: NA
Standby: NA
Harmonic suppression: Better than 50 dB
Frequency tolerance: Within 0.005%
Modulation capability: 80% minimum
Receiver
Sensitivity: 0.5 µV for 10 dB (S+N)/N
Selectivity: 10 dB at 6 KHz, 60 dB at 30 KHz
Image rejection: 10 dB
Audio output: 3 watts
Speaker location: Front

FIELD TEST RESULTS

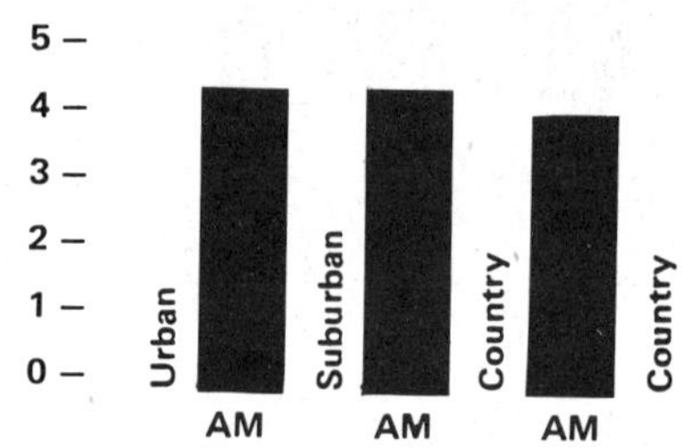

Measured AM Transmitter Power:
As base station on AC: 3.40 watts
As mobile station on DC: 3.15 watts
Test Engineer's Comments: Transceiver furnished with "hot" lead only, for DC operation. Requires separate additional lead to chassis ground.

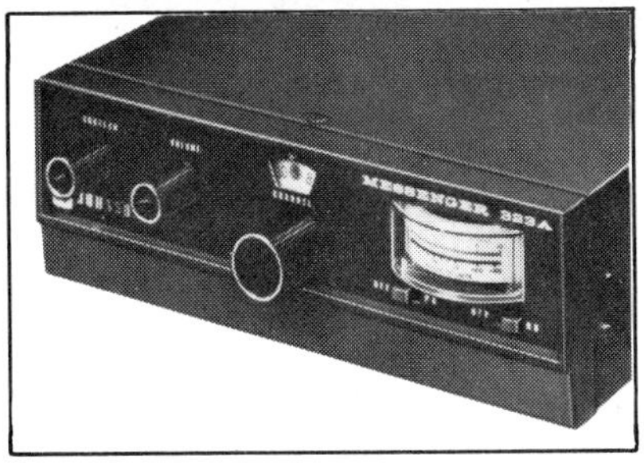

E.F. Johnson MESSENGER 323A

Manufacturer Says: Mobile suitable for portable operation.
1976 Price: $249.95, suggested retail
Features: Special "tapered" automatic noise limiter, switchable noise blanker, large S/RF meter. Handle power pack complete with center-loaded antenna and rechargeable batteries converts transceiver into portable.

MANUFACTURER'S SPECS
Dimensions: 8"W x 2-1/2"H x 9-1/2"D
Weight: 5.3 lbs.
Channels: 23
Power supply: 13.8 V DC positive or negative ground
Circuitry: Frequency synthesizer, single conversion superheterodyne
Operating environment: –22°F to +140°F (–30°C to +60°C)
Transmitter
Power output: 4 watts
Output impedance: 50 ohms
Power consumption:
 Transmit: NA
 Receive: NA
 Standby: NA
Harmonic suppression: Better than 50 dB
Frequency tolerance: Within 0.005%
Modulation capability: 70% minimum
Receiver
Sensitivity: 0.5 μV for 12 dB (S+N)/N
Selectivity: 6 dB at 6 KHz, 60 dB at 20 KHz
Image rejection: 45 dB
Audio output: 4 watts
Speaker location: Bottom

FIELD TEST RESULTS

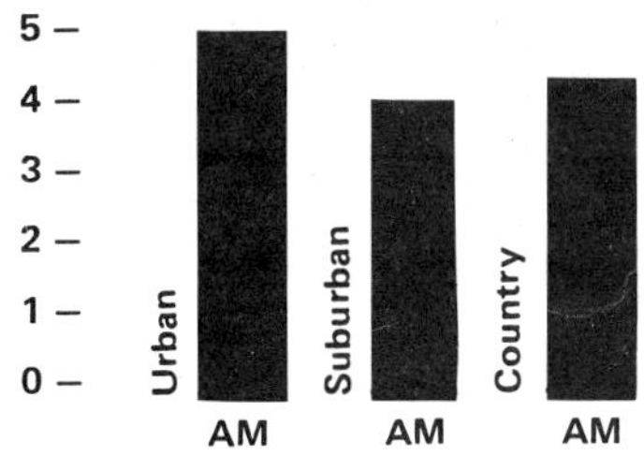

Measured AM Transmitter Power: 3.20 watts
Test Engineer's Comments: Tapered automatic noise limiter is self-adjusting to reduce noise limiting with stronger signals thereby improving intelligibility. Switch for negative or positive ground. Single "hot" lead furnished, additional lead required for chassis ground.

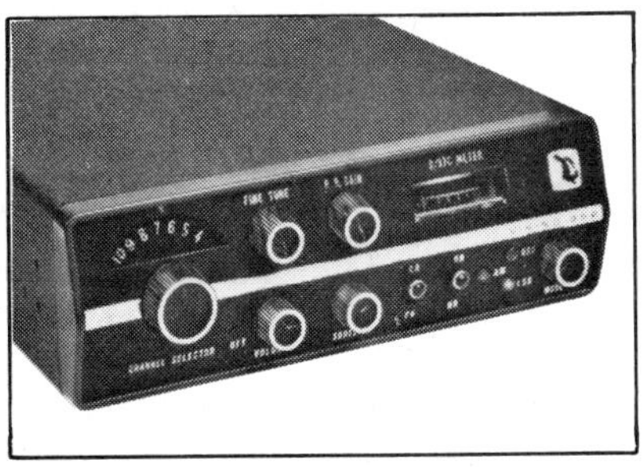

E.F. Johnson VIKING 352

Manufacturer Says: Best performer.
1976 Price: $359.95, suggested retail
Features: AM/SSB mobile transceiver adaptable to base station use with custom pedestal mount containing power supply. Automatic noise limiting built-in, switchable noise blanker, speech compression, RF gain control, mode indicator always lit when set is on. S/RF meter.

MANUFACTURER'S SPECS
Dimensions: 7.5"W x 2.4"H x 10.7"D
Weight: 6 lbs.
Channels: 23
Power supply: 13.8 V DC negative or positive ground
Circuitry: Frequency synthesizer
Operating environment: –22°F to +140°F (–30°C to +60°C)
Transmitter
Power output:
 AM: 4 watts at 13.8 V DC
 SSB: 12 watts PEP at 13.8 V DC
Output impedance: 50 ohms
Power consumption:
 Transmit: 1.8 amp fully modulated
 Receive: 0.4 amp squelched
Harmonic suppression: Better than 50 dB
Frequency tolerance: Within 0.005%
Modulation capability: 80% minimum AM
Receiver
Sensitivity: AM: 0.5 μV at 10 dB (S+N)/N; SSB: 0.35 μV at 10 dB (S+N)/N
Selectivity: 6 dB at 4.5 KHz, 60 dB at 30 KHz
Image rejection: 50 dB
Audio output: 2 watts
Speaker location: Bottom

FIELD TEST RESULTS

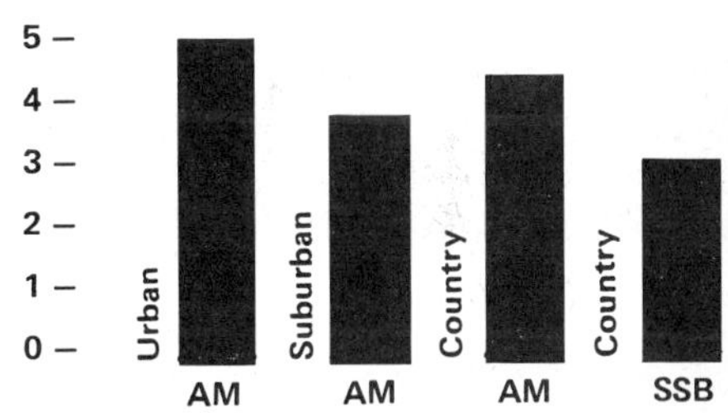

Measured AM Transmitter Power: 2.80 watts
Test Engineer's Comments: Only "hot" lead provided. Be sure to ground chassis when connecting set.

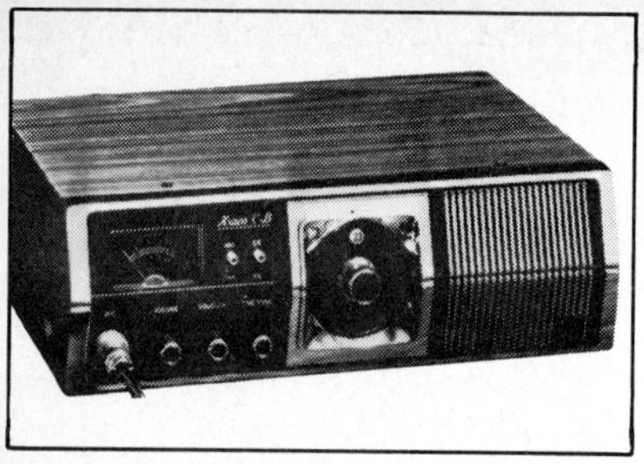

Kraco Enterprises, Inc. KB 2345

Manufacturer Says: Base/mobile best performer.
1976 Price: $279.95, suggested retail
Features: Base/AM transceiver suitable for mobile operation. Switchable automatic noise limiter, large S/RF meter, fine tuning control, mode indicator lights.

MANUFACTURER'S SPECS
Dimensions: 2-1/4"W x 4-1/8"H x 10"D
Weight: 11 lbs.
Channels: 23
Power supply: Base: 105-120 V AC 50-60 Hz Mobile: 11-16 V DC
Circuitry: Frequency synthesizer, dual conversion superheterodyne
Operating environment: NA
Transmitter
Power output: 4 watts
Output impedance: 50 ohms
Power consumption:
- Transmit: NA
- Receive: NA
- Standby: NA

Harmonic suppression: 50 dB
Frequency tolerance: Within 0.005%
Modulation capability: NA
Receiver
Sensitivity: 0.7 μV for 10 dB (S+N)/N
Selectivity: NA
Image rejection: 50 dB
Audio output: 3 watts
Speaker location: Front

FIELD TEST RESULTS

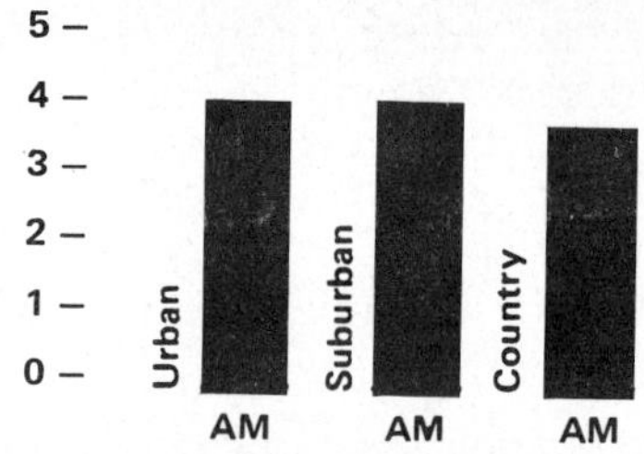

Measured AM Transmitter Power:
As base station on AC: 3.75 watts
As mobile station on DC: 3.35 watts
Test Engineer's Comments: Noise limiter moderately effective; reception generally good even with ANL off.

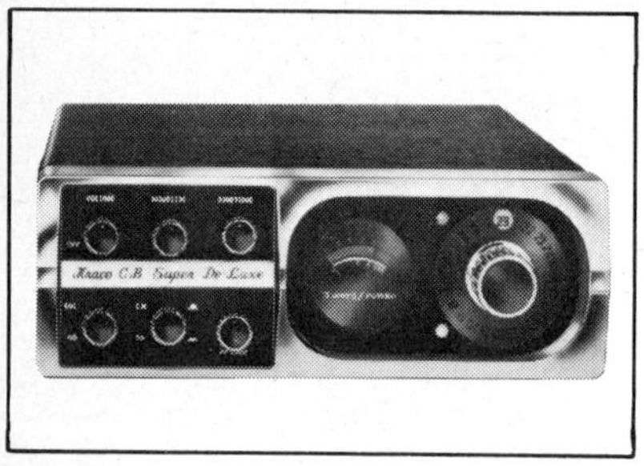

Kraco Enterprises, Inc. KCB 2330

Manufacturer Says: Best bang for the buck.
1976 Price: $219.95, suggested retail
Features: Switchable automatic noise limiter, switchable noise blanker. Large S/RF meter. Fine tuning control, RF gain control.

MANUFACTURER'S SPECS
Dimensions: 7-1/8"W x 2-3/8"H x 9-1/4"D
Weight: 5 1/2 lbs.
Channels: 23
Power supply: 13.8 V DC negative or positive ground, operating range 11 to 16 V DC
Circuitry: Frequency synthesizer, dual conversion superheterodyne
Operating environment: NA
Transmitter
Power output: 4 watts
Output impedance: 50 ohms
Power consumption:
- Transmit: NA
- Receive: NA
- Standby: NA

Harmonic suppression: 50 dB
Frequency tolerance: Within 0.005%
Modulation capability: NA
Receiver
Sensitivity: 0.7 μV for 10 dB (S+N)/N
Selectivity: 45 dB at 10 KHz
Image rejection: NA
Audio output: 3 watts
Speaker location: Bottom

FIELD TEST RESULTS

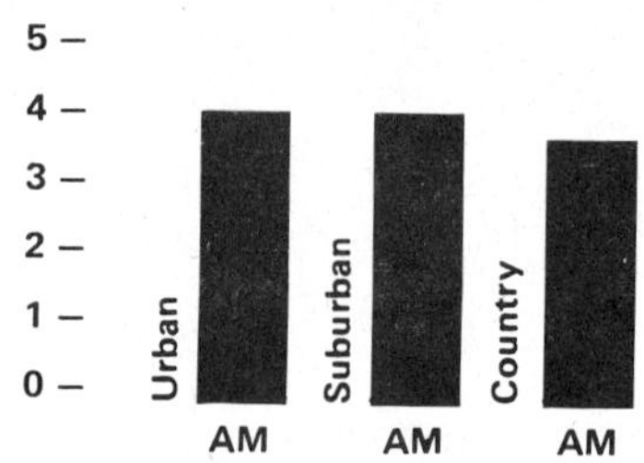

Measured AM Transmitter Power: 3.40 watts
Test Engineer's Comments: Noise controls moderately effective. Large meter easy to read head-on and from right, fair to good from left because of recessing.

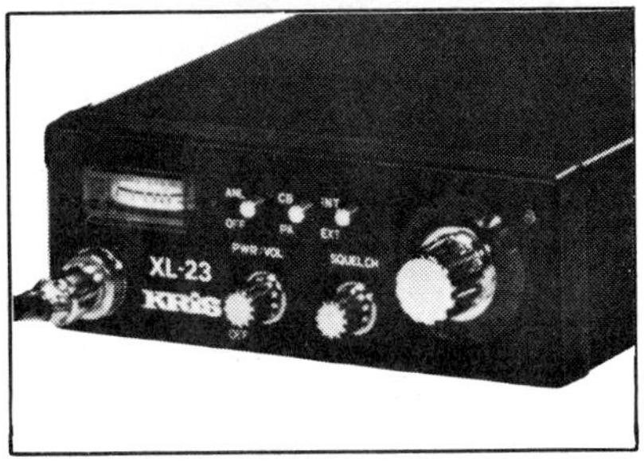

Kris XL-23

Manufacturer Says: Best performer, best bang for the buck.
1976 Price: $349.95, suggested retail
Features: Switchable automatic noise limiter, S/RF meter with jack for large external S-meter, TVI trap.

MANUFACTURER'S SPECS
Dimensions: 6.50"W x 2.25"H x 8.375"D
Weight: 4-1/2 lbs.
Channels: 23
Power supply: 13.8 V DC negative or positive ground
Circuitry: Frequency synthesizer, dual conversion superheterodyne
Operating environment: –22°F to +122°F (–30°C to +50°C)
Transmitter
Power output: 4 watts
Output impedance: 50 ohms
Power consumption:
Transmit: NA
Receive: NA
Standby: NA
Harmonic suppression: More than 50 dB
Frequency tolerance: Within 0.005%
Modulation capability: 95%
Receiver
Sensitivity: 0.5 µV for 10 dB (S+N)/N
Selectivity: 6 dB at 5 KHz minimum
Image rejection: 60 dB
Audio output: 3 watts
Speaker location: Bottom

FIELD TEST RESULTS

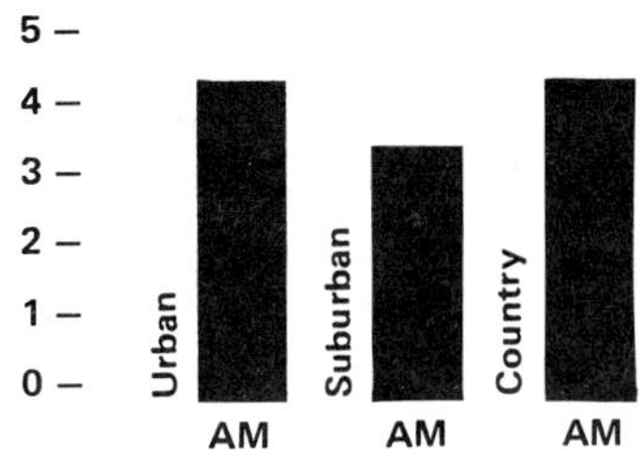

Measured AM Transmitter Power: 3.60 watts
Test Engineer's Comments: Transmit relay contacts wired to permit use of broadcast-band radio or tape over external speaker system shared with the CB transceiver. Other system's audio will be off during transmit mode.

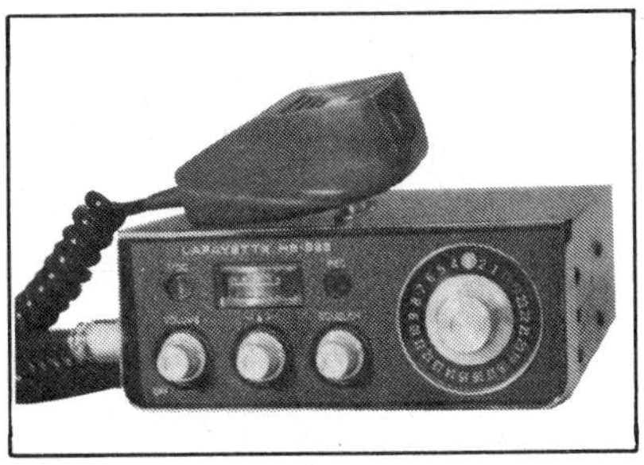

Lafayette Radio Electronics Corp. HB-525F

Manufacturer Says: Best bang for the buck.
1976 Price: $179.95, suggested retail
Features: S/RF meter, delta tune switch, TVI trap, antenna load adjustment, lighted indicators for receive and transmit modes. Optional portable power pack. Charger for power pack also serves as power supply for transceiver when used as base station.

MANUFACTURER'S SPECS
Dimensions: 6-1/4"W x 2-3/8"H x 8"D
Weight: 7 lbs.
Channels: 23
Power supply: 12.6 V DC nominal; negative or positive ground
Circuitry: Frequency synthesizer, dual conversion superheterodyne
Operating environment: NA
Transmitter
Power output: 4 watts
Output impedance: 50 ohms
Power consumption:
Transmit: NA
Receive: NA
Standby: NA
Harmonic suppression: Exceeds FCC requirements at time of type acceptance
Frequency tolerance: Within 0.005%
Modulation capability: 90% or better
Receiver
Sensitivity: 0.5 µV for 10 dB (S+N)/N
Selectivity: 6 dB at ±2.5 KHz; 45 dB at ±8 KHz
Image rejection: NA
Audio output: 2.7 watts into external speaker jack
Speaker location: Bottom

FIELD TEST RESULTS

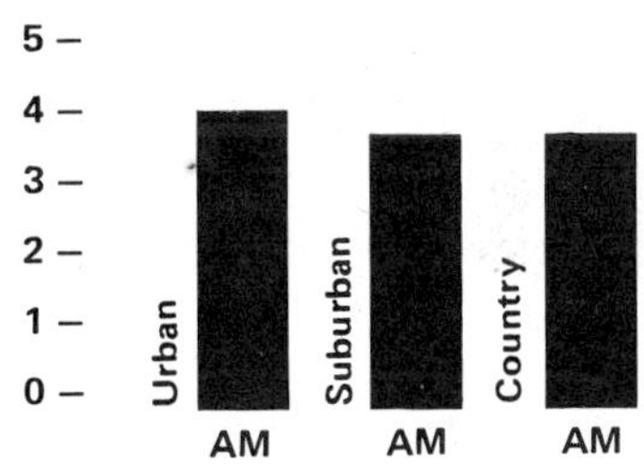

Measured AM Transmitter Power: 4.00 watts
Test Engineer's Comments: Polarity switch for negative or positive ground; locks in position. TVI trap easy to adjust. Built-in switching circuit for burglar alarm. Antenna loading adjustment.

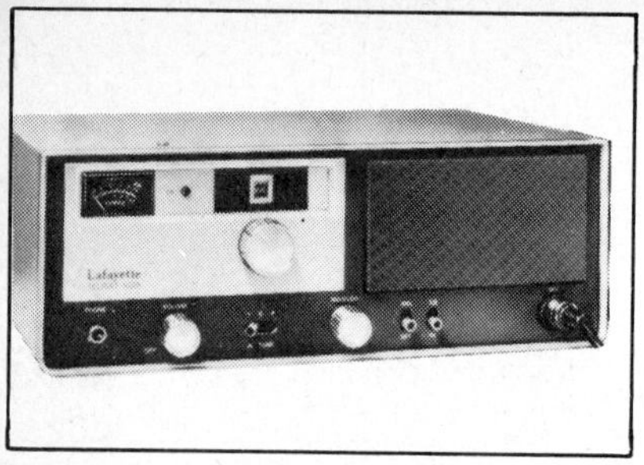

Lafayette TELSAT 1023

Manufacturer Says: Best performer.
1976 Price: $179.95, suggested retail
Features: AM base station with mobile or emergency operating capability on nominal 12-V battery, negative ground system. Delta tune, switchable automatic noise limiter. S/RF meter.

MANUFACTURER'S SPECS
Dimensions: 12-3/16"W x 4-11/16"H x 9-7/16"D
Weight: 7.7 lbs.
Channels: 23
Power supply: 105-120 V, 50/60 cycles AC, or 11.5-14.5 V DC, negative ground
Circuitry: Dual conversion superheterodyne
Operating environment: NA
Transmitter
Power output: 4 watts
Output impedance: 50 ohms
Power consumption:
Transmit: NA
Receive: NA
Standby: NA
Harmonic suppression: Meets FCC requirements at time of type acceptance
Frequency tolerance: Within 0.005%
Modulation capability: 90% typical
Receiver
Sensitivity: 1 μV for 10 dB (S+N)/N
Selectivity: 40 dB at 10 KHz
Image rejection: 55 dB
Audio output: 3 watts maximum
Speaker location: Front

FIELD TEST RESULTS

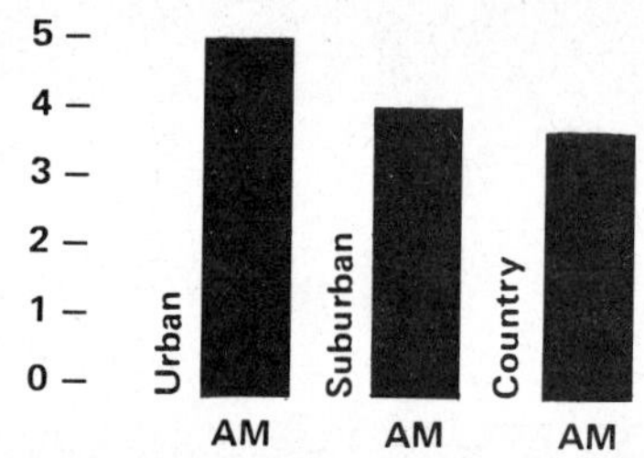

Measured AM Transmitter Power:
As base station on AC: 3.85 watts
As mobile station on DC: 3.60 watts
Test Engineer's Comments: Large S/RF meter judged easily readable. Would have preferred modulation indication on meter instead of adjacent light. 12 V DC power cable must be purchased as accessory.

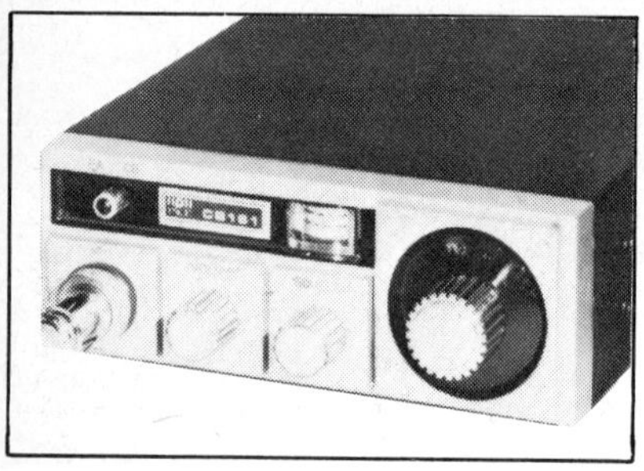

Pace, Division of Pathcom, Inc. PACE CB 161

Manufacturer Says: Best bang for the buck.
1976 Price: $149.95, suggested retail
Features: Phase lock loop, S/RF meter, public address, squelch control.

MANUFACTURER'S SPECS
Dimensions: 6-5/8"W x 1-7/8"H x 9"D
Weight: 4 lbs.
Channels: 23
Power supply: 12 V DC nominal negative or positive ground
Circuitry: Frequency synthesizer, dual conversion superheterodyne
Operating environment: NA
Transmitter
Power output: 4 watts
Output impedance: 50 ohms
Power consumption:
Transmit: NA
Receive: NA
Standby: NA
Harmonic suppression: 50 dB
Frequency tolerance: Better than ±0.005%
Modulation capability: 85% minimum, typically 95%
Receiver
Sensitivity: 0.5 μV for 10 dB (S+N)/N
Selectivity: NA
Image rejection: 50 dB minimum
Audio output: 4 watts
Speaker location: Bottom

FIELD TEST RESULTS

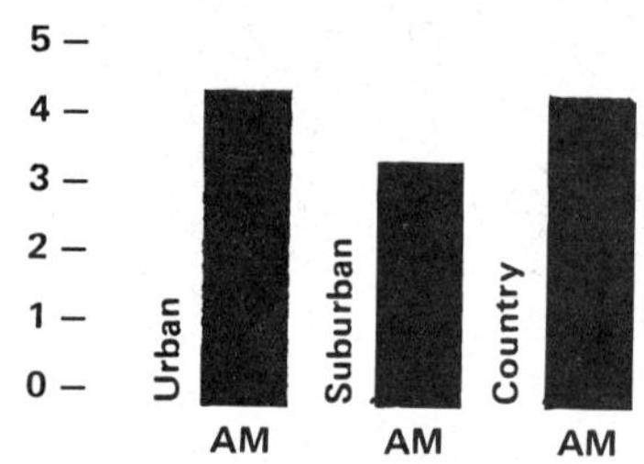

Measured AM Transmitter Power: 4.00 watts
Test Engineer's Comments: Meter judged somewhat small for easy reading.

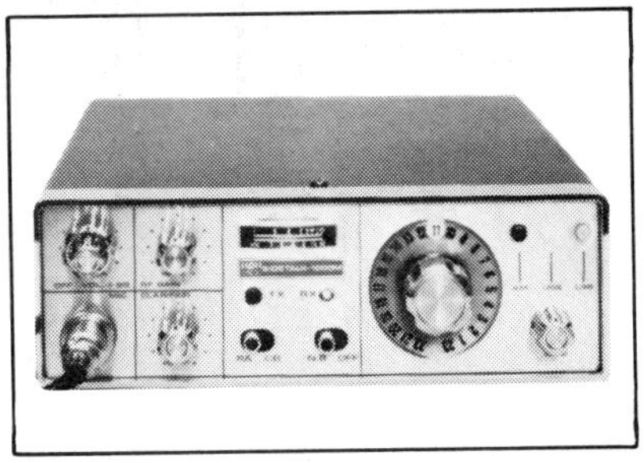

Pace, Division of Pathcom, Inc.
PACE SIDETALK 1000M

Manufacturer Says: Best performer.
1976 Price: $369.95, suggested retail
Features: AM/SSB with S/RF meter, switchable noise blanker, RF gain control, indicator lights for all operating modes.

MANUFACTURER'S SPECS
Dimensions: 7-1/2"W x 2-1/4"H x 10"D
Weight: 10 lbs.
Channels: 23
Power supply: 12.5 V DC nominal, range 10-16 V DC, negative or positive ground
Circuitry: Frequency synthesizer, dual conversion superheterodyne
Operating environment: NA

Transmitter
Power output: AM: 4 watts; SSB: 12 watts PEP
Output impedance: 50 ohms
Power consumption:
- Transmit: 1.4 amp
- Receive: 0.75 amp with audio output
- Standby: 0.1 amp squelched

Harmonic suppression: 50 dB minimum
Frequency tolerance: Within 0.005%
Modulation capability: 100% AM average speech

Receiver
Sensitivity: AM: 1 μV for 10 dB (S+N)/N; SSB: 0.5 μV for 10 dB (S+N)/N
Selectivity: 6 dB at ±2.1 KHz; 40 dB at ±10 KHz
Image rejection: 50 dB minimum
Audio output: 3 watts
Speaker location: Bottom

FIELD TEST RESULTS

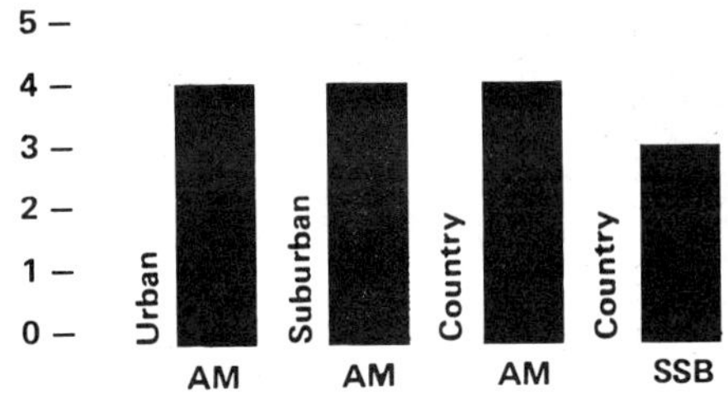

Measured AM Transmitter Power: 3.55 watts
Test Engineer's Comments: A consistent performer. Power line noise filter is effective only in negative ground installations.

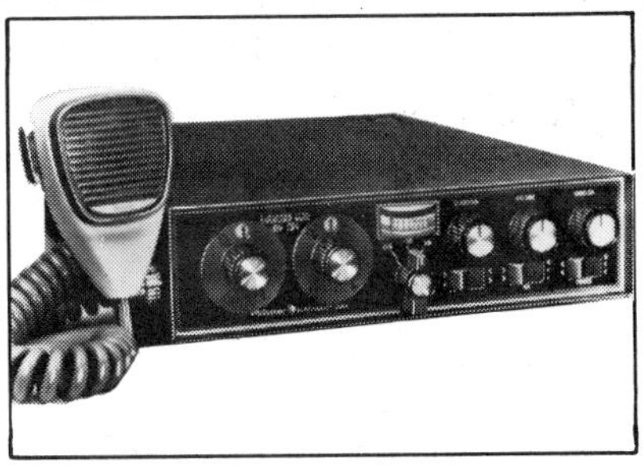

Palomar Electronics
DIGICOM 100

Manufacturer Says: Best performer.
1976 Price: $495.00, suggested retail
Features: AM/SSB mobile transceiver, switchable noise blanker. Speech compressor. Plug-in modular circuits. Dual channel-selectors and programmable digital synthesizer.

MANUFACTURER'S SPECS
Dimensions: 10-1/2" x 2-1/2" x 15"
Weight: 8-1/2 lbs.
Channels: 23
Power supply: 13.5 V DC negative or positive ground (with internal modification by dealer)
Circuitry: Programmable digital synthesizer
Operating environment: NA

Transmitter
Power output: AM: 4 watts; SSB: 12 watts PEP
Output impedance: 50 ohms
Power consumption:
- Transmit: NA
- Receive: NA
- Standby: NA

Harmonic suppression: 60 dB
Frequency tolerance: Within 0.005%
Modulation capability: 90% maximum AM

Receiver
Sensitivity: AM: 1.0 μV for 10 dB (S+N)/N; SSB: 0.5 μV for 10 dB (S+N)/N
Selectivity: NA
Image rejection: 70 dB
Audio output: 2 watts
Speaker location: Bottom

FIELD TEST RESULTS

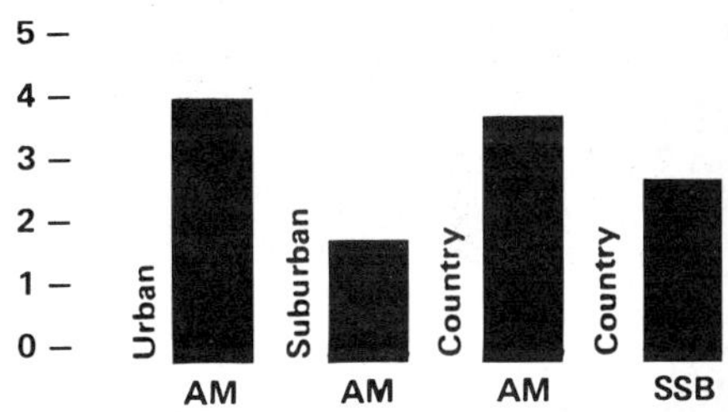

Measured AM Transmitter Power: Varied 3.8-4.0 watts
Test Engineer's Comments: Meter found very inaccurate. Dual selector-knobs judged difficult to comprehend when setting. Chassis unusually large for mobile installation.

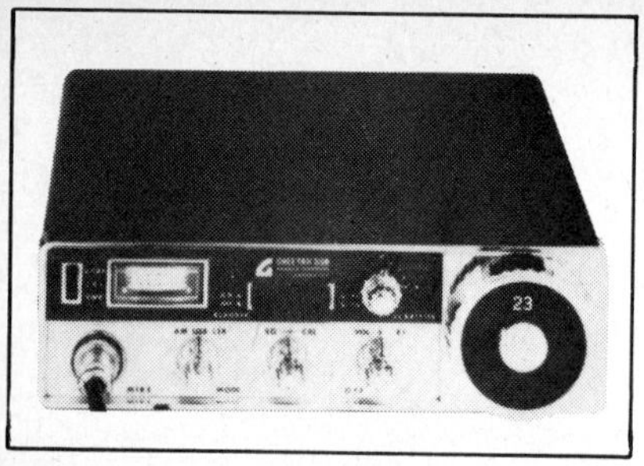

Pearce-Simpson CHEETAH SSB

Manufacturer Says: Best bang for the buck.
1976 Price: $419.95, suggested retail
Features: Dual conversion superheterodyne. Noise blanker. Crystal lattice SSB filter. S/RF/SWR meter. SWR meter reads both forward and reflected power. RF gain control. Fine tuning control with range of ±600 Hz. Transmitter has automatic load control for SSB. TVI trap.

MANUFACTURER'S SPECS
Dimensions: 8-5/8"W x 2-1/2"H x 10-1/8"D
Weight: 7 lbs.
Channels: 23
Power supply: 13.8 V DC negative or positive ground
Circuitry: All-transistor frequency synthesized
Operating environment: −5°F to +122°F (−20°C to +50°C)
Transmitter
Power output: AM: 4 watts; SSB: 12 watts PEP
Output impedance: 50 ohms
Power consumption:
Transmit: NA
Receive: NA
Standby: NA
Harmonic suppression: 50 dB minimum
Frequency tolerance: Within 0.005%
Modulation capability: 100%
Receiver
Sensitivity: AM: Less than 0.6 μV for 10 dB (S+N)/N
SSB: Less than 0.2 μV for 10 dB (S+N)/N
Selectivity: AM: 6 dB at 5 KHz, 50 dB at 20 KHz
Image rejection: 60 dB minimum
Audio output: 3.5 watts
Speaker location: Left side

FIELD TEST RESULTS

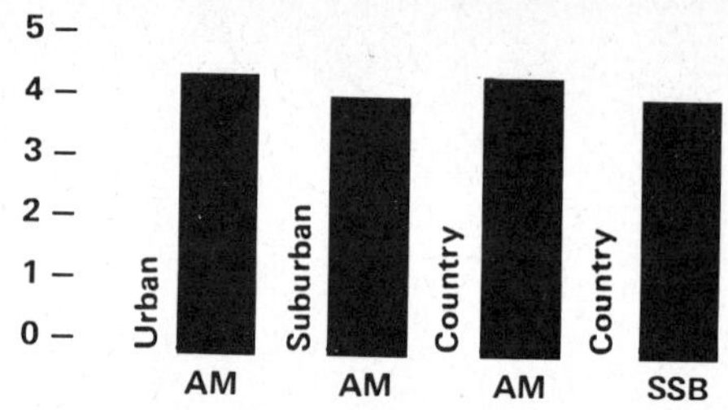

Measured AM Transmitter Power: 3.05 watts
Test Engineer's Comments: Meters judged especially useful. TVI trap very desirable in light of new FCC interference standards announced July, 1976. Transmitter audio-frequency response range more restricted on SSB.

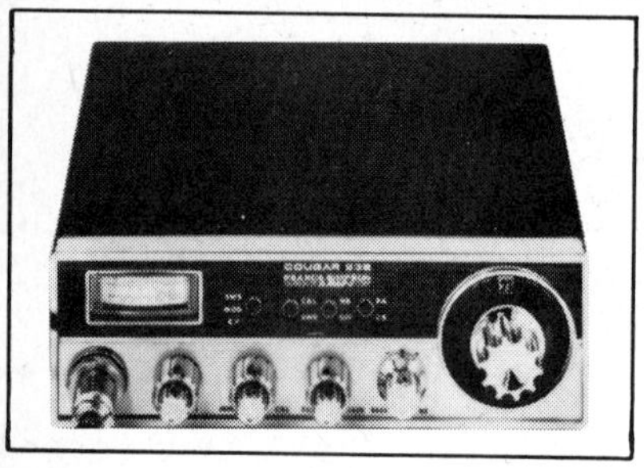

Pearce-Simpson COUGAR 23B

Manufacturer Says: Best bang for the buck.
1976 Price: $239.95, suggested retail
Features: Delta tuning. S/RF/SWR meter with SWR calibration, built-in automatic noise limiter, mike gain control, tone control, switchable noise blanker.

MANUFACTURER'S SPECS
Dimensions: 6-7/8"W x 2-1/4"H x 8-3/4"D
Weight: 4 lbs.
Channels: 23
Power supply: 13.8 V DC negative or positive ground
Circuitry: Frequency synthesizer, dual conversion superheterodyne
Operating environment: −22°F to +122°F (−30°C to +50°C)
Transmitter
Power output: 3.5 watts
Output impedance: 52 ohms
Power consumption:
Transmit: NA
Receive: NA
Standby: NA
Harmonic suppression: 55 dB minimum
Frequency tolerance: Within 0.005%
Modulation capability: 90%
Receiver
Sensitivity: 0.5 μV for 10 dB (S+N)/N using 1000 Hz, 30% modulation
Selectivity: 6 dB at 5 KHz
Image rejection: 60 dB minimum
Audio output: 5 watts
Speaker location: Bottom

FIELD TEST RESULTS

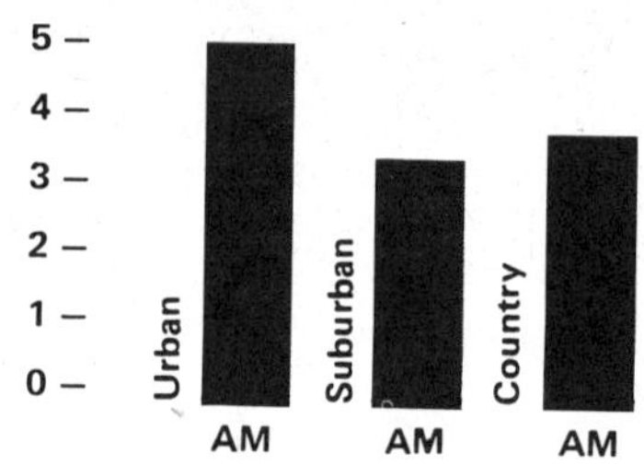

Measured AM Transmitter Power: 2.90 watts
Test Engineer's Comments: Meter functions especially useful but meter judged too small for accurate reading at a glance.

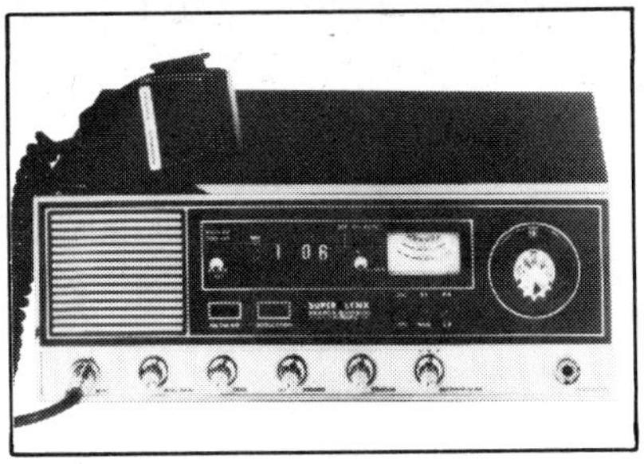

Pearce-Simpson SUPER LYNX

Manufacturer Says: Base station suitable for mobile operation.
1976 Price: $284.95, suggested retail
Features: AM base station equipped for mobile use on negative or positive ground. Switchable automatic noise limiter. Lighted digital clock with automatic turn-on. S/RF/Modulation meter. Mike gain control. Tone control. Fine tuning.

MANUFACTURER'S SPECS
Dimensions: 13-1/4"W x 4-1/2"H x 10"D
Weight: 8 lbs.
Channels: 23
Power supply: 13.8 V DC negative or positive ground and 117 V AC
Circuitry: Frequency synthesizer, dual conversion superheterodyne
Operating environment: –22°F to +168°F (–30°C to +65°C)
Transmitter
Power output: 3.5 watts
Output impedance: 50 ohms
Power consumption:
Transmit: NA
Receive: NA
Standby: NA
Harmonic suppression: 55 dB minimum
Frequency tolerance: 0.003%
Modulation capability: 100%
Receiver
Sensitivity: 0.5 μV for 10 dB (S+N)/N using 1000 Hz, 30% modulation
Selectivity: 6 dB at 5 KHz
Image rejection: NA
Audio output: 3.5 watts
Speaker location: Front

FIELD TEST RESULTS

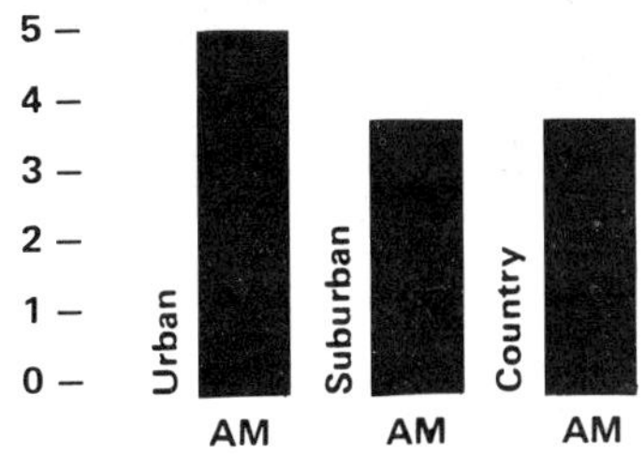

Measured AM Transmitter Power:
As base station on AC: 3.55 watts
As mobile station on DC: 3.10 watts
Test Engineer's Comments: Large meter judged easy to read. Manufacturer offers pedestal to tilt set upward for more convenient operation as base station in home or RV.

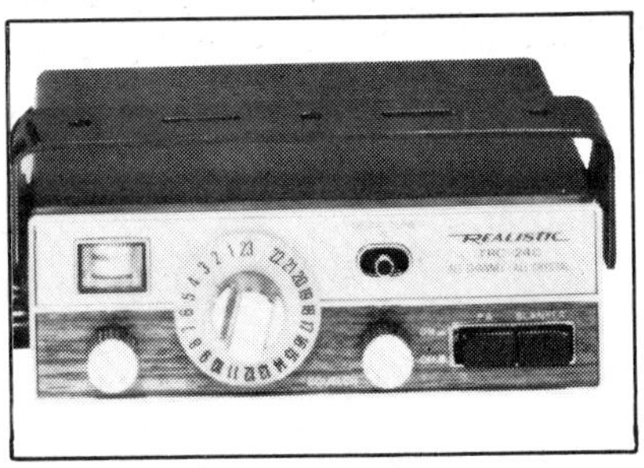

Radio Shack REALISTIC TRC-24C

Manufacturer Says: Best bang for the buck.
1976 Price: $159.95, suggested retail
Features: Automatic noise limiter built-in. Switchable noise blanker. Modulation indicator light in S/RF meter. Delta tuning switch.

MANUFACTURER'S SPECS
Dimensions: 6"W x 1-3/4"H x 7"D
Weight: NA
Channels: 23
Power supply: 13.8 V DC negative or positive ground
Circuitry: Frequency synthesizer, dual conversion superheterodyne
Operating environment: NA
Transmitter
Power output: 4 watts
Output impedance: 50 ohms
Power consumption:
Transmit: NA
Receive: NA
Standby: NA
Harmonic suppression: NA
Frequency tolerance: Within 0.005%
Modulation capability: 100%
Receiver
Sensitivity: 0.5 μV for 10 dB (S+N)/N
Selectivity: 6 dB at ±3 KHz
Image rejection: 50 dB or more
Audio output: 3 watts
Speaker location: Bottom

FIELD TEST RESULTS

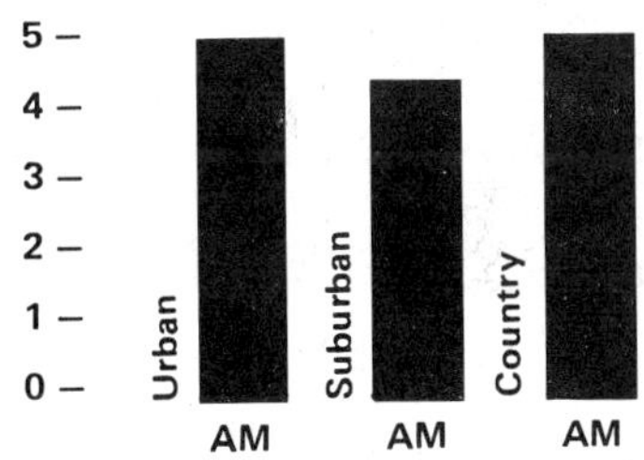

Measured AM Transmitter Power: 3.35 watts
Test Engineer's Comments: Noise controls especially effective. Meter judged too small for easy reading.

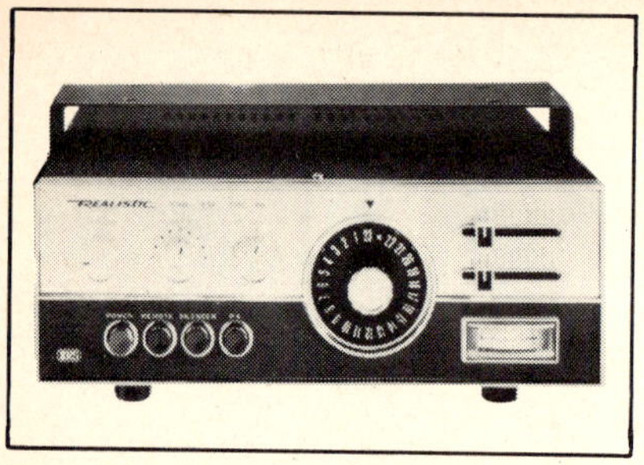

Radio Shack REALISTIC TRC-48

Manufacturer Says: Best performer.
1976 Price: $329.95, suggested retail
Features: Base/Mobile AM/SSB transceiver. Clarifier. Remote control on mike adjusts speaker volume. PA useable while monitoring CB receiver. Switchable automatic noise limiting and noise blanker. RF gain control.

MANUFACTURER'S SPECS
Dimensions: 8-7/8″W x 2-3/4″H x 10-1/2″D
Weight: NA
Channels: 23
Power supply: 12 V DC nominal, negative ground
Circuitry: Dual conversion superheterodyne
Operating environment: NA
Transmitter
Power output: AM: 4 watts; SSB: 12 watts PEP
Output impedance: 50 ohms
Power consumption: As mobile station on DC: 2.5 amp maximum
As base station on AC: 100 watts maximum
Harmonic suppression: 55 dB
Frequency tolerance: Within 0.005%
Modulation capability: NA
Receiver
Sensitivity:
AM: 0.5 μV or better for 10 dB (S+N)/N
SSB: 0.2 μV or better for 10 dB (S+N)/N
Selectivity: AM: 6 dB at 5 KHz; SSB: 6 dB at 2.1 KHz
Image rejection: 50 dB or better
Audio output: 4 watts
Speaker location: Bottom

FIELD TEST RESULTS

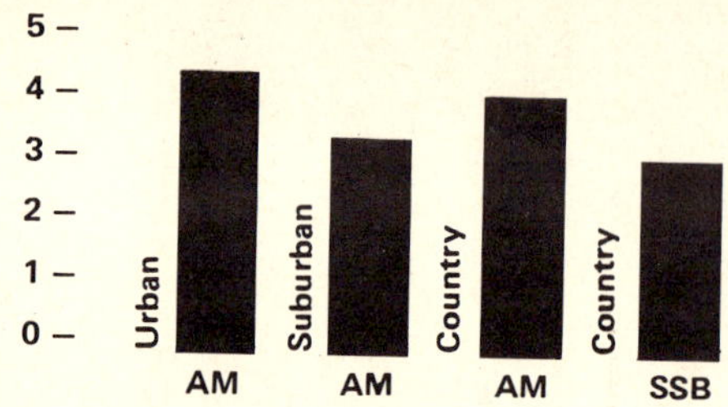

Measured AM Transmitter Power:
As mobile station on DC: 3.20 watts
As base station on AC: 3.45 watts
Test Engineer's Comments: Clarifier control shifts both receiver *and* transmitter on AM/SSB by a small amount, a convenient feature for operation within a CB network. ANL *and* noise blanker switched simultaneously by *Silencer* button; separate controls preferred.

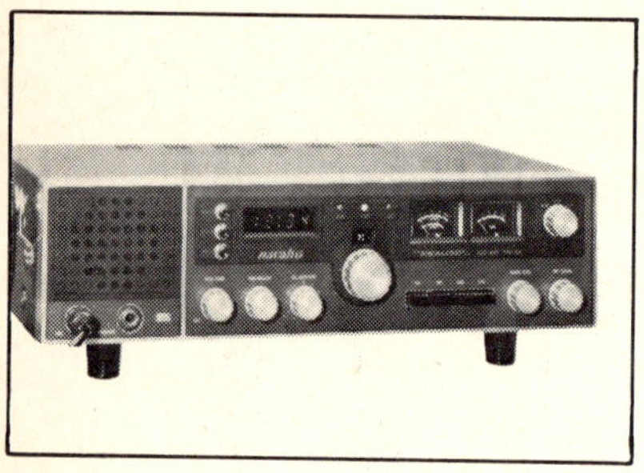

Radio Shack REALISTIC TRC-57 NAVAHO

Manufacturer Says: Base station suitable for mobile operation.
1976 Price: $399.95, suggested retail
Features: Digital phase lock loop. LED digital clock shows 12-hr. or 24-hr. time. Switchable automatic noise limiter and noise blanker. Meters for S/RF and SWR. SWR calibration control. RF gain control. Automatic modulation control. Fine-tuning control.

MANUFACTURER'S SPECS
Dimensions: 14-7/8″W x 3-7/8″H x 10-1/2″D
Weight: 16 lbs.
Channels: 23
Power supply: 120 V AC, 13.8 V DC negative or positive ground
Circuitry: Frequency synthesizer, dual conversion superheterodyne
Operating environment: –4°F to +122°F (–20°C to +50°C)
Transmitter
Power output: AM: 4 watts; SSB: 12 watts PEP
Output impedance: 50 ohms
Power consumption:
Transmit: On DC: AM: 1.7 amp; SSB: 1 amp; with no modulation
AM: 2.2 amp; SSB: 2.2 amp; at maximum output power
On AC: AM: 65 watts; SSB: 45 watts; with no modulation
Receive: On DC: 0.7 amp squelched, 1.6 amp full audio
On AC: 40 watts squelched, 60 watts full audio
Harmonic suppression: 55 dB
Frequency tolerance: Phase lock loop within 100 Hz
Modulation capability: 100%
Receiver
Sensitivity: AM: 0.5 μV for 10 dB (S+N)/N; SSB: 0.125 μV for 10 dB (S+N)/N
Selectivity: AM: 60 dB; SSB: 70 dB
Image rejection: 50 dB
Audio output: 6 watts
Speaker location: Front

FIELD TEST RESULTS

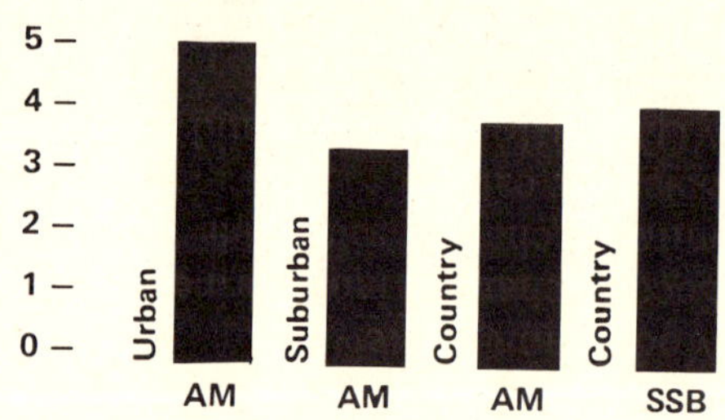

Measured AM Transmitter Power:
As base station on AC: 3.65 watts
As mobile station on DC: 3.55 watts
Test Engineer's Comments: Good fine-tuning control. Dual meters easy to read. LED clock operates on AC only. Fast clock setting judged especially useful.

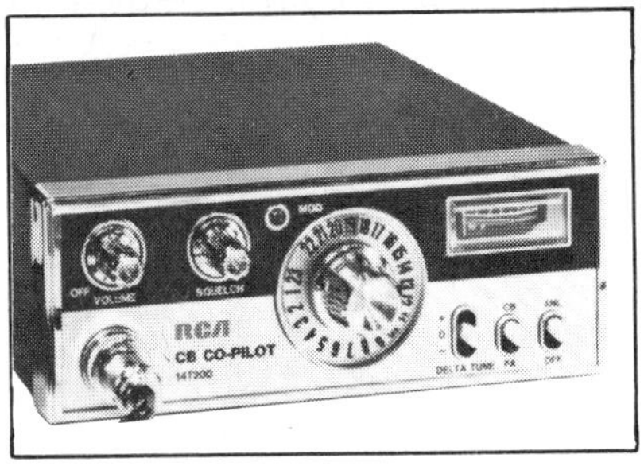

RCA 14T200 CO-PILOT

Manufacturer Says: Best performer/best bang for the buck.
1976 Price: $169.95, suggested retail
Features: Switchable automatic noise limiter, delta tune switch, S/RF meter, LED modulation indicator, automatic modulation level control.

MAUFACTURER'S SPECS
Dimensions: 6.5"W x 2.4"H x 7.75"D
Weight: 3 lbs. 5 oz.
Channels: 23
Power supply: 13.8 V DC negative or positive ground, operating range 12 to 15 V DC
Circuitry: Frequency synthesizer, dual conversion superheterodyne
Operating environment: −22°F to +122°F (−30°C to +50°C)

Transmitter
Power output: 4 watts
Output impedance: 50 ohms
Power consumption:
  Transmit: 1.4 amp nominal
  Receive: 1.1 amp full audio
  Standby: 0.24 amp squelched
Harmonic suppression: 50 dB minimum
Frequency tolerance: Within 0.005%
Modulation capability: NA

Receiver
Sensitivity: 0.7 μV for 10 dB (S+N)/N
Selectivity: 6 dB at 6 KHz, 50 dB at 20 KHz
Image rejection: 60 dB
Audio output: 3 watts minimum
Speaker location: Bottom

FIELD TEST RESULTS

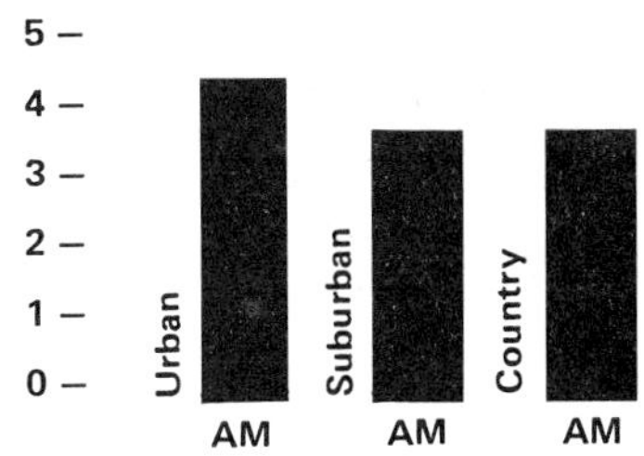

Measured AM Transmitter Power: 3.55 watts
Test Engineer's Comments: Small meter surprisingly readable.

Sharp Electronics Corporation CB-800

Manufacturer Says: Best performer/best bang for the buck.
1976 Price: $139.95, suggested retail
Features: Large LED channel indicator, switchable automatic noise limiter, delta tuning switch, automatic modulation level control. Channel 9 flashing indicator.

MANUFACTURER'S SPECS
Dimensions: 5-3/4"W x 2-1/4"H x 7-7/8"D
Weight: 3.6 lbs.
Channels: 23
Power supply: 12 V DC nominal, negative or positive ground
Circuitry: Frequency synthesizer, dual conversion superheterodyne
Operating environment: NA

Transmitter
Power output: 4 watts
Output impedance: 50 ohms
Power consumption:
  Transmit: NA
  Receive: NA
  Standby: NA
Harmonic suppression: Better than 50 dB
Frequency tolerance: Within 0.005%
Modulation capability: 100%

Receiver
Sensitivity: 0.7 μV for 10 dB (S+N)/N
Selectivity: 6 dB at ±3 KHz, 50 dB at ±10 KHz
Image rejection: NA
Audio output: 3 watts
Speaker location: Bottom

FIELD TEST RESULTS

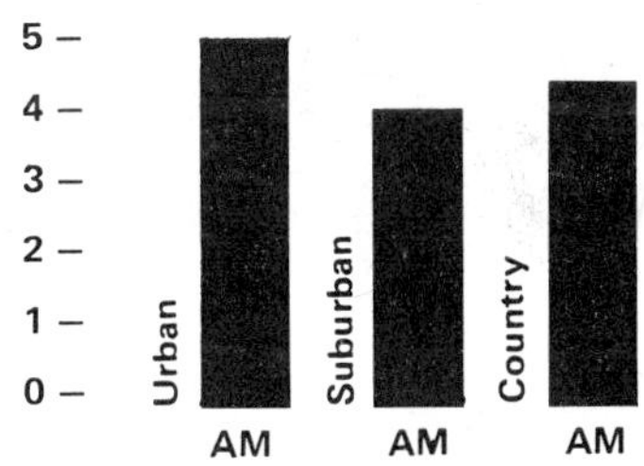

Measured AM Transmitter Power: 3.05 watts
Test Engineer's Comments: Digital channel indicator large and clear in most ambient light levels. Recommend shaded mounting position.

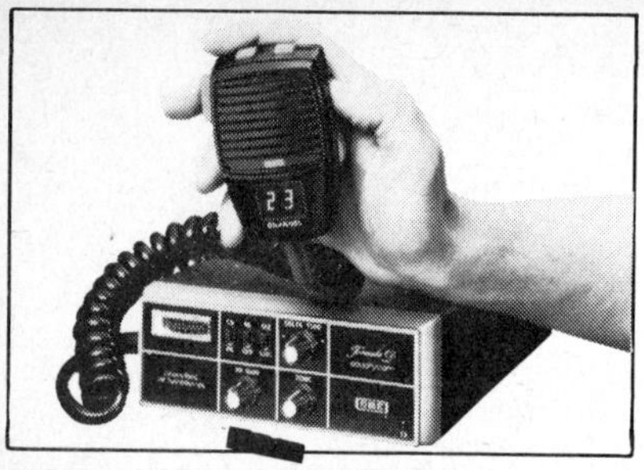

SBE, Inc. 32CB FORMULA D TOUCH/COM

Manufacturer Says: Best performer
1976 Price: $289.95, suggested retail
Features: Major controls on microphone, two-speed channel changing, LED channel indicator, switchable noise limiter, delta tune, switchable rf gain control, tone control.

MANUFACTURER'S SPECS

Dimensions: 6.7"W x 2.4"H x 9.4"D
Weight: 6 lbs.
Channels: 23
Power supply: 13.8 V DC, negative or positive ground. Operating range 11.7 to 15.9 V DC
Circuitry: Frequency synthesizer, dual conversion superheterodyne, phase lock loop
Operating environment: –4°F to +122°F (–20°C to +50°C)

Transmitter

Power output: 4 watts
Output impedance: 50 ohms
Power consumption:
Transmit: 2.2 amp at 13.8 V DC with 95% modulation
Receive: 1.5 amp full audio
Standby: 1 amp squelched
Harmonic suppression: 57 dB at ±500 KHz, 64 dB down at all other frequencies
Frequency tolerance: Within 0.003%
Modulation capability: to 100%

Receiver

Sensitivity: 0.05 μV for 10 dB (S+N)/N
Selectivity: 50 dB at ±10 KHz. 60 dB at ±20 KHz. 65 dB at ±30 KHz
Image rejection: 40 dB or better
Audio output: 3.5 watts
Speaker location: Bottom

FIELD TEST RESULTS

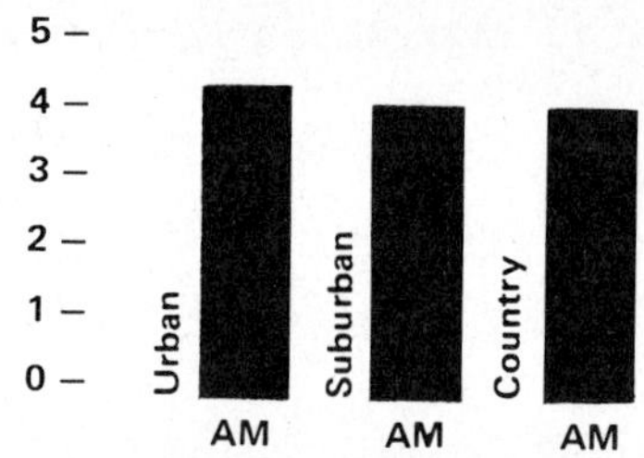

Measured AM Transmitter Power: 3.6 watts
Test Engineer's Comments: Remote control-head switches and controls easily manipulated in one-handed operation without juggling. Speech-compressor circuit good. Polarized window for channel indicator appears to improve LED visibility in daylight.

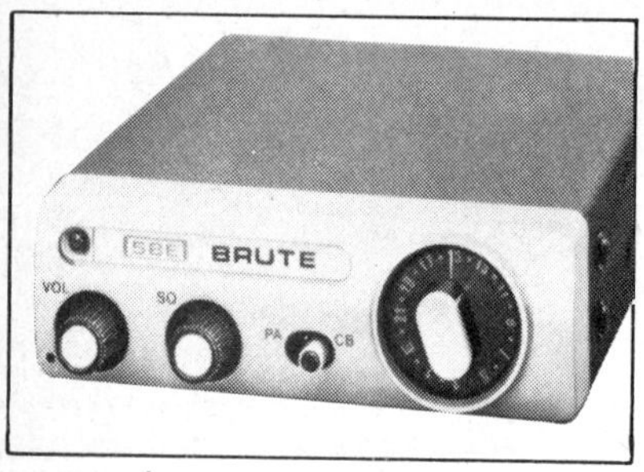

SBE, Inc. 34CB BRUTE

Manufacturer Says: Best bang for the buck.
1976 Price: NA
Features: Extremely compact mobile AM transceiver. Built-in automatic noise limiter. RF indicator lamp. Phase lock loop.

MANUFACTURER'S SPECS

Dimensions: 4-1/2"W x 1-3/4"H x 5-3/4"D
Weight: 2.2 lbs.
Channels: 23
Power supply: 13.8 V DC negative or positive ground, operating range 11.7 V DC to 15.9 V DC
Circuitry: Frequency synthesizer, dual conversion superheterodyne, PLL
Operating environment: –4°F to +122°F (–20°C to +50°C)

Transmitter

Power output: 4 watts maximum
Output impedance: 50 ohms unbalanced
Power consumption:
Transmit: 1.4 amp at 95% modulation
Receive: 0.2 amp squelched, 0.35 amp at 2 watts audio
Standby:
Harmonic suppression: NA
Frequency tolerance: Within 0.003%
Modulation capability: 59-100%

Receiver

Sensitivity: 0.7 μV for 10 dB (S+N)/N
Selectivity: 6 dB at ±5 KHz, 45 dB at ±10 KHz, 55 dB at ±30 KHz
Image rejection: NA
Audio output: 2 watts
Speaker location: Bottom

FIELD TEST RESULTS

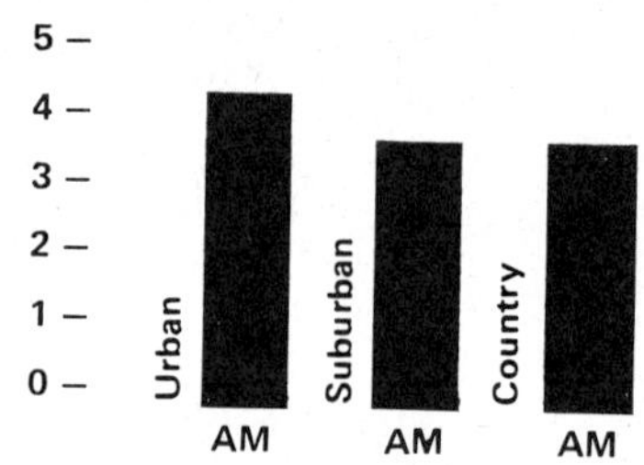

Measured AM Transmitter Power: 3.45 watts
Test Engineer's Comments: Channel selector difficult to read in high ambient light levels. Extreme compactness and light weight make for easy installation almost anywhere, especially in subcompact vehicles. Permanently attached microphone would be problem in glove box unless lid is open during operation.

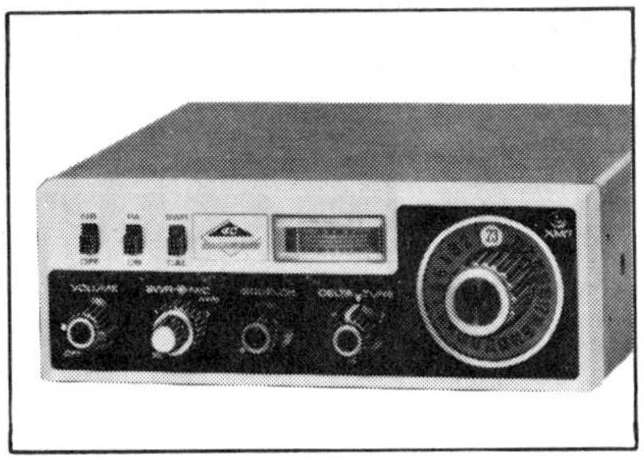

Tram Diamond Corporation
DIAMOND 40

Manufacturer Says: Best bang for the buck.
1976 Price: $229.95, suggested retail
Features: Fail-safe transmitter final-stage circuitry. Microphone gain control. Delta tune. S/RF/SWR meter with SWR calibrate. Switchable noise blanker. Automatic noise limiter.

MANUFACTURER'S SPECS
Dimensions: 6-57/64"W x 2-23/64"H x 8"D
Weight: 6 lbs.
Channels: 23
Power supply: 13.8 V DC negative or positive ground
Circuitry: Frequency synthesizer, dual conversion superheterodyne
Operating environment: –22°F to +122°F (–30°C to +50°C)
Transmitter
Power output: 4 watts
Output impedance: 50 ohms
Power consumption:
  Transmit: 1.8 amp
  Receive: 1.3 amp full audio output
  Standby: 0.5 amp squelched
Harmonic suppression: Meets FCC requirements at time of type acceptance
Frequency tolerance: Within 0.005%
Modulation capability: 100%
Receiver
Sensitivity: 0.3 μV provides more than 1 watt audio, with gain control full and noise blanker off. Signal mod. 30% at 1 KHz sine wave
Selectivity: 6 dB at 4 KHz, 60 dB at 20 KHz
Image rejection: More than 50 dB
Audio output: NA
Speaker location: Bottom

FIELD TEST RESULTS

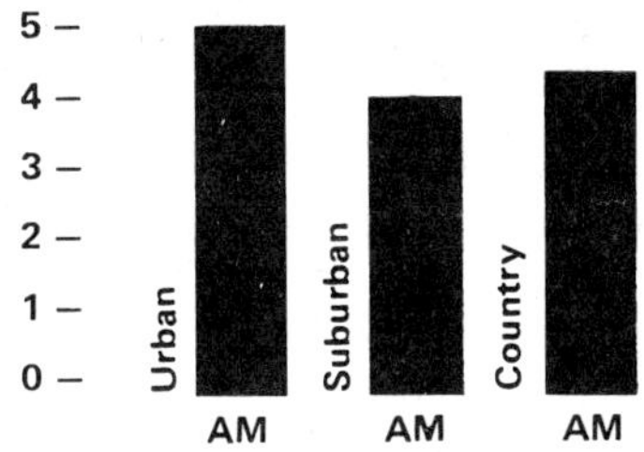

Measured AM Transmitter Power: 4.05 watts
Test Engineer's Comments: SWR meter judged especially useful but panel meter illumination too dim.

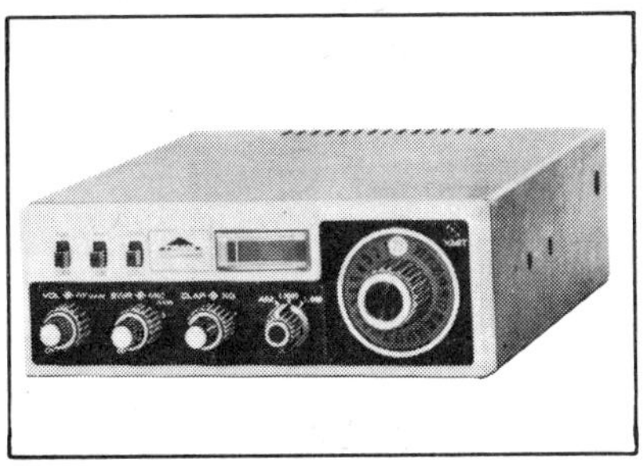

Tram Diamond Corporation
DIAMOND 60

Manufacturer Says: Best performer.
1976 Price: $419.95, suggested retail
Features: AM/SSB mobile transceiver. Fail-safe automatic protection circuitry for final transmitter transistors. Microphone gain control. S/RF/SWR meter with calibration control. Switchable noise blanker. Automatic noise limiter. RF gain control.

MANUFACTURER'S SPECS
Dimensions: 6-57/64"W x 2-23/64"H x 9-23/32"D
Weight: 6.5 lbs.
Channels: 23
Power supply: 13.8 V DC negative or positive ground
Circuitry: Frequency synthesizer, dual conversion superheterodyne
Operating environment: –22°F to +122°F (–30°C to +50°C)
Transmitter
Power output: AM: 4 watts; SSB: 12 watts PEP
Output impedance: 50 ohms
Power consumption:
  Transmit: AM: 2 amp full modulation; SSB: 2.2 amp for 7 watts PEP
  Receive: 0.5 amp squelched, 1.3 amp full audio output
Harmonic suppression: Meets FCC requirements at time of type acceptance
Frequency tolerance: Within 0.005%
Modulation capability: 100%
Receiver
Sensitivity: AM: 0.3 μV provides more than 1 watt audio, with gain control full and noise blanker off. Signal mod. 30% 1 KHz sine wave. SSB: 0.1 μV provides more than 1 watt audio output.
Selectivity: AM: 6 dB at 4 KHz, 50 dB at 20 KHz SSB: 6 dB at 2.2 KHz, 50 dB at 5.5 KHz
Image rejection: More than 50 db
Audio output: NA
Speaker location: Bottom

FIELD TEST RESULTS

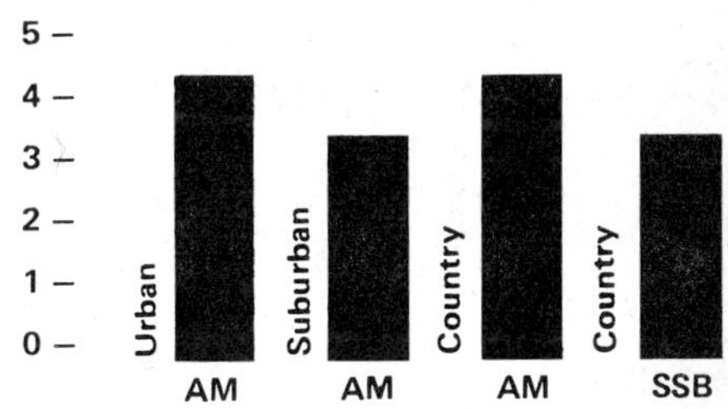

Measured AM Transmitter Power: 3.55 watts
Test Engineer's Comments: SWR meter judged especially useful but illuminated panel meter too dim. RF gain control helpful on SSB and for nearby stations that could overload this sensitive receiver.

CHAPTER 9

PORTABLE CB SETS

If you've gotten the impression all rigs are variations on a standard theme, welcome to the new world of hand-held portables—where little is standard among sets that *appear* to be the same.

Hand-held portables are completely self-contained CB broadcasting stations. The essential package consists of a built-in microphone, speaker, transceiver, telescoping antenna and batteries.

When mass-production methods for miniaturized electronics made it possible for every kid on the block to have a walkie-talkie, there was a proliferation of flea-powered toy transceivers with outputs up to 100 milliwatts—the maximum allowed by the FCC for unlicensed operation. A husky hog-caller can get more range with an enthusiastic shout. Nevertheless, many of these toys are currently on CB frequencies. Don't confuse them with the well-designed industrial 100 milliwatt transceivers for the CB band. These are intended for short-range communication around construction sites, industrial areas, forestry camps and the like. These have better sensitivity than toys—1 microvolt or less for 10 dB quieting—so similar units can communicate up to about 1/2 mile with reasonable reliability.

A practical way to use these short-range hand-held portables is with a good base station or mobile transceiver as a control center. Then, if outlying portable-to-portable communication becomes difficult, the superior transmission and reception capability of the central control station can be used by its operator to relay messages. To do this legally, the portable sets must be licensed.

By now, you will not be surprised to learn the FCC develops a legal twitch at the thought of licensed base or mobile stations talking to unlicensed portables.

There are two kinds of portable sets. Those with 100 milliwatts or less power don't *require* a license but they may qualify for a license anyway. Those with power greater than 100 milliwatts on CB channels must have a CB license and must not supply more than 4 watts rf power to the antenna.

Communication among 100-milliwatt units does not require a license, but it is illegal for a licensed CB station to communicate with unlicensed stations.

Any radio set licensed by the FCC must be type accepted. This means the FCC checks the manufacturer's specs on a particular model and approves it for sale with the understanding that all other sets with that model designation will be made the same way.

Type acceptance by the FCC isn't really a guarantee of good quality because the FCC only checks certain things like rf power output and frequency drift. Manufacturers who build apparatus to comply with FCC rules are a lot different than toy makers, so an FCC type-approved set is probably of good quality and should give satisfactory performance.

We are not quite through with

This footloose CBer is operating his Midland 13-861 on rechargeable batteries. When he hikes back to his car, he can operate mobile from the vehicle's electrical system.

our legal twitch about portables. It happens that the FCC approves portables for more than one kind of operation. If they have approved one model for calling the chickens, you can't legally use it for baying at the moon. So, you have to check two things: Is it type approved? If so, for what?

Each CB portable must be FCC type accepted for operation as a Class D Citizens Band station under Part 95 of the regulations. You must attach a transmitter identification card to each station and observe all operating rules and regulations.

Many electronic catalogs call *all* hand-held portables *walkie-talkies* whether they can be licensed or not. Read the fine print for any unit in the 100-milliwatt category even if the advertisement says it is a CB set. If it does not have FCC type approval under Part 95 you won't be able to use it for licensed CB operation when necessary. Read the fine print for more powerful portables, too. Many transceivers are made for operation on industrial channels squeezed in among the CB channels. They carry type approval under FCC Part 91. There are also transceivers for transportation use approved under FCC Part 93. These operate at the edge of certain CB channels. Don't become an illegal channel-jumper by purchasing the wrong rig just because an advertisement offers *CB Walkie-Talkies.*

No matter what it's called, a good 4-watt hand-held CB portable can have all the legal power and almost all the features of a good mobile rig. If you feed the signal into a more efficient antenna than the whip attached to the set you can get a good range, depending on antenna height. You can also improve range with the existing whip by getting it higher above terrain. So go climb a tree! It's legal. Just don't exceed the antenna height limit of 20 ft. above the top-most branch, even if you figure a way to climb up there!

The variety of portable designs

Portables of 100 miliwatts or less can communicate among themselves but can't be used to communicate with licensed CB transceivers unless the portable is FCC type-accepted as a CB set.

is like the march of transceiver technology; your problem is you'd never know it from outward appearance of the sets. There are lots of single-conversion superheterodyne receivers, though the trend is toward double-conversion for better selectivity. Don't worry about it in the deep woods unless there is a large population of CBers.

Here are the important items to check in hand-held portables.

READING SPECIFICATIONS OF PORTABLE CB SETS

Channels—You can get all-channel capability. Many sets come with 3 to 6 channel *capability*—a tricky word. Be sure the transceiver is equipped with crystals for the channels *you* want; the set may be furnished with just one pair of crystals, good for only one channel. You'll have to buy the others—in pairs—at a cost of $5 per pair or more.

I advise getting full-band portables. The Citizens Band is getting busier and you will need to be able to choose among 40 channels in hope of finding a quiet frequency. If you settle for less, be sure to include emergency channel 9.

Transmitter Power—Unlike mobiles and base transceivers, all of which are designed for a full 4 watts into the antenna terminals—and usually come close to this goal—hand-held portables may be designed for lower rf outputs. Among the sets requiring a CB license, you'll find units putting out a puny 1/4 watt as well as full 4-watt rigs. Much advertising for portables gives the power *into* the final transmitting stage, *not the output* into the antenna—a misleading numbers game. Read the fine print. For less-than-full-power sets, be especially wary about transmitter power specs; it is better to compare power *from the final* stage to the antenna rather than *power to the final* because the efficiency of this stage may differ among various set designs.

Some transceivers have a switch for reducing transmitter power in short-range operations, conserving the batteries. You'll also find sets that offer alternative power supplies. Check the transmitter power with each type of power supply—it may be different on batteries than on AC.

Power Supply—Nothing is standard. You'll find sets operating with 8 or 10 AA penlite cells, 10 or 12 rechargeable AA nickel cadmium batteries, 9-volt transistor-type batteries, nominal 12-volt vehicle batteries with 13.8-volt generators, and 115-volt AC through an adapter which converts to nominal 12 volts DC. A battery-condition indicator is essential. This is a light or meter with a test switch or button which tells you if the batteries are still OK.

Some portables use disposable one-time-only batteries, usually size AA flashlight cells. Some use rechargeable Nicad cells. Some give you the option of using either

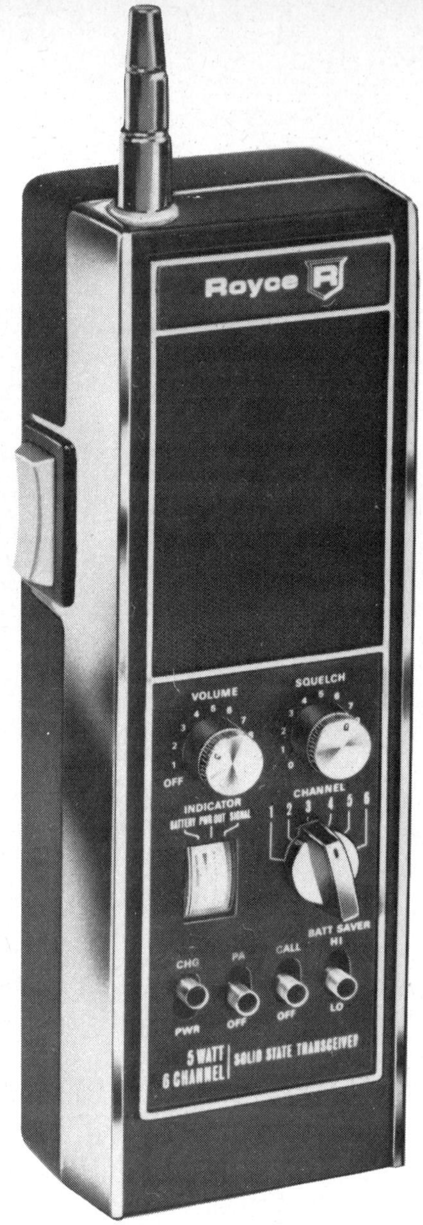

BATT SAVER **switch on this Royce hand-held 6-channel set cuts battery drain when switched to** LO **by reducing rf power to 1 watt.** HI **is 3 watts output for longer range. Don't be confused by the numbers game all portable manufacturers seem to play. The 5-watt label means 5 watts *into* the final stage. Output is limited to 4 watts just like any other CB set.**

BATTERIES FOR PORTABLES

When you depend on batteries to power your CB rig, the major problems are getting enough supply voltage to put out a good signal, and battery life.

Flashlight batteries of zinc-carbon provide 1.5 volts each when they are fresh. You'll need 8 to 10 of these for a transceiver, depending on the rig's make and model. The nice part of disposable zinc-carbon batteries is their general availability everywhere.

Zinc-carbon batteries can be stored for long periods under refrigeration. High temperatures cause rapid deterioration, so don't toss your portable into the vehicle glove box or leave it in the sun. You may find it helpful to rotate sets of batteries in your rig to allow recovery time after each use. Zinc-carbon batteries are relatively inexpensive per unit—but turn out to be costly in terms of useful operating hours.

Alkaline batteries cost about twice as much as zinc-carbon batteries but can give you up to 10 times the useful life. Current drain can be continuous with less harm to the battery. Ambient temperatures of $-20°F$ to $+140°F$ do not affect performance significantly. You can buy a recharger for *some* alkaline batteries but it may take 12 to 15 hours. Caution: Most alkaline cells are for one-time use. You can only recharge those cells specifically sold as rechargeable. Labels aren't very informative. When in doubt, don't recharge: *Batteries can leak or explode* if not made for recharging.

Nickel-cadmium batteries are popularly known as Nicads. You get only 1.2 volts per unit so you need more of them to provide proper power input. Nicad battery contacts must be kept scrupulously clean; oxidation, no matter how invisible, robs energy. Polish to a bright finish with a pencil eraser—if you rotate sets of Nicads, polish each time you install a set in your rig. Some battery packs use sets of Nicads welded together for better contact.

You can recharge Nicads up to 500 times. Use only a charger specifically marked for Nicads. Generally you can't overcharge Nicads so you don't need a meter on your charger. Follow the recommended charging times or make it a practice to recharge overnight.

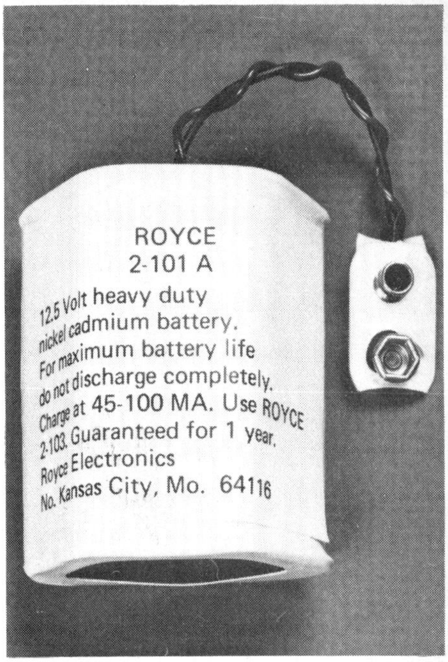

Some manufacturers package Nicad cells in one plug-in assembly so you don't have problems with oxidation of the individual battery contacts. These ***battery packs*** **need a special charger.**

type.

Because disposable and rechargeable cells don't have the same voltage, the battery compartment is larger on a set that lets you use either type and you have more flexibility. Disposable batteries furnish 1.5 volts per cell, so 8 cells in series gives a total of 8 x 1.5 = 12 volts. To get the same voltage from Nicads with only 1.2 volts per cell, you have to use 10 cells. Therefore the battery compartment for a set using either type must have room for 10 cells. If you are using Nicad, use all 10. If you are using disposable batteries, use only 8 and fill the empty spaces with 2 dummy batteries which come with the set.

Getting more power output on batteries is easy with a set designed to work with 12 volts of internal batteries of either type *plus* work on a vehicle generating 13.8 volts at fast idle or faster engine speed. This means the set can really use 13.8 volts all the time and will probably operate with reduced power if you install only 12 volts worth of batteries. You can improve power by installing one extra disposable 1.5-volt battery to get 13.5 volts. Use only one of the two dummy batteries.

Weight—Portables seem to get heavier the farther you carry them. For a small set, about half the weight is in batteries. Specifications quote weight without batteries. When figuring the total weight, add half a pound for 8 flashlight cells

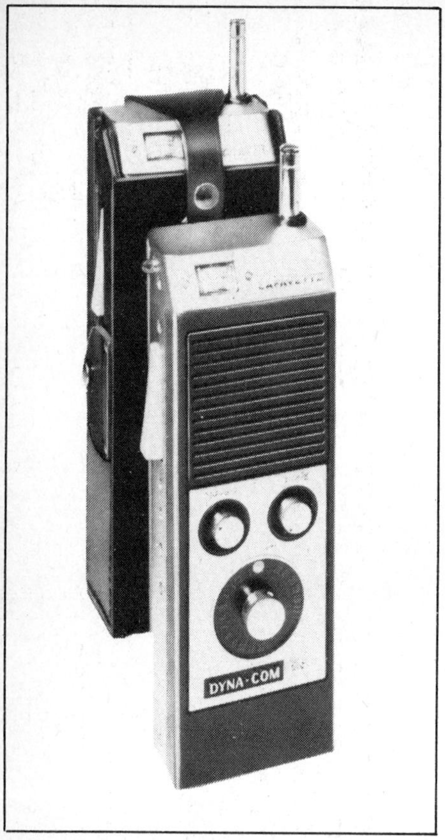

Portable sets should be protected from dirt and moisture which can enter through exposed plug-in connectors on set. This Lafayette Dynacom 23 has a flap which closes over the connectors for protection.

and 2 dummies, or about 9 ounces for 8 alkaline cells and 2 dummies, or 10 ounces for 10 Nicads, or proportionately more for additional batteries.

Antenna—You'll get a telescoping base-loaded or center-loaded antenna with most sets. Look for an option to connect a different antenna such as a fiberglass dipole, a shorter-than-standard whip for tight quarters, or a marine, mobile or base-station antenna.

Some portables come with a built-in swivel at the base of the antenna. For best communication range, CB antennas should be vertical. A swivel allows you to position the antenna that way, no matter how you hold the set.

Controls—The position of controls varies from set to set. You'll find them on the top, front or both. Consider their position carefully for convenience in the ways you'll be operating the set.

You'll need to get at the channel selector switch, squelch and volume controls, and the press-to-talk switch without gymnastics or loss of time. If there is provision for an external microphone-speaker combination you'll have more freedom because you won't have to bring the whole set up to your mouth to transmit, or to your ear to hear.

FEATURES

Speech compressor circuitry will assure full modulation. A jack for external *public address* is handy, but you'll need to carry an extra speaker which ought to be mighty efficient—meaning heavy—because you'll be lucky to get more than 2.5 watts audio output. Automatic *noise limiting* is important, particularly around vehicles and construction sites. *Automatic gain control* is a must if you value your hearing and your peace of mind. An *external mike* may be an optional accessory. In most portables it is not an ordinary mike but a microphone and speaker combined like the one in the transceiver. Don't expect top-flight voice quality, although some of these units are surprisingly good—make a listening test. A *meter* usually combines several functions. You should have: Battery condition, power output and received signal strength.

Connectors—I shudder at the exposed mini-connectors on most hand-held portables. They are an open invitation to dirt, moisture and corrosion—and the first place to look for trouble when your set becomes intermittent. If you're not going to use all those little accessory jacks, cover them with a piece of duct tape—found in any hardware store. Carrying cases are carefully cut out to provide access to such jacks and furnish no protection.

Carrying Cases—Hand-held portables are very convenient for people with three hands. A good case is important to free your hands for other activities and to protect the set against the inevitable bumps and knocks. Cases are sometimes included in the selling price of the rig; otherwise you get only a wrist or shoulder strap and an opportunity to cough up more money. Cough! While you're at it, fold up a small plastic bag and tuck it in the case. No case is waterproof and you will later use the plastic bag to protect the set when working in the rain or in salt spray at the beach. Slide the set into the plastic bag and close it snugly around the antenna. Then you can put the whole thing back in the carrying case.

Battery Chargers—Some sets recharge Nicads simply by being plugged into an external charger while the batteries remain in the set. Of course, your set is out of service during the recharging period—about 3 to 4 hours. It's better to have an extra set of Nicads, expensive as they are, so you can substitute them when fresh batteries are needed. You'll need a separate battery charger capable of charging about 10 Nicads. Caution: Nicads should never be stored while fully discharged because they lose their ability to take and hold a charge. Keep them charged.

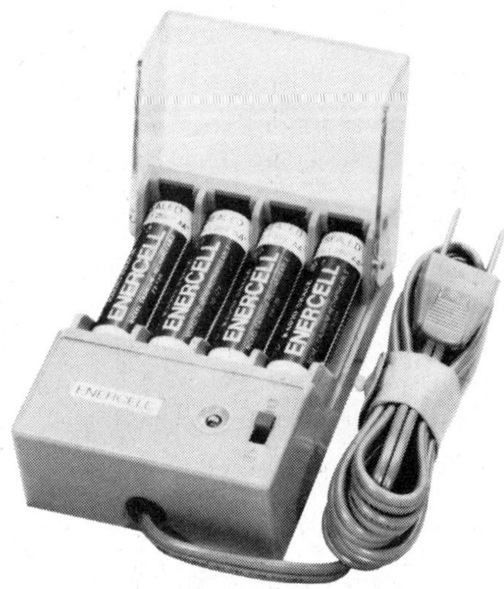

Recharge individual Nicad batteries with an accessory like this Radio Shack unit if your portable doesn't have its own charger.

CHAPTER 10

HOW TO SET UP YOUR BASE STATION

A base station can be a CBer's pride and joy. Most are loaded with knobs, dials, lights, meters, special controls and even digital clocks for 12-hour or 24-hour time. A deluxe base station creates a sense of professionalism. You expect it to be more powerful because it is larger than a mobile rig. Right? Well, *almost* right.

The maximum power permitted into the antenna is the same as for mobile rigs and portables. Four watts is four watts. But, if you're not going to drive it around or carry it with you, you can do a lot of tricks with the antenna. Four watts into a gain antenna reaching 60 ft. above the ground is a lot of *effective* radiated power compared to four watts fed into a shortened, loaded whip antenna bobbing around on somebody's car-trunk lid. Four watts into a gigantic directional-beam antenna 20 ft. above your roof puts your signal way beyond anything you can do with your trusty hand-held portable set or any kind of mobile rig.

It's true, any CB set can feed these super antennas. You can hook up a mobile or portable set and get the same range. But base-station transceivers are more convenient, sometimes better made, and always more fun. A base station's gadgets seem to take on more meaning to CB operators and often they turn out to be enormously useful in wringing out the last mile of communication range.

WHO AND WHERE

There's more to getting set up than just buying an attractive rig and bringing it home or to your business location. I recommend you do some shopping and a lot of heavy thinking before laying out any money.

Who is going to use the base station and where should it be located for maximum convenience? If you are going to use your mobile CB and a home base for family communications, consider putting the base station in or near the kitchen—that's where mama is much of the time. If the kids are going to use it after school, mama probably doesn't want it in her kitchen. If papa is going to use it after work and during the evening, he'd probably like to have it in his study, home office or "shack." If you plan to join a CB emergency-monitoring team, think about where you should locate it for maximum convenience around the clock. Who knows when you may be operating from midnight to 5 AM?

Consider privacy and the noise environment. If you put the base station where there is a lot of noise from other sources—the back room at a boiler factory or near the home hi-fi set when the kids want to play records—it isn't going to work out. Don't forget *you* will be a source of noise when you are operating.

UTILITIES

Besides a proper location and environment, your base station

Put your personal communications center in a room where nobody will bother your equipment and your station operations won't bother anybody. This radio shack is from Radio Shack—a CB transceiver, police-band scanner and short-wave receiver.

needs some connections with the outside world. The transceiver is connected to an outside antenna by coax cable. If you locate the CB against an outside wall, or near one, it's easier to run the cable inside the building and the installation will look better.

You need electrical power for the CB set and maybe for other receivers or other transmitters, depending on how far you get into radio communications. Also, maybe a coffee pot for late night vigils, a desk lamp, electric clock, electric typewriter, tape recorder and whatever else you end up finding to be necessary.

Most household and business-building circuits are wired and fused for a total load of about 1500 watts. This is a 15-ampere circuit. You can check the fuses or read the switches on a circuit breaker to see how many amps each circuit is intended to furnish.

To locate the fuse or circuit breaker for the area where you plan to put in radio equipment, plug a table lamp into a wall socket. Turn on the lamp. Then have somebody flip off circuit breakers in the fuse box, or unscrew fuses until the lamp goes off. That's it. Read the ampere rating and note which fuse or circuit breaker it is.

Next, you want to find out what else is on the same electrical circuit in your house. Each fuse or circuit breaker in the electrical panel feeds one circuit in the building.

By moving the table lamp from outlet to outlet while switching the circuit on and off at the main fuse panel, you can trace the circuit and find whatever else is connected to it.

When you have figured that out, you have two things to worry about. You don't want noisy electrical appliances on the same circuit as your CB rig. "Noisy" means electrical noise, not acoustical noise you hear in the room. Anything with an electric motor—such as the blender in the kitchen—will feed electrical noise into a CB set on the same circuit. The power supplies in CB sets are supposed to filter out such noises, but they don't do it perfectly.

The other thing to worry about is overloading the circuit, which is why you checked the ampere rating of the fuse or circuit breaker. Every electrical appliance should be marked with the number of watts it uses—usually the marking is on the back, sometimes in raised letters cast into plastic. Light bulbs are marked on the bulb. Add up all the wattage ratings of everything you can find connected to the circuit you intend to use. Then add in the wattage ratings of the CB set and anything else you plan to connect for use with the CB. If you haven't picked the CB yet, use a relatively high wattage rating in your calculations so you have some margin to spare. If you figure on 100 watts for the CB, you allow for a deluxe unit with all features.

As a rule of thumb, you can estimate the total wattage a circuit can supply by multiplying the amperes by 100. A circuit fused for 15 amps will supply 1500 watts. This number is conservative by 10% or so, but why not have a little in reserve?

If the total wattage demand of everything on the circuit will be less than the capacity of the circuit, and you don't have kitchen appliances or your furnace motor on the same circuit, you are making progress.

You may have to use plug-in "double sockets" or "triple sockets" to make all the connections for your shack. This will normally be OK if you run the electric cords neatly and don't group a lot of high-wattage units onto the same outlet. If you are getting close to the limit, banish the coffee pot—you'll be surprised how much power they take.

How About a Land Line?—Many uses of CB require a phone at hand. Emergency monitors, for example, grab the phone immediately and relay info to the proper authorities.

It won't work if your CB is upstairs and the nearest phone is downstairs. Of course the phone company will put a phone or a phone plug anywhere you want it, but you may save the fee by thinking it over and possibly making your plan so the CB ends up where the phone already is. Don't laugh! We get smarter by profiting from those little mistakes—preferably made by somebody else.

Let's assume you have taken a step at a time and solved the problems of where to locate your equipment to get power, telephone and antenna connections with minimum fuss and expense. Now pick your base station if you don't already have it.

DREAM A LITTLE

Even if your use of CB is strictly utilitarian, there is something rewarding about operating a base station that does just about everything. You get a feeling some smart electronics engineer put it all together just for your convenience. A fine base station can have a lot of surprisingly thoughtful features.

Imagine a base station that turns itself on so you don't miss scheduled contacts, listens for private calls, monitors channel 9 while you're working other channels, automatically keys the transmitter to turn it on when you talk, tailors the transmitter audio circuits to improve the quality of your voice on the air and helps tune your antenna. You get the idea: Convenience, flexibility and high precision.

I confess I am hooked on all these features of first-class base transceivers. So were our professional communication engineers who did the performance field-testing in Chapter 8. They chose the Tram D 201 for the test bases and found every meter and control on the rig served some useful purpose and they used them frequently in setting up and running the tests.

In the next few pages features are discussed which you might like to have in your base transceiver. Make sure you buy not only capability but also performance. A meter can be a toy or a measurement device. A voice-operated transmitter—VOX—can make life miserable if you can't adjust it to your manner of speaking. A built-in clock control may add to service headaches. No feature is worth the investment unless it does its job well, and you need it.

BASIC FEATURES OF EVERY BASE-STATION RIG

Don't be surprised to find bottom-of-the-line base stations looking like stripped-down mobile rigs. Some are made for a price. And some are made for simplicity of use, so the relative absence of controls has nothing to do with price. Don't be misled by the term bottom-of-the-line; the least-expensive base station by one manufacturer may still be more costly than the least-expensive model of another manufacturer. Shop for combinations of features you want before considering price.

Some manufacturers make base stations for every price bracket, adding features as the price increases. You may find sleek-looking transceivers with fancy cabinets, large dials and meters but no more control flexibility and operating features than a budget-priced mobile.

I don't recommend spending much extra for appearance just so you have a rig that looks like a base station but works like a mobile. It's better to select a good mobile set from the more competitive market and install it in an accessory cabinet with a good, regulated power supply so you can operate it on house current. I'll tell you how to pick the right supply shortly.

Here are the minimum features of inexpensive base stations:

- AM only
- Squelch
- Fine tuning by a three-position switch
- Built-in automatic noise limiter
- Headphone jack
- S/RF meter
- Volume control
- Modulation-indicator light
- Front-mounted speaker

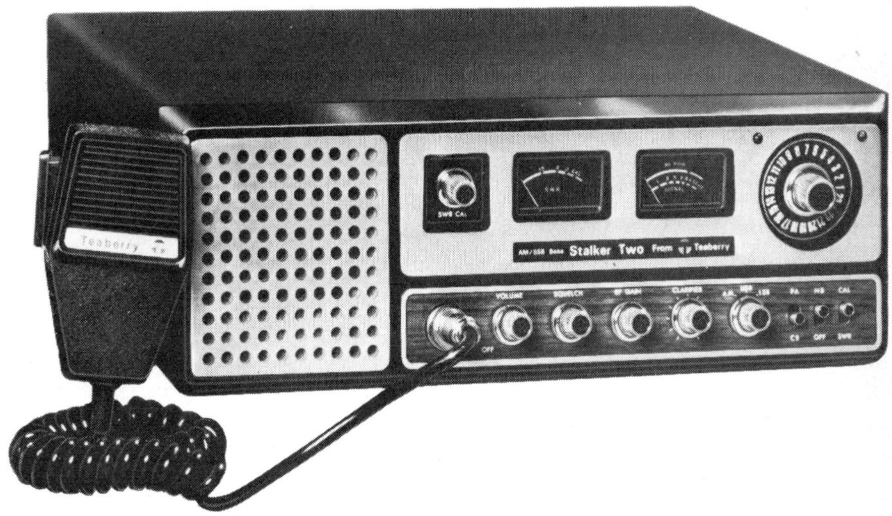

Serious CBers head for the sidebands when they put in base stations. This is Teaberry's Stalker Two with the rock-steady phase-lock-loop which is best for SSB.

Most sets have a DC power cord so you can also use them as mobile sets with negative- or positive-ground vehicle power.

At higher prices you'll get rf and microphone-gain controls, switchable noise controls, continuous fine tuning, receiver tone control, single-sideband and other features in larger cabinets.

BASE-STATION FEATURES YOU WON'T MIND PAYING EXTRA FOR

Some sets use different circuitry for certain AM functions than they do for SSB because the circuits give better performance that way. For example, two kinds of noise limiters may be used—one that works best on AM, another that works best on SSB. If you are paying for an AM/SSB set, you won't mind the extra cost of circuits that make each system work at its best.

In AM/SSB sets, look for two ranges on the fine-tuning control—a fairly wide range on AM, such as 3,000 Hertz; and a narrower range on SSB, such as 1,200 Hertz. Because the control knob turns the same number of degrees on either scale, the reduced tuning range of 1,200 Hertz has the effect of "expanding" the scale. This means you must turn the control knob farther for the same frequency change when receiving SSB, therefore the control is easier to get precisely at a particular spot. This is important in SSB receivers because fine tuning must be exactly right or voices are garbled.

Some sets have continuous receiver tuning across the band so you can hear a station that is so far off frequency you couldn't fine tune it with crystal-controlled circuits and a fine-tuning control or switch.

A few sets are claimed by their manufacturers to combine the best advantages of transistors and vacuum tubes by using some of each. I've used these sets and think they are great performers so it may be worth giving them the additional ventilation they require. For reliability, the choice is probably all solid-state. If you're considering a base station that might have to operate on a battery for prolonged periods of emergency, rigs with any vacuum tubes will drain the battery faster.

Better Meters—Meters are on nearly every base station. There is room for large ones in most cases. However, there are differences in the way they perform. Some meters show only relative power—a large number is better than a small one, but neither tell you how much *actual* power there is. The best meters give actual AM carrier power—the output in watts of your unmodulated signal. On SSB, a good rf power meter will be of the averaging type. Otherwise, because of voice characteristics, it would be jumping wildly. Such a meter tends to read below mid-scale when you are talking because of the gaps between spoken words. For *test* purposes only when reading PEP output, announce that you are testing, then give a short sharp whistle into the microphone. This gives continuous modulation and a more accurate power reading.

Another difference in meters on base stations is calibration of the S meter for received signal strength. Typically, these meters read from 1 to 9 or S-1 to S-9 and then read higher signal strengths on a dB scale in steps of 10 dB. A received signal strength of S-1 may be unintelligible. A received signal strength of S-9 is very strong and signals above that may cause you to turn down the rf gain control.

CBers often read the S scale in "pounds"—another form of slang. The main use of an S meter is to tell another operator how strong his signal is at your receiving point. *I'm reading you at 8 pounds, good buddy,* means S-8.

Because of this casual or "gossipy" use of S-meter readings, and because there is no agreed standard among CB manufacturers today, a meter that reads *relative* signal strength is normally OK. You can pay more for a set that reads actual received signal strength—if you can find one—but so many other things affect field strength that an accurate reading is usually not worthwhile in ordinary applications. Intervening buildings or hills between transmitter and receiver make exact S-meter readings meaningless because the building or hill can significantly reduce the received-signal strength. At a slightly different location, the same signal may jump from S-3 to S-9 just because you got out of a building's "shadow."

A built-in SWR meter with calibration controls should be part of a base transceiver. You can buy an external meter as an accessory, but you're much more likely to check SWR periodically, as you should, if you can do it by flipping a switch. Built-in meters measure standing waves at the most critical point—the final output stage of the transmitter. At best, an external accessory meter measures a few inches away.

Talk Power—This phrase used in manufacturer's literature is a catch-all referring to any features that help you get full modulation or close to it.

A good base station should have a mike gain control. This lets you adjust for full modulation with your own voice and your own way of speaking—loud or soft. To prevent overmodulating, the FCC now requires a modulation-limiting circuit in transmitters equipped with a mike gain control.

A good base station should have speech compression. This increases the modulation percentage of the transmitter, on the average, by reducing the difference between the loudest parts of your speech and the softest parts. With your speech compressed or made more uniform in loudness, the transmitter can be driven harder and modulated at a higher percentage without overmodulating. Therefore a speech compressor gives you more talk power.

You also want a transmitter tone control which affects your voice quality as it goes out onto the air. Very deep or very high-pitched voices can be made more intelligible by suppressing the extreme voice frequencies and bringing up the middle tones. Improved intelligibility is another contribution to that ambiguous but desirable feature called *talk power.*

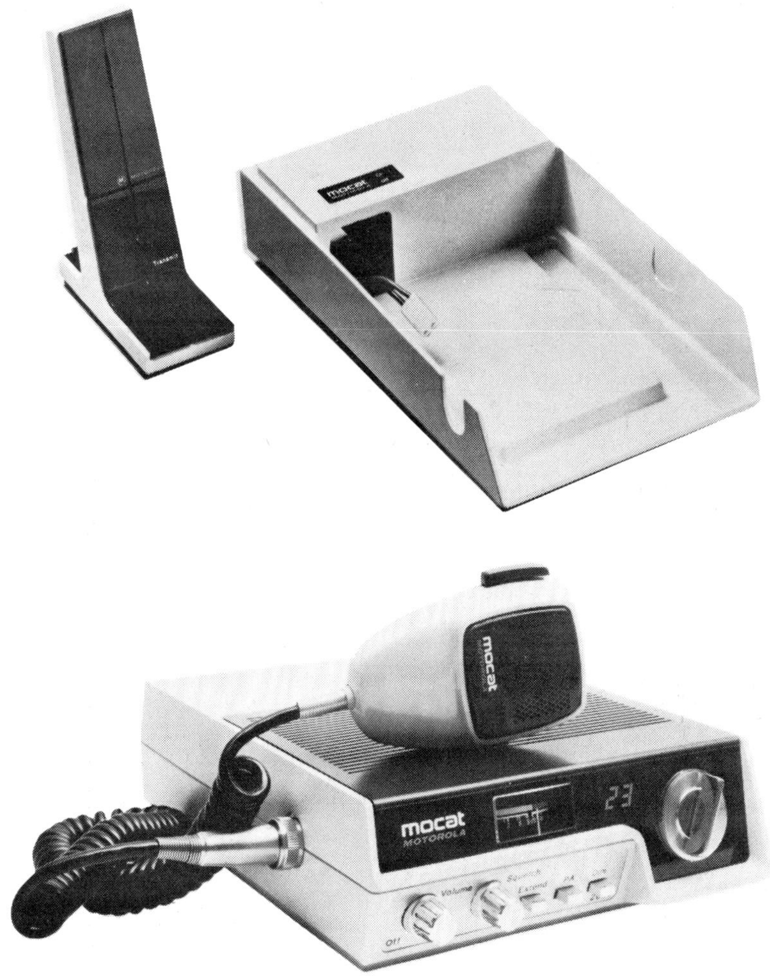

A good mobile rig works fine as a base station. This is Motorola's base-station adapter for mobiles like the Mocat 2020. You can also get a desk microphone for a more convenient base station use.

MY CHOICE

In addition to the basic features listed earlier in this chapter, I choose the following for convenience of operation and improved talk power:

Modulation meter instead of modulation-indicating light

Switchable noise limiters instead of non-switchable.

RF gain control

Mike gain control

Continuous fine tuning rather than a switch

Two-range fine tuning

Receiver tone control

Transmitter tone control

SSB operation

RF power meter reading actual watts

Speech compression

Modulation limiter

Second-channel monitoring with monitored channel selectable

Selective calling

VOX with sensitivity and delay adjustments

HOW TO INSTALL A BASE STATION

You may find my safety recommendations for installing a base station are more cautious than those in the instruction manual packed with your set. Play it safe and you won't have any shocking anecdotes to tell.

Where you bring coax in from the outside, use a feed-through bushing through the wall or window sill to make the entry weathertight. Feed-throughs to work with walls up to 14-inches thick are available in electronics parts stores. You can saw a long one to fit your wall. Drill the hole slanting upward from outside to keep water from running along the cable into your house. Some feed-throughs have an external overhang for additional drip protection. Make a drip-loop in the cable so the coax slants upward for the last few inches before going into the bushing.

Locate near an AC power outlet. If you anticipate operating on battery power in an emergency, buy one of the new sealed batteries which never needs water or ventilation. You can put one on a shelf with a small trickle charger and practically forget about it. Because they are sold with what appears to be a true life-time warranty, chances are good you'll never have to buy another one. Older style batteries with caps for venting and filling cost slightly less, but you can anticipate buying new ones every two years or so. And, you'd have to provide for ventilation to the outside or plan to live with the unpleasant odor they create. Caution: Do not charge an unsealed battery in a room without lots of ventilation.

Charging batteries give off explosive gases.

Accessory Power Supply—If you intend to use a mobile rig as a base station with an accessory AC power supply it is best to use the one recommended by the manufacturer for your particular transceiver. If you buy a general-purpose power supply, be sure it is adequate for the power consumption of your set. Read the specs for maximum current drain in the transmit mode.

Don't buy just any power supply marked "for CB use." Get a unit that provides 1 or 2 amps more than the current drain given for "transmit:full modulation" on your rig's specification sheet. A typical power supply for base station use of a mobile transceiver should provide a *minimum* of 3 amps regulated DC current with a surge of 5 amps. It's good to have circuit breakers and an indicator light to show when it's on.

You may find it cheaper to buy a sealed battery with a trickle charger. This combination also gives you emergency operating capability when power fails—something you don't get with an accessory power supply.

Ground the System—If you place the set on metal such as a table or filing cabinet, and put the microphone on another metal object, with perhaps other accessories on additional metal surfaces, be sure to bond all cabinets and tables together with a common ground. Fasten braided metal bonding straps or #12 copper wire between them; clean away paint at the connection points to make contact with bare metal. Connect to ground as described in the next step.

Ground the transceiver and metal furniture to a metal cold-water pipe or an earth ground. First connect a #12 wire of copper or electrical-grade aluminum to the metal chassis of your set, under a screw head. Some base stations have ground-connection terminals specifically marked. Use a star washer on the screw to bite into the metal for a better connection.

Run the other end of the wire to a grounding clamp on the cold-water pipe. Buy the clamp at your CB store, an electrical supply or building supply store. Be sure to sandpaper or file a clean spot on the pipe before tightening the clamp. If you connect to an earth ground instead of a cold-water pipe, drive a copper groundstake into the earth at least 5 ft. or whatever your local electrical code requires. In the desert you may have to go down 12 ft. or more, which is difficult. *Ground your transceiver even if it is not on a metal surface.* The antenna should also be grounded through a spark-gap lightning arrester described in the section on lightning protection in Chapter 13.

Installing a Telephone Patch—Some CBers connect their transceivers to home or office telephone systems. Their rigs can send or receive calls by land line over CB. When somebody with a mobile rig wants to contact another person in the area who is without a CB but has a phone, a base-station operator communicating with the mobile station can dial the third party and connect that person to the base transceiver through a phone patch. When the phone patch is on, the third party has two-way communication with the mobile station. It's mighty handy for certain business calls, if you don't mind being overheard—and plenty useful when you're late for an appointment or trying to make reservations as you drive toward a destination. The burden, though, is on the base operator.

Phone patches are sold as accessories for base stations. You'll find instructions packed in the carton that tell you it can be connected directly to a telephone terminal block—that little square box on the baseboard to which your telephone is already connected—or to a wall jack, the kind used to plug in extension phones.

I suggest you use the wall jack and connect the phone patch to a diplexer—a plug-socket combination that goes between the wall jack and your extension phone plug. You plug the phone into the diplexer and the diplexer into the wall jack. Then connect the phone patch unit into the diplexer. You can use either phone or phone patch with this set-up. Most phone patches sold today have a protection circuit to isolate the phone line from any DC voltages of your transceiver. The phone patch should be connected to your transceiver according to the instructions furnished by the accessory manufacturer.

Phone patches are legal under FCC rules only if they provide a way for the base-station operator to hear the communication going out over the transmitter and a switch that can cut off illegal transmissions.

CHOOSE AND INSTALL AN ANTENNA

As indicated at the beginning of this chapter, the antenna you choose makes the difference. This is another decision that looks easy but isn't.

Start with the material in the following chapter which tells what you need to know about antenna types, radiation patterns, gain and other basic theory so you can understand sales brochures and sales clerks.

Chapter 13 tells you how to install a base station antenna without falling off the roof, or watching the antenna fall off the roof in the next strong breeze.

CHAPTER 11

WHAT YOU REALLY NEED TO KNOW ABOUT ANTENNAS

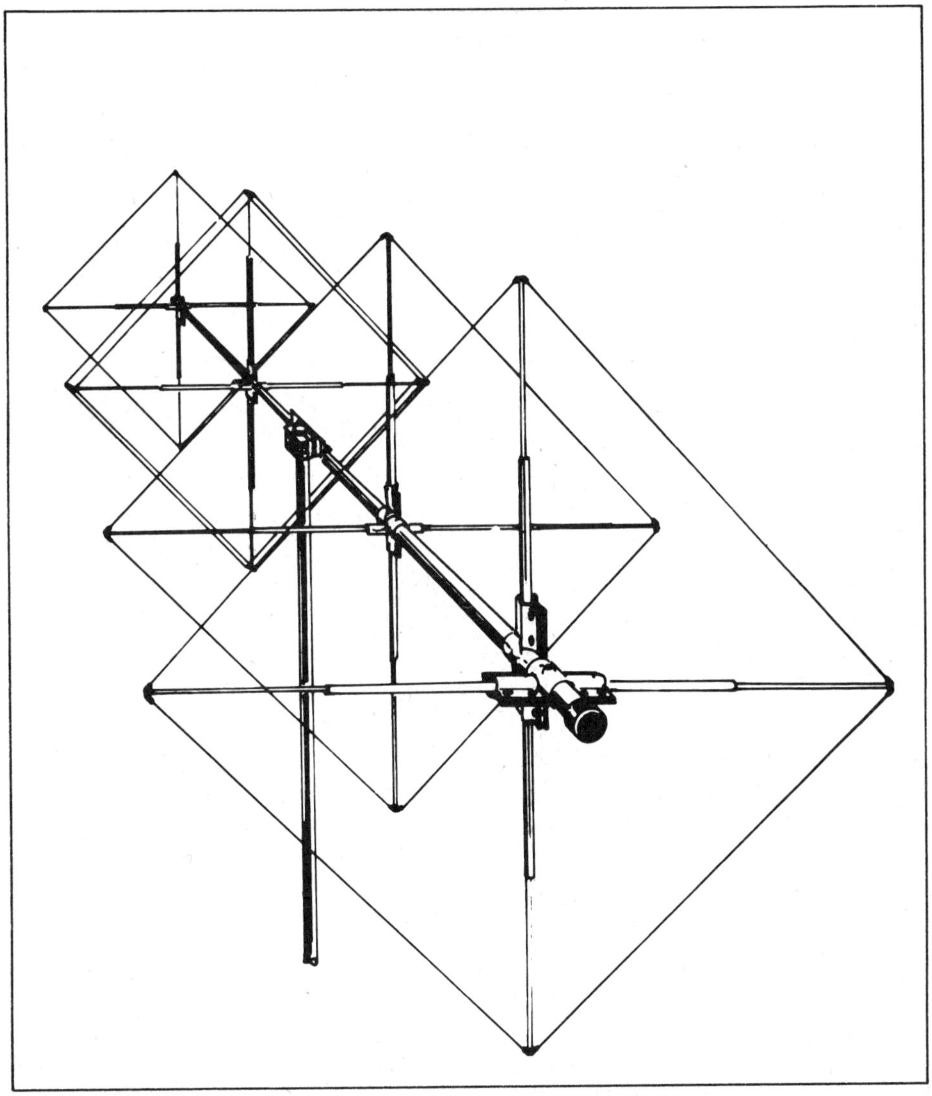

This *Big Gun II* by Hy-Gain is a quad antenna.

The antenna is the best way to improve performance because the rest of your CB system is a sealed box with a legal limit of 4 watts of radio-frequency (rf) power output fed to the antenna.

Disregard marketing jargon like "power packed" or "super space-age." What you need to know are the technical terms that tell you about antenna performance and their practical significance.

WAVES

An antenna radiates electrical energy into space, or receives energy from another antenna. Your CB uses the same antenna for both purposes, switching the antenna connection between transmitter and receiver.

If you could see energy radiated from an antenna as you view it from above, it would look like waves of water expanding in rings, similar to waves in a pool where you have just dropped a stone.

The waves grow weaker as they recede, which is why your radio signal grows weaker with increasing distance from the antenna. Waves bouncing against an object are reflected and distorted, which explains some of the uncertainties of the paths radio waves take among buildings and vehicles in big cities.

WAVELENGTH

When viewing waves in water, you can see or measure the length of each wave. The distance from the crest of one wave to the crest of the next is called *wavelength.* Radio waves also have wavelength. In the CB band, the wavelength is approximately 36 ft. Higher radio frequencies have shorter wavelengths.

FREQUENCY

The number of waves created in one second is called the *frequency.* Because the waves repeat the same pattern over and over again, they are considered as *cycles.* One complete wave is one cycle. To honor a pioneer in radio science named Hertz, we now use the word *Hertz* to mean *cycles per second.* If a wave makes 10 cycles per second, we say the frequency is 10 Hertz (Hz).

Radio engineers use the prefixes *kilo* and *mega* when dealing with high frequencies. Kilo means one thousand. Mega means one million.

The frequency of your voice waves is thousands of cycles per second, or *kilohertz* (KHz). The frequency of the radio waves in the CB band is near 27 million cycles per second, or 27 *megahertz* (MHz). The abbreviations KHz and MHz are commonly used on CB spec sheets.

FREQUENCY BAND

A group of frequencies with upper and lower limits is called a band. The CB band extends from 26.96 MHz to 27.41 MHz. Within this band, each CB channel operates on a different frequency. By tuning your CB set to the assigned frequency for a particular channel, you broadcast and receive on that frequency. If you are communicating with others on the same channel, they are using the same frequency. A list of CB channel frequencies is in Chapter 20.

RADIATION

We often call antennas *radiators* because they radiate rf energy. Radiation from a CB antenna travels upward into space as *sky waves* and along the surface of the earth as *ground waves.* CB stations communicate using ground waves. Antennas with a lower *angle of radiation* send stronger ground-wave signals and give you more communication range.

ANTENNA LENGTH

There is a vital relationship between the length of an antenna and the length of the radio wave it must radiate or receive. An antenna is most efficient when it is exactly as long as the rf wavelength it is transmitting or receiving. But a full-wavelength antenna of about 36 ft. would be mighty tall for your vehicle, boat or hand-held portable.

Fortunately, an antenna is also efficient at certain fractions of the exact wavelength. So you'll find 5/8-wavelength and 1/2-wavelength antennas on base stations and boats, and 1/4-wavelength antennas on vehicles.

Electrical Length—The physical length of an antenna—as you would measure it with a tape measure—is not the same as its apparent length when "viewed" by radio waves. CB antennas are physically a little shorter than their electrical wavelength, depending on the material and construction of the antenna.

Loading Coils—Even a quarter-wavelength antenna, about 9 ft., is often too long for some mounting locations. To shorten it physically while maintaining the same

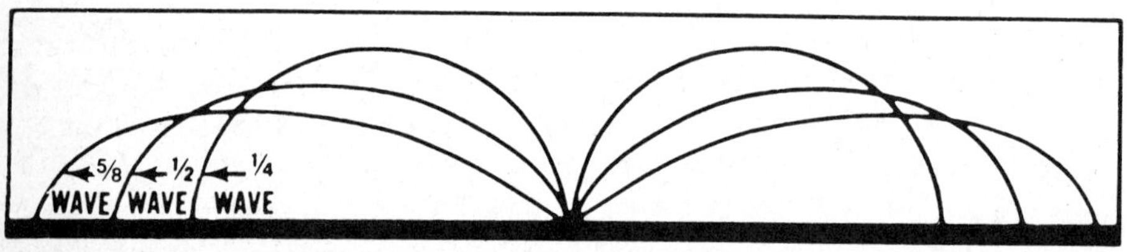

Depending on length, CB antennas have different angles of radiation. As you can see, 5/8-wave antennas have the lowest angle and therefore put more signal where your good buddy can receive it.

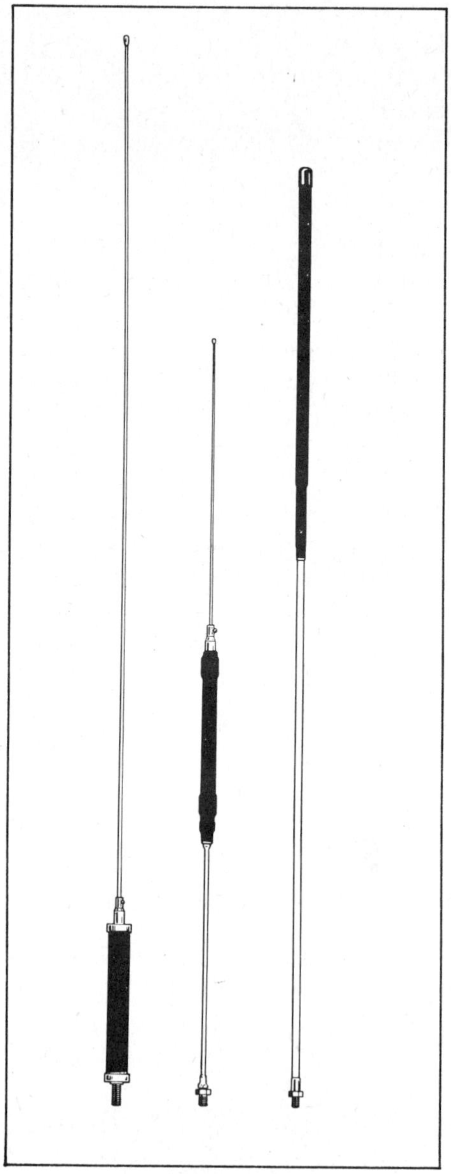

Bulges in CB antennas are loading coils to make them act longer electrically than they really are in inches. Here are CPD Blue Streak bottom-loaded, center-loaded and top-loaded antennas. Bottom-loaded antennas on a car don't move around as much due to wind currents.

electrical length, manufacturers use a *loading coil* as part of the antenna. Think of this as a part of the antenna length coiled up somewhere along the antenna—at the base, near the center, or at the top. This makes the antenna physically shorter because part of its length is coiled, but it continues to have the same electrical wavelength as before.

Loading coils don't radiate rf energy and, in fact, use up some of it. Loaded antennas are less efficient as radiators. The shorter they are made by use of antenna loading, the less efficient they become.

ANTENNA MATCHING

The efficiency of an antenna also depends on how well it is matched to the transmission line and transmitter. If there is proper *matching*, maximum power will flow from the transmitter to the antenna and be radiated.

Think of this problem as connecting several lengths of garden hoses. Using different diameters will impede the maximum flow of water. If we were actually using hoses, we would match them by getting the same diameter, measured in inches. In the rf world of transmitters and antennas, what we match is called *impedance* and its measurement is in *ohms*.

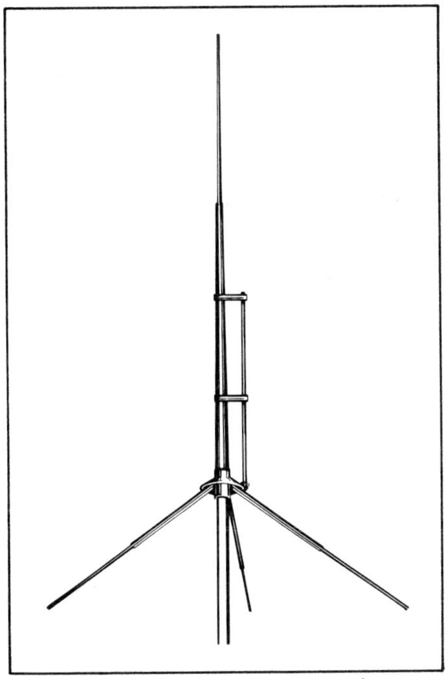

Antenna groundplanes are a metal car body or wires or radial rods as shown on this Turner SK-22B.

Impedance is technically complicated. You need only check to be sure that impedance specifications for transceiver, coaxial cable and antenna are all the same and then install the antenna so it works at its specified impedance. In CB the standard impedance is 50 ohms.

GROUNDPLANES

Some antennas have 50-ohms impedance only when they are located above an electrically conducting surface called a *groundplane*. It is an artificial ground. Antennas shorter than a half-wave need a groundplane.

Groundplanes can be the metal roof or body of a vehicle, or metal rods or wires extending outward from the base of the antenna which extend like spokes of a wheel and are called *radials*. A groundplane is sometimes called a *counterpoise* in antenna literature and in marine applications.

A CB antenna, vertical above its groundplane should have an impedance of 50 ohms. If the antenna is bent over by wind or the weight of a loading coil, or not mounted properly in the first place, the impedance will be wrong and antenna efficiency reduced.

Even though antennas shorter than a half-wave need groundplanes for correct impedance, their use is not restricted to shorter antennas. Using a groundplane with 1/2-wave or 5/8-wave antennas lowers the radiation angle so there is more ground wave and less sky wave. This puts more of your signal along the surface of the earth and extends communication range.

STANDING WAVES

When there is a mismatch of antenna impedance, the antenna does not accept all of the power flowing from the transmitter. Some of it bounces back or is *reflected* just as water waves reflect off a rock.

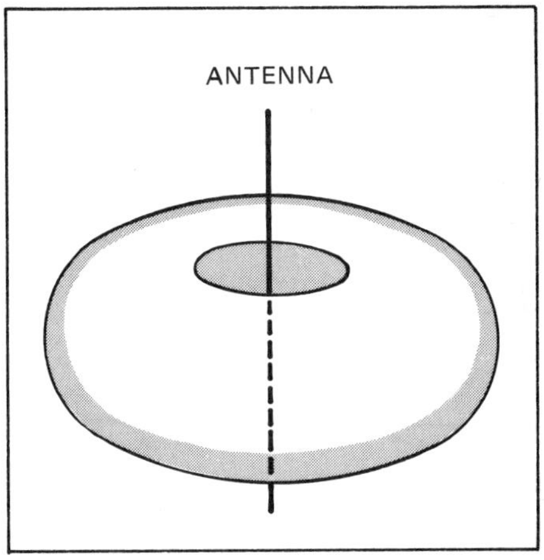

Omni antennas radiate equally in all directions of the compass, but not straight up or straight down.

When that happens, rf waves are flowing along the cable between transmitter and antenna, in both directions simultaneously. These waves combine with each other to form stationary peaks and troughs called *standing waves*. The peaks represent higher-than-normal voltages which can burn out the final power stages of your transmitter. The height of the standing-wave peaks is expressed by a measurement called *standing-wave ratio* (SWR). You want this ratio to be as low as possible.

Impedance matching is important for two reasons: To increase radiated power and to reduce the size of standing waves.

ANTENNA GAIN

A theoretical antenna, called *isotropic*, radiates a spherical ball of energy—equally in all directions. That means as much rf signal would go straight down into the earth as straight up into the sky. Neither of these directions is useful for getting Smokey reports.

Omni Antennas—An improvement is a radiation pattern shaped like a doughnut with no signal going straight up or down. This pattern is called *omnidirectional*. The word means all-directional but it is intended to mean all directions in a *horizontal* plane.

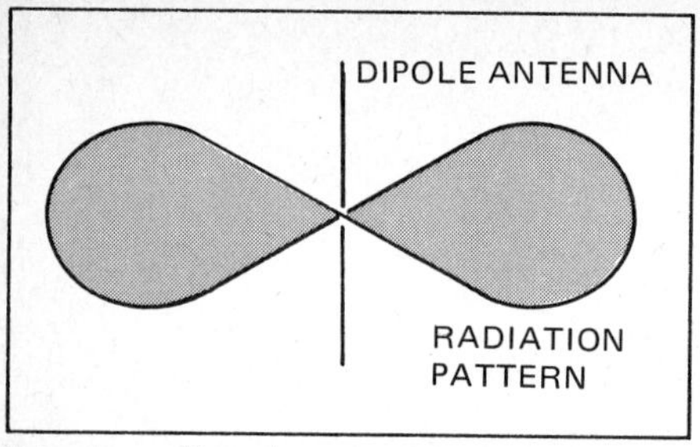

Dipoles radiate front and back in two lobes, but not to either side.

Because an *omni* antenna doesn't waste energy in the wrong directions, it puts more energy where it can be useful and gives more communication range along the surface of the earth. Therefore we say the omni has *gain* compared to an isotropic—sometimes we say it has *power gain*. Obviously gain didn't result from amplification in the antenna, it resulted from controlling the radiation pattern so there is a gain in energy where we want it—accompanied by a reduction in energy where we don't want it to be anyway.

An omni makes a doughnut pattern. A *directional* antenna has even less horizontal coverage, therefore it has more gain than an omni —in the direction it is sending energy.

Dipole Antennas—A *dipole* antenna is a 1/2-wavelength radiator divided in half. Its radiation pattern is bi-directional—shaped like a figure 8, viewed from above. Folding each half a dipole into an elongated loop is favored by some antenna manufacturers because it offers a better impedance match over the frequency band. *Folded dipoles* may become more popular because of the need for broader-band antennas to cover 40 channels rather than just 23. Twin folded dipoles crossed at right angles make good omnidirectional antennas.

A simple dipole has 2.1 dB gain over an isotropic radiator.

Beam Antennas—By adding more pieces to a dipole antenna it can be made into a directional *beam.* The back lobe of the dipole radiation pattern is no longer present and the front lobe is narrower. To use a directional beam antenna, you have to point it very accurately at the other antenna you are communicating with.

Beam antennas are made by mounting the dipole crossways on a boom and using additional elements in front and behind the dipole. Longer elements, called *reflectors* are behind the dipole. Shorter elements, called *directors,* are in front. The more directors and reflectors there are, the narrower the beam and the longer the antenna is along its boom. This design is commonly called a *Yagi* antenna, named after its inventor.

Gain of Beam Antennas—Because beams put more rf power in a certain spot, they have gain over a dipole and *also* gain over an isotropic. When you compare specs on antenna gain, be sure all are to the same reference antenna.

A three-element Yagi has 8 dB gain over a dipole and 10 dB gain over an isotropic.

Because only the dipole is connected to the transmitter, the other elements in a beam antenna are sometimes called *parasitic* elements. The improvement in directivity due to adding parasitics to a dipole is expressed by a measurement called *front-to-back ratio.* A high ratio means more front radiation and less to the back. You may also see a *side-rejection ratio* which expresses the relative *insensitivity* to unwanted signals from the side of a beam antenna used as a receiver.

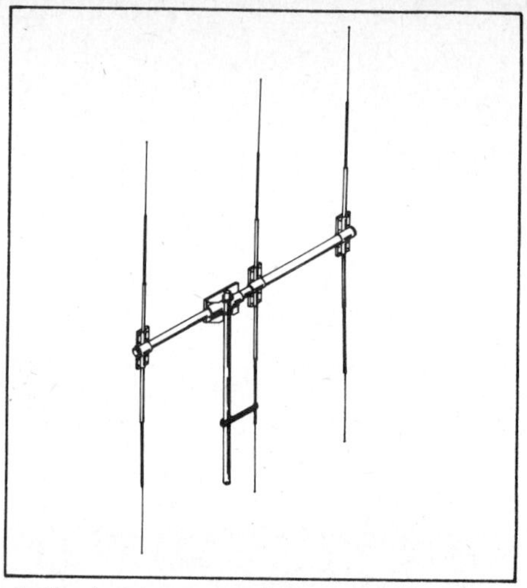

A dipole is converted to a beam antenna, or Yagi, by adding director elements in front and reflector elements in back. This is a Hy-Gain CB 3.

Quad Antennas—The *quad* is an array of rectangular loops, each with a perimeter of about one wavelength. They operate on the same general principle as the Yagi, but a quad has about 2 dB more gain than a Yagi with the same number of elements. A 3-element quad has about the same gain as a 4-element Yagi—useful in installations where you are short of space. Remember a beam antenna must be

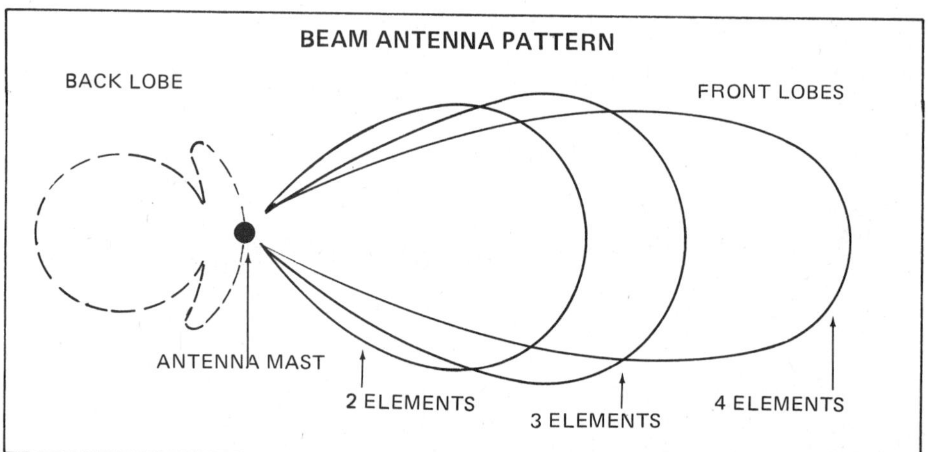

The number of antenna elements determines beam antenna patterns. More elements give a stronger front lobe and a smaller back lobe.

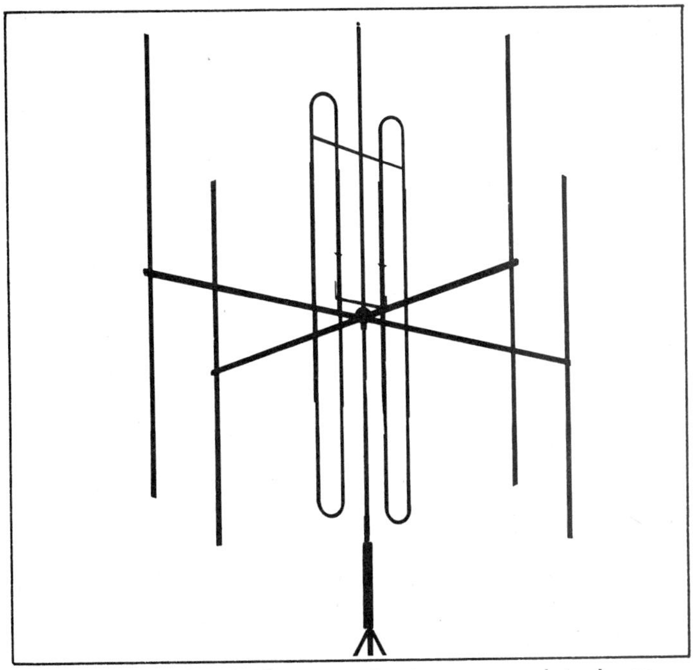

This Turner electrically scanned antenna can point a beam by switching among the four vertically polarized antennas extending outward from the mast. Some similar antennas use three of these elements instead of four.

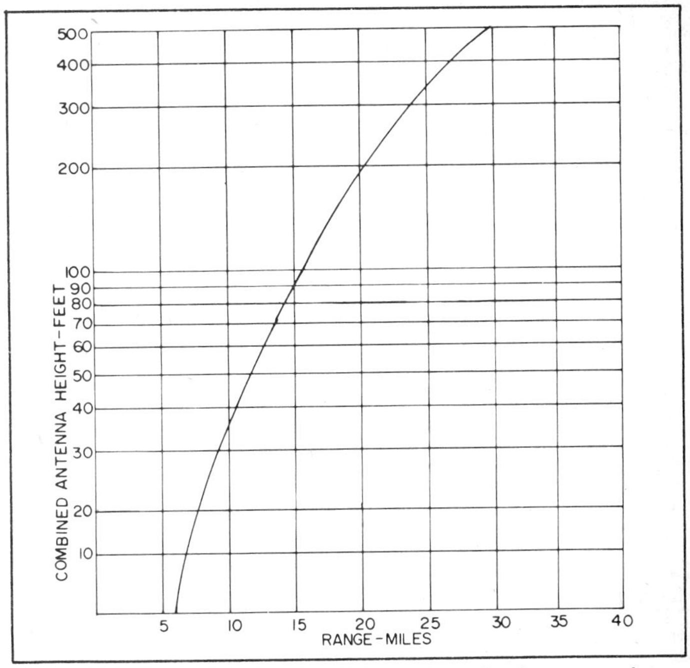

If nothing prevents ground-wave communication at maximum range, add the heights of the two antennas and find communication range from this curve. Courtesy of E. F. Johnson Co.

pointed, physically or electrically, at its target. If you point it physically, you need room to swing it in a circle.

Y-quads and V-quads combine quad and Yagi techniques on the theory that a quad reflector lowers the angle of radiation and improves front-to-back ratio.

Stacking beam antennas in the same plane produces another 3 dB of gain.

EFFECTIVE RADIATED POWER

An isotropic antenna would spray out your powerful 4 watts equally in all directions of a sphere. Any antenna with some directionality puts more power where you want it. Suppose a directional antenna makes a signal which is twice as strong as it would be with an isotropic. There is an effective increase in power at that location because the signal is actually stronger—even though it is weaker somewhere else. That improvement is called *effective radiated power* (ERP). All gain antennas have higher ERP than an isotropic.

POLARIZATION

A whip antenna standing vertically is said to be vertically polarized. Beam antennas are polarized according to the orientation of the dipole or *driven element* of the beam. Because the parasitic elements and the driven element of a simple beam antenna all point the same direction, you can determine the polarity of a beam antenna just by looking at it. If all of the elements on the boom point up and down, it has vertical polarization.

Two antennas communicating with each other should have the same polarization—both vertical or both horizontal. Because of the many vehicular vertically polarized antennas in use, most base-station antennas are also vertically polarized. If you attempt to communicate without the same polarization, there is a signal loss of 20 dB.

There are advantages in working with horizontal polarization: Much noise and interference in the CB band is vertically polarized, so if you and your good buddy are both horizontally polarized, you aren't receiving much of the noise. Also most communications are vertically polarized, so you have less interference and clutter from other stations. Of course, both parties must be horizontally polarized which usually limits this trick to base stations.

Dual Polarization—Some base-station antennas use two full sets of elements—one set vertical and one horizontal. This looks like an array of crosses along the boom. One polarization or the other is selected electrically by a switch at the base station.

Don't confuse this with the small radial wires you see on top of some vertical radiators. This "top hat" dissipates static charges and helps reduce noise in the receiver but they don't affect the vertical polarization.

SCANNING ANTENNAS

Some special beam antennas are built to scan or "look around" in any direction without physically pointing or rotating the beam. They are 4 beam antennas mounted

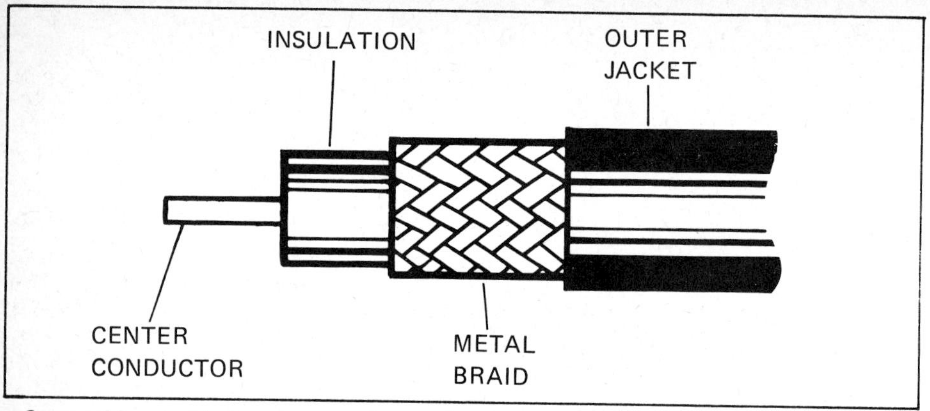

Coax cable, used between transceiver and antenna, has minimum loss and maximum shielding. Pick the right type and install it carefully.

together, one aimed north, one south, one east, one west. By electrically switching the 4 beams and using them in combinations the antenna can be used as an omni or pointed as a beam in steps of 45° around the circle. This means there is one "phantom" aiming point halfway between each pair of beams.

The gains of this antenna type are: Used as an omni, 3 to 6 dB over isotropic. Used as a beam, 8 dB along the individual beam antennas and 7 dB along the phantom intermediate beams, compared to isotropic.

HOW FAR CAN YOU TALK

If interference, noise, or an intervening building or mountain don't prevent communications, two antennas can "talk" as far as they can "see" each other. When one antenna is below the horizon, you can still communicate because the actual CB range is about 1/3 longer than line-of-sight.

The line-of-sight distance between two antenna tips is determined by the height of each. It will be the same with two antennas that are each 20-ft. high, or one that is 40 ft. and the other at ground level. Therefore line-of-sight is determined by the combined heights of the two antennas. Height means above nearby terrain so if you are on a hill, count the altitude gain due to the hill.

The accompanying graph shows practical communication ranges for various combined antenna heights. Think of these as ideal—the best you can expect to get from groundwave communications. Very often you will get much less due to noise and interference.

HOW TO PICK THE RIGHT COAXIAL CABLE

The best connection between your antenna and a transceiver is coaxial cable. It consists of an inner conductor enclosed in an insulating material surrounded by metallic braid. The braid is called the outer conductor and also serves as electrical shielding for the inner conductor. The entire cable is wrapped in a vinyl jacket.

No cable is a bargain if it deteriorates rapidly and degrades the signal. A lot of work goes into a good antenna installation so you don't want to replace the coaxial cable frequently.

Coax comes in white and black. White makes a neat, non-marking installation for boats, but a white jacket alone is no guarantee you are buying the best coax for marine use. The chemical composition of the vinyl cable jacket may contaminate the polyethelene insulation surrounding the inner conductor, causing losses at radio frequencies.

There are only two cable types you should use: RG-8A/U has lower signal losses in a length of cable because it is larger. It is also more rugged. Where there is room and a lot of sharp bends are not required, use RG-8A/U—such as down the inside of a mast or inside a vehicle from one point to another. RG-8A/U should also be used in base stations for lower losses in cable runs over 50 ft.

In tight spaces or where sharp bends are needed, use RG-58C/U. This has more electrical loss than the other type and is less rugged, but it is smaller. It is relatively easy to bend because the inner conductor is stranded. Stranded conductors are less liable to fracture due to vibration. The copper-braided shield is tinned which reduces corrosion and makes soldering easier.

Both of the listed cable types have outer covers which will not contaminate the insulator.

Caution: Beware of the word *type* as in *RG-58C/U type.* It means the cable is *similar* to one that meets military specs, but you never know *how similar.* MIL-spec cable is the best, provided it is fresh. Enough high-technology wars have been fought to flood the market with surplus cable. Examine the cut end of coax before buying. The inner polyethelene insulation should be milky white; the braided shielding silvery and fresh-looking. Any discoloration of these materials is an indication of old age.

CHAPTER 12

HOW TO SELECT AND INSTALL MOBILE ANTENNAS

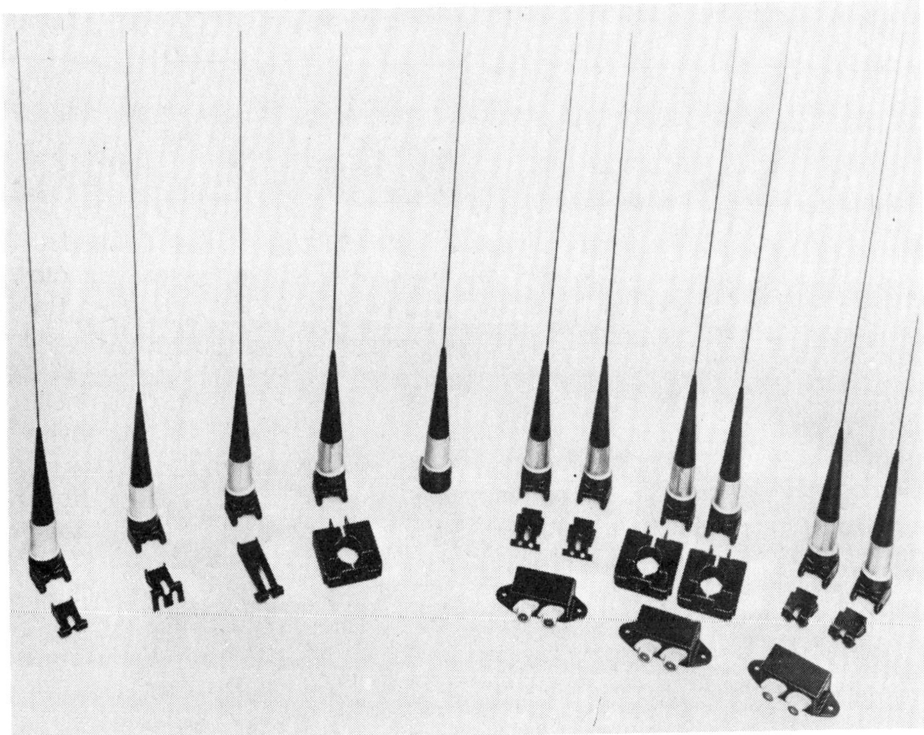

Most of the differences among antennas of a given type are in the mounts. These are all base-loaded Motorola antennas.

I prowled some shopping-center parking lots, eyeballing CB-antenna installations on autos, vans, station wagons and trucks. Ninety percent of them violated good mounting and hook-up practices described in this chapter. No matter how much money you've invested in CB, I think it's a shame not to get the talk power you paid for. You get it by paying attention to *all* the little things. Most make a small difference by themselves but add up to a large improvement when you go on the air with the right combination.

With a bewildering array of antennas displayed in plastic-wrapped packages imprinted with lightning bolts and potent-sounding names, it's easy to be confused. Often the package doesn't tell you very much about the antenna. But the names sure are powerful. It's hard to choose between a Cosmic Zapper and a Space Blaster.

This chapter begins with a refresher course on different kinds of antennas and mounting locations.

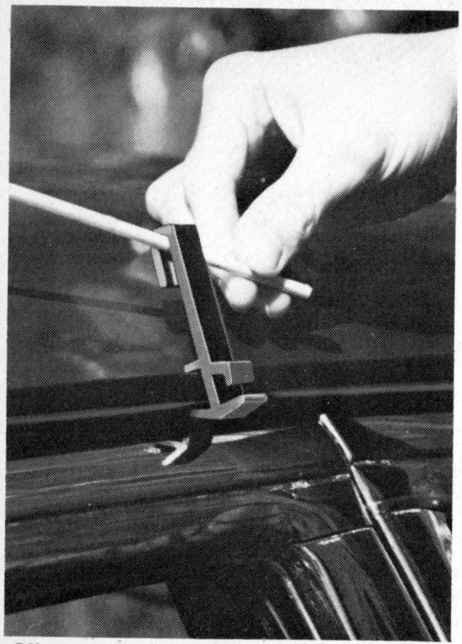

Clip your whip. Accessory clips fasten the antenna tip to the rain gutter so you can garage your vehicle.

Then I tell you how to do things the antenna instruction booklet assumes you already know.

ANTENNA LENGTH

Full-length quarter-wave whips—without loading coils—are astonishingly tall when you mount them on a vehicle. Nine feet up from a bumper is a long way. Nine feet from a car roof or truck cab is higher than the usual 13.5-ft. height limitation for vehicles.

Don't assume I am against full-length quarter-wave antennas. They are more efficient than anything shorter that is made *electrically* into a quarter-wave by insertion of a loading coil. But it really isn't that simple because antenna performance depends greatly on where and how you mount it. An inefficient antenna in the best location on your car may outperform an efficient antenna that is poorly mounted.

Location of the antenna affects its radiation pattern. If you want to talk up and down the highway and aren't much concerned with side-to-side, there are some things you can do—discussed a little later in this chapter.

Location of the antenna affects the groundplane the antenna "sees" under it and therefore the impedance it shows the transmitter. If you don't get a 50-ohm antenna connected to your 50-ohm transmitter, you lose radiated power and the SWR measurement tells you to do something about impedance matching.

All quarter-wave antennas are not the same physical length. It depends on material and shape. There are thin steel whips measuring inches longer than thicker fiberglass whips of the same *electrical* length. You might solve a clearance problem, saving critical inches through careful choice among fiberglass whips.

Caution: Fiberglass whips bend quite a bit. There is a limit in cold weather. If your fiberglass whip strikes an object when the antenna is brittle with cold, it may snap. Don't transmit with a broken antenna.

Whips and the Trucker—This advice applies primarily to full-length quarter-wave whips. They are so tall it seems right to mount them low along the side of the cab. Don't! This solves the clearance problem but causes a radiation problem. Too much of the antenna runs closely alongside the metal of the truck which absorbs signal and detunes the antenna. Go for height, paying attention to clearances. Better still, look over the alternatives: Short antennas and better locations to take advantage of the vehicle as a groundplane.

Whips and RV's—If your RV has a fiberglass body or roof (or other plastic) you may need to install a groundplane for the antenna. Adaptations of marine antennas are sold for RV use and claimed not to require a groundplane. Read the specs carefully and be sure the manufacturer says very plainly, "You don't need a groundplane to use this antenna on a plastic roof."

If you use an antenna that needs a groundplane, install one according to this chapter's instructions.

Because full-length quarter-wave antennas on RV's give the same height problems as on any other vehicle, here's a compromise you may consider. While traveling, use a shorter antenna. When parked, switch to a fold-down roof-mounted full-length antenna for better efficiency and longer communication range. You'll need a coaxial switch and a fold-down mount designed for RV use.

SHORT ANTENNAS

Using a loading coil to make an antenna physically short but electrically long has several disadvantages. If a heavy loading coil is in the center, the antenna wobbles around more in respect to the groundplane and changes tuning.

A short antenna trades signal strength for compactness. You can get loaded antennas as short as 18 in. for the CB band. Typically they radiate only about 50% as well as a full-length quarter-wave—down 3 dB in output.

A compensating advantage is the compactness of the shorter units. You can mount them higher without worrying about clearance; you can get them up on the roof where they have a good groundplane; and you can place them on your vehicle so they have some directivity.

ANTENNA RADIATION PATTERNS

Depending on where you place the antenna in respect to its groundplane, the radiation pattern can be changed from circular. You can improve transmission and reception in some directions at the cost of having less in other directions.

Think of the groundplane as a reflecting mirror which captures rf signal leaving the antenna in a downward direction and bounces it up again so it zooms along over the surface of the earth and reaches folks out in radioland. With that way of thinking, you can often predict the antenna pattern from various mounting locations. The accompanying drawings show the effect of common mounting locations.

HOW TO INSTALL A GROUNDPLANE IN AN RV

You can line the inside of the vehicle roof with copper mesh. If you're lucky, you may find copper screening in a hardware store. Some industrial electronic parts suppliers stock it. I don't think it is worth the search even though it's the old-timer's favorite and makes an effective groundplane. Copper mesh is hard to fasten to the roof and rarely looks neat because you wrinkle it when you handle it.

It's simplier and neater to buy a groundplane kit for non-metallic vehicle bodies.

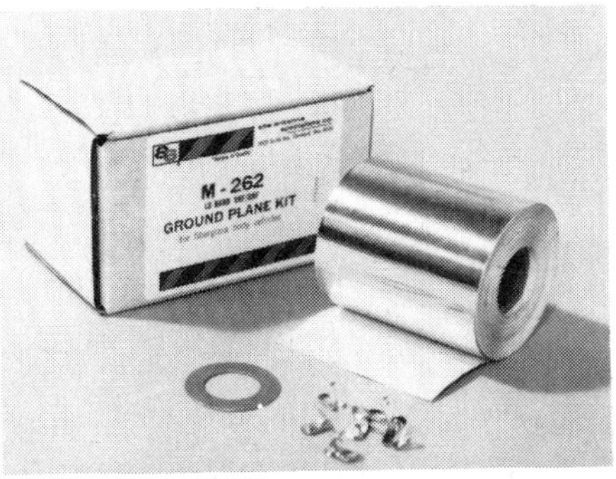

This groundplane kit by *Antenna Specialists* has everything you need to install a groundplane on a non-metallic roof.

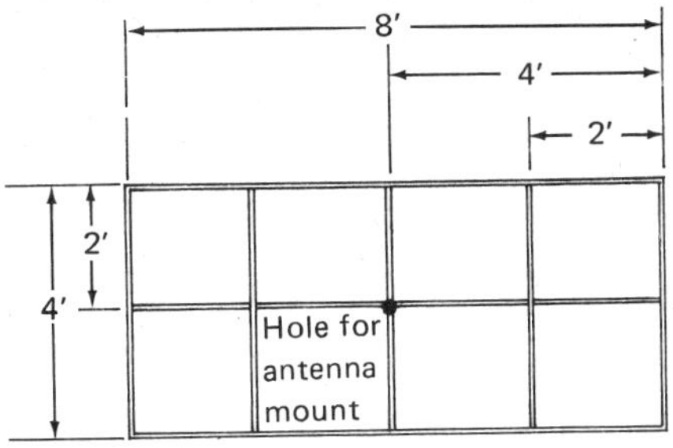

Lay out the groundplane pattern and mark the location of each strip.

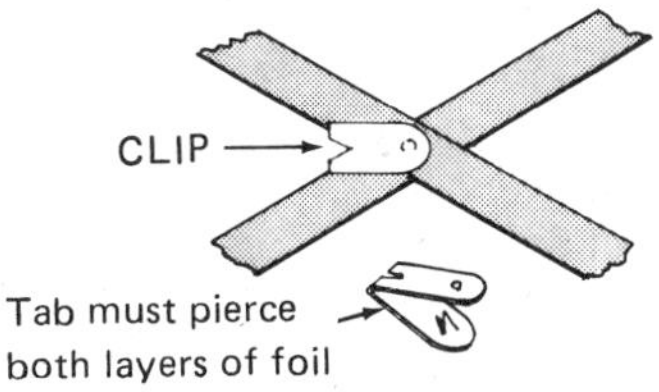

Peel backing off foil and fasten *lightly* to roof surface. Then fasten clip tabs through *both layers* of foil at each intersection to assure electrical continuity. Then press foil tightly in place.

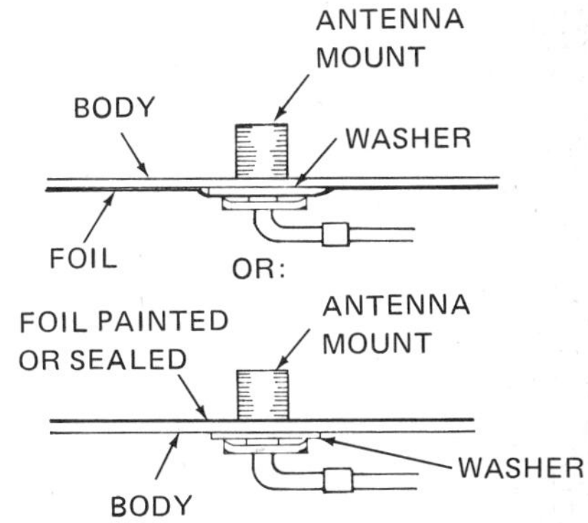

Drill hole for antenna base in the center intersection, starting from the foil side so the drill bit doesn't peel away foil. Use large washer on inner surface of roof. If tape is installed on outside of roof, paint with clear acrylic to prevent deterioration.

Clean the mounting surface. Let dry thoroughly.

Cut three 8-ft. and five 4-ft. sections of tape.

Mark antenna location as center-point for the grid pattern. You'll need to put the antenna close to the middle of the roof for a groundplane pattern of this size.

Remove the paper backing from the foil as you put each strip in place in the pattern shown in the accompanying drawing.

Drill the hole for the antenna in the center intersection of the grid and follow antenna manufacturer's instructions for through-the-roof mounting *except* use the wide, flat metal washer contained in the groundplane kit. Put this washer under the antenna-mounting nut on the underside of the roof. This prevents damage to the fiberglass. Seal around the hole to prevent water leaks. Use non-corrosive sealer.

To prevent rusting of the metal clips used at the intersections, cover them with a non-corrosive marine silicone sealer such as Boatlife.

If you don't use a through-the-roof antenna mount, solder a wire to the center intersection clip of the grid and run this wire to the ground side of your antenna. Naturally the antenna must be above the groundplane for it to do any good.

DUAL ANTENNAS

Dual antennas have a legitimate purpose besides looking good. If you want to get some directional effect from a pair, they must be mounted correctly. In CB lingo this is called *co-phased.* Mounting one on each side of the vehicle forms the signal into two beams, one reaching forward along the highway and the other reaching back where you came from. This is great for use on the highway because your ears are where the action is. But the disadvantage is reduced communication range from side to side.

OTHER FACTS

Higher is better because elevation means increased range and less distortion due to adjacent metal of the vehicle body. Don't forget elevation also gets your signal over *other* vehicles a bit better, too. So why not pick the most efficient antenna, place it in the highest location for the radiation pattern you want, and go on the air?

Mounting complexities are associated with each location and for various antenna types. They are sometimes compounded by the shape and construction of the vehicle. I'll try to take all the facts into consideration for you in the following discussion and give you my recommendations.

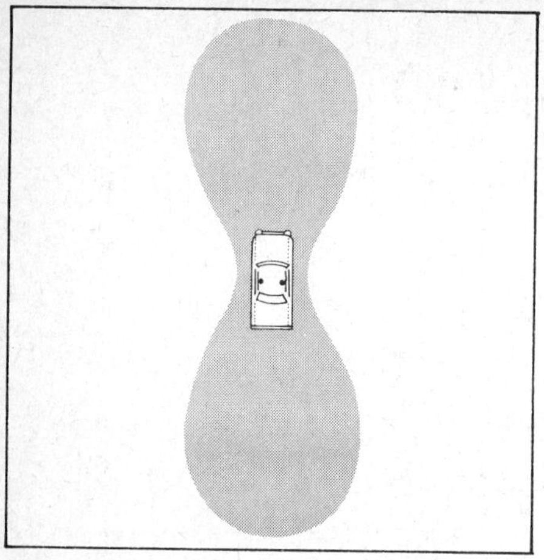

Co-phased antennas give a figure-8 pattern with most signal reaching forward and backward. To be effective, antennas must be 8 or 9 feet apart.

Center-roof mounting gives a nearly omnidirectional pattern. This magnetic base mount by Antenna, Inc. adheres to vinyl-topped metal roofs. I tested it against my garage door and at high speeds.

CENTER ROOF MOUNTING

Do you want to put a hole in the roof for permanent mounting? If your answer is no, here is a simple solution but it has limitations.

Magnetic Base Mount—The easiest, fastest mounting arrangement you can find. You don't need tools. You can pop in on the roof dead center or even move to the side, front or rear depending on what direction you want your signal to radiate. You can remove the antenna faster than you can say *10-7* and hide it in the vehicle. These are base-loaded because center or top loading would bobble and work the magnet loose. They are about 3 ft. high so be sure to remove yours before entering the family garage. They scratch paint when they get knocked off.

A magnetic base does not hold to a fiberglass roof, nor does it hold well to metal surfaces that aren't flat. The only one I've found for vinyl-covered metal roofs is made by Antenna, Inc,: it really works, even on thick vinyl.

Naturally, holding power is reduced with thicker paint or vinyl between the magnet and the metal surface—and by dirt between magnet and roof. A wet roof, even though clean, will not hold as well as a dry one. Keep a rag handy to wipe the spot where you zonk the magnet base down. Caution: Wind-blown rain may still get under the magnet to reduce its holding power.

You'll see magnetic mounts with labels saying, "Wind Survival: 100 MPH." When you wrench off the plastic wrapping – which has a survival strength to frustrate King Kong—you may find the instruction sheet inside admits to only 80 MPH tenacity. You wouldn't really exceed 80, would you?

Ground speed of your vehicle is not-wind speed. If you are doing a legal 50 MPH into a 30 MPH headwind, the antenna up there is fighting an 80 MPH gale. Don't forget wind gusts or turbulence caused by other vehicles passing you in the opposite direction.

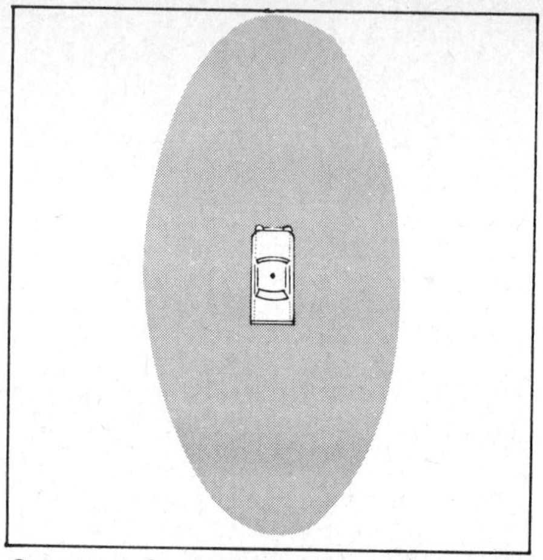

Center-roof mounting gives an egg-shaped antenna pattern, but it's about as close to omni as you can get on a vehicle.

Trucks and buses can produce a wall of wind with very high velocities. You'll also find turbulence in tunnels, underpasses, around buildings and on bridges.

Recommendation—Use a magnetic base mount—a handy removable device for city and suburban use. It's a good way to rotate one CB among several vehicles. Removal is one way to foil theft. Don't forget to put the metal keeper bar or disc in place on the magnet to retain its strength when the mount is stored. Keep a pair of pliers handy to remove the keeper—prying it off can damage the gasket on the mount base. For very windy highway use, choose an antenna that mounts more securely.

Through-the-Roof Mount—Solid and substantial, these make a good electrical ground. Caution: Steel roofs are strong enough, but an aluminum or fiberglass roof requires a reinforcing plate or washer.

With a permanent antenna mount, you can still add a quick-disconnect accessory for the antenna itself—a necessity for garaging and car-washes.

TRUNK MOUNTS

These are two types: Trunk lid and trunk groove. The trunk-lid mounting bracket slips over the edge of the lid and fastens with set screws. The coax cable is squeezed between the lid and the rubber gasket. You can loosen the screws, or use fold-over brackets I'll describe shortly and quickly stow the antenna in the trunk *if the antenna is short enough.* Caution: The antenna moves when you open the trunk. Choose a mounting spot where it won't spear your rear window.

The trunk groove mount attaches to the groove around the trunk opening. It remains stationary when you open the trunk. Drill two small holes in the groove to fasten the mount. The coax squeezes between the lid and the rubber gasket. Make sure there is an adjustment on the mounting bracket to position the antenna so it doesn't interfere with opening and closing the trunk.

Trunk-lid mounts like this Turner fasten to the forward metal lip with a set screw.

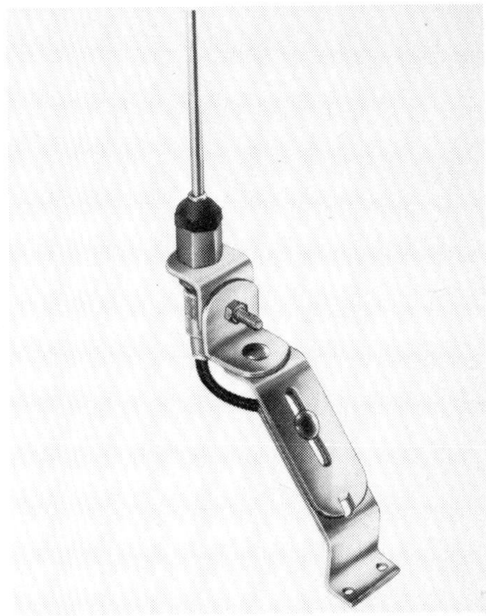
This Turner trunk-groove mount has adjustment so it won't interfere with opening of trunk-lid.

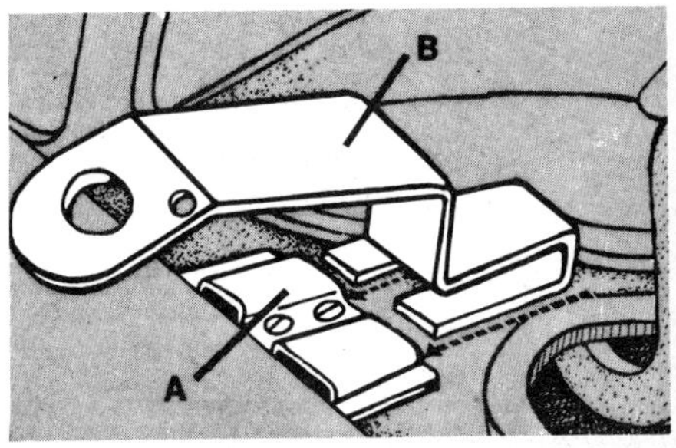

Trunk-groove mounts usually require tools for removal if you're gonna make the antenna disappear. This Stowaway mount by Commando Communications works without tools. Slide part B into part A, close the trunk lid. The wedge-shaped track makes a good ground every time you put the antenna in place. To remove, just slide the antenna out of the mount.

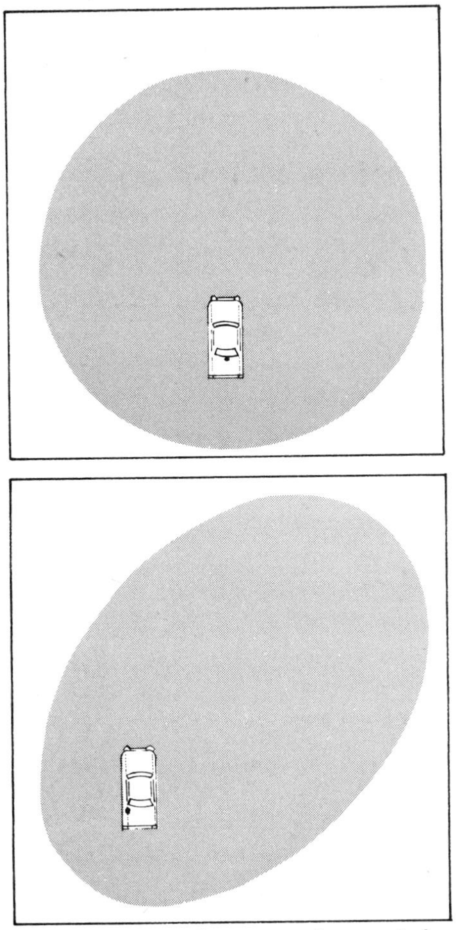
Antenna pattern from a trunk-mounted antenna favors forward direction. Trunk groove mounting forces you to put antenna to one side or another, resulting in angled pattern.

Not all autos were created equal by the stylists. Be sure a trunk mount can be adjusted so the antenna is vertical regardless of the angle at which the base is fastened.

Trunk mounts of both types are used for dual co-phased antennas but this spaces them only about 5 ft. apart which adds only a small increase in performance.

For best performance, antennas need a good ground. Trunk lids connect to the car body only through hinges which may not be a good electrical ground. If there is any doubt, *bond* the lid to the car frame by flexible copper-braid connection to bare metal at each end.

As with all mounts, trunk-lid and trunk-groove types should be attached so at least one screw contacts bare metal. If you remove and replace the mounts, be sure that screw goes in the same hole and reaches bare metal each time. If your CB performance drops suddenly, check this.

BUMPER MOUNTS

These are normally used with physically long antennas. They usually come with metal straps or chains; stainless steel won't leave ugly stains. If you have a compact car, you'll need a mount designed to fit a smaller bumper so check this before buying the mount. Some clamp-type mounts work best on pre-1975 cars which don't have fat, protective bumpers. You can also drill holes in a bumper and bolt a mount in place.

This Stowit mount fastens to the inside of a trunk lid and pivots back under the lid when you want to conceal it. Don't forget to leave enough slack in the coax cable for the trunk lid to swing up.

Run a good ground from antenna base or bumper metal to the car, using flexible copper braid. You can get braid in electronic parts stores and good auto parts stores.

Tall whip installations tend to be semi-permanent, so you are advertising your CB. Quick-disconnect accessories let you remove the whip from the mount, but you'll be hard put to stow it away except in vans, campers and trucks that accommodate its great length.

A long whip is bound to whomp something and also bend with wind pressure on highways, so get one with a shock spring, or add one. They are sold as accessories–stronger springs for longer, heavier antennas. They add height to your whip, and contribute to clearance problems. When you add a shock spring, rather than buy an antenna that comes with one, you lengthen the antenna and will have to retune it.

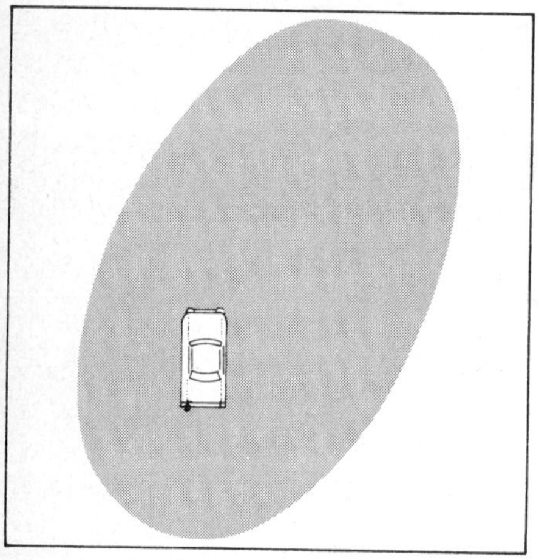

Rear-bumper mount sends signal forward. If antenna is off-center, pattern is angled in opposite direction.

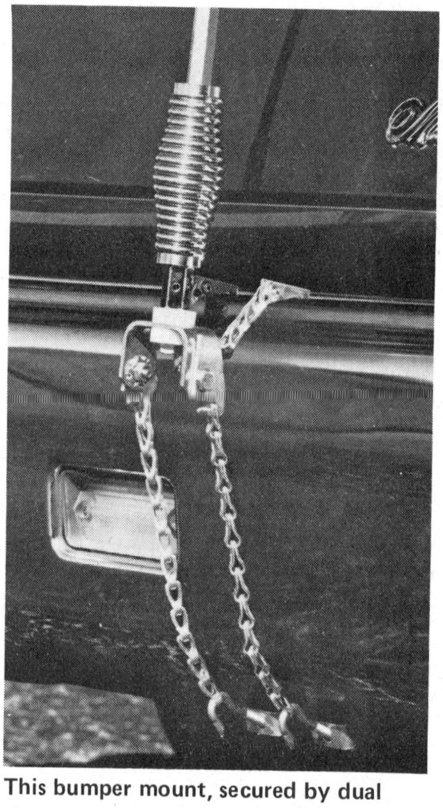

This bumper mount, secured by dual chains, exposes the antenna base to damage by somebody else's bumper.

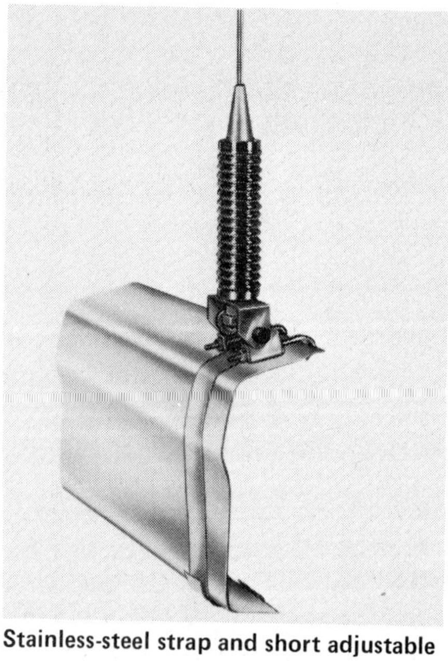

Stainless-steel strap and short adjustable hooks on this Turner bumper mount allow positioning antenna base where it's less vulnerable.

Caution: Not all bumper mounts keep the antenna base out of the way of another vehicle's bumper. Make sure the mount you select does not project beyond the bumper or you'll invite damage the first time you leave your car parked on the street.

Newer vehicles have resilient bumpers designed for 5 MPH crashes. They may not make a good electrical ground to the vehicle.

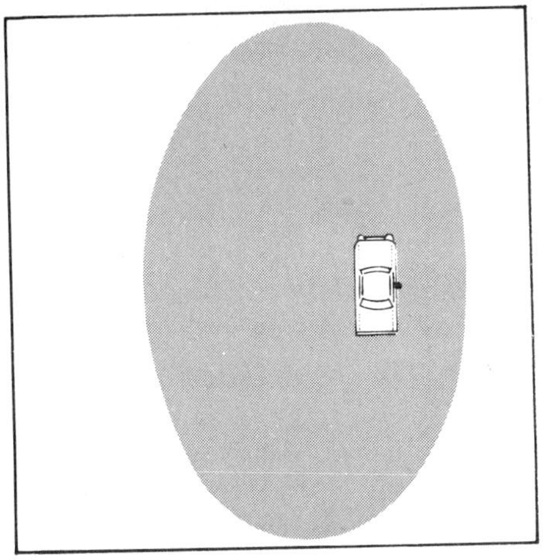

Rain-gutter mount angles antenna pattern across opposite side of vehicle.

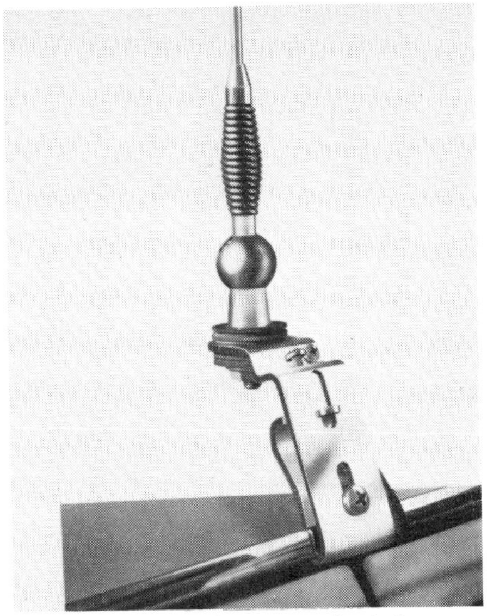

This Turner gutter mount adapts to nearly all rain gutters.

RAIN-GUTTER MOUNTS

Attach with one or two screws, or with a small spring-loaded clamp. The coax cable passes into the vehicle through one of several possible openings:

Door lead-in takes the cable between door and doorframe, squeezing it through the pliable gasket. Too tight a squeeze can pinch the cable and eventually break the center lead. A door gasket may leak wind and rain around the cable.

Window lead-in is poor. You can't close the window all the way without pinching the cable, so you'll have air and rain leakage in bad weather. If you pass the cable through a door window, the cable may get yanked when the door is opened before loosening the window to allow slack.

Trunk-groove lead-in is best. The wider rubber gasket doesn't damage the cable. There's no interference with door and window operation. And, a bit of rain or wind leakage here probably won't hurt anything.

Now let's get to the mounts themselves. With the screw type you can force one set screw through the paint to get a good electrical ground. Installation or removal is quick, but you have to keep a screwdriver in the vehicle.

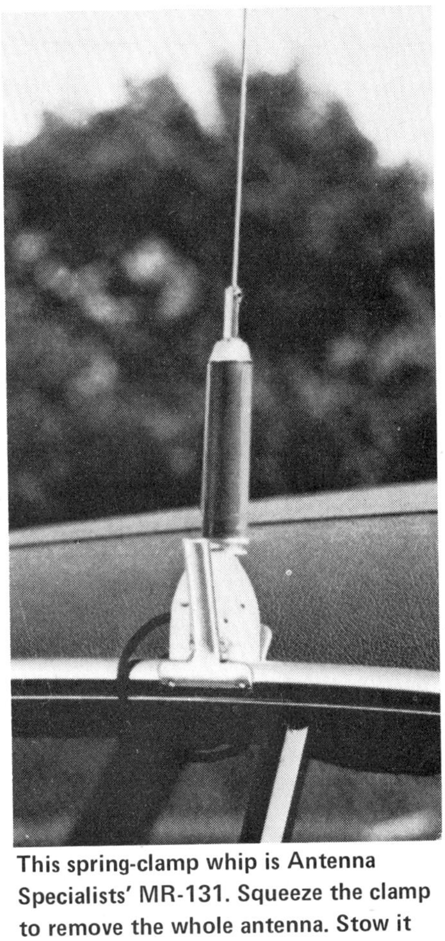

This spring-clamp whip is Antenna Specialists' MR-131. Squeeze the clamp to remove the whole antenna. Stow it under the seat where it won't show.

The spring-clamp mount doesn't require tools. You'll need a spot of bare metal for a good ground.

Caution: Not all rain gutters are strong enough to hold a CB antenna. Any weakness may be aggravated by a bobbing center-loaded antenna. Inspect your gutters before deciding to hang an antenna there.

Not all rain gutters are the same depth, nor is there standard spacing above vehicle doors. Select a mounting bracket adjustable for gutter depth. Check clearance to be sure the door will open. Don't use a mount where screws project downward to scrape you or your passengers getting in or out of the vehicle.

MIRROR MOUNTS

Good for trucks, tractors, RV's, station-wagon luggage racks, or wherever there's a solid hunk of tubing to hook on to. The coax may be passed through door gasket—at some risk of pinching—or through a hole drilled in the door or body.

If you move your set from vehicle to vehicle, there are easy on-and-off brackets.

Some mirrors vibrate, which can crack the base of all but the best mounts. This vibration, plus rust and corrosion, impairs the electrical bond between antenna and vehicle ground. If your installation was made some time ago, this is a good place to look for losses in efficiency.

Some mirrors are *insulated* from vehicle ground by rubber or plastic gaskets at the mounting points. Sometimes the door to which the mirror is mounted is electrically insulated from vehicle ground. Use copper braid between the antenna mount and the vehicle to avoid these problems.

COWL OR TOP DECK MOUNTS

To avoid drilling, consider replacing your present broadcast-band antenna with a disguise antenna to handle your regular radio *and* CB. Cowl mounts require access underneath to install and make connections. Retractable units require

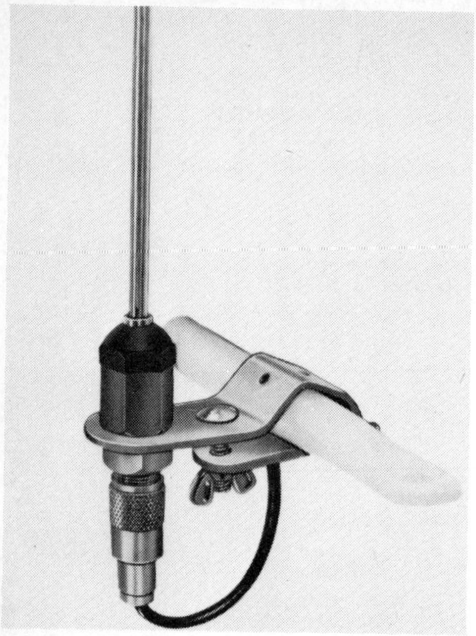

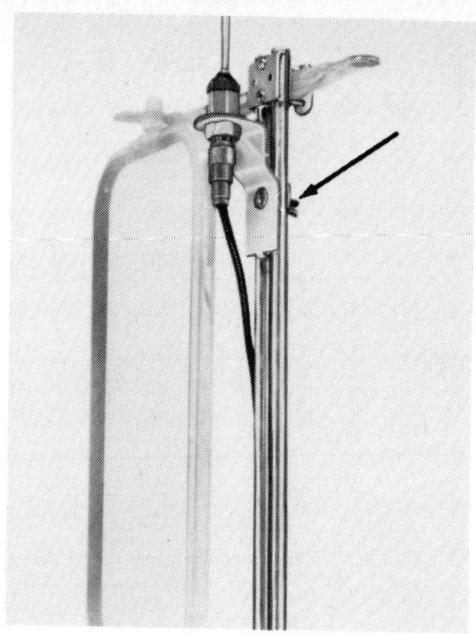

Mirror mounts put the antenna up above the body. Sometimes the top strut of a mirror bracket makes an antenna too high for clearance. With sliding clamp like this Turner SK 811, a wing-nut positions antenna.

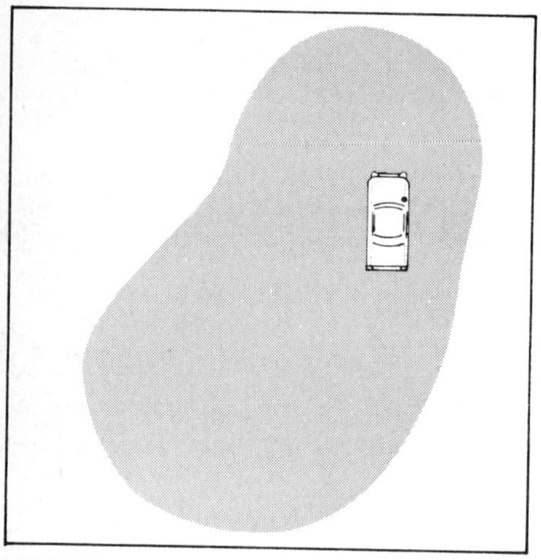

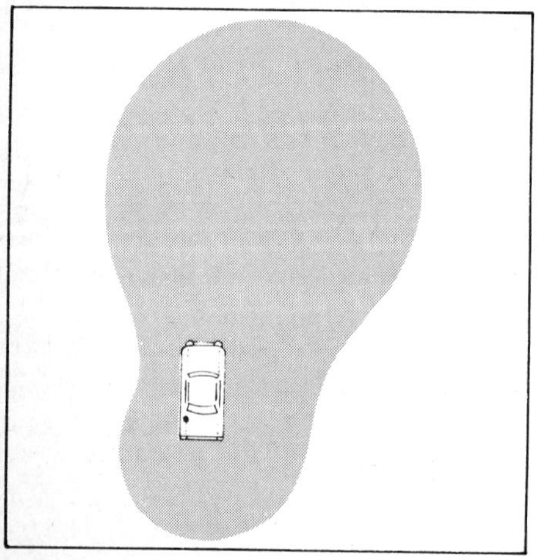

Cowl mounting puts more signal where you've been. Rear fender mounting gives you a bit more range to where you are going. Both patterns are slightly angled by off-center antenna.

space beneath for the mechanism.

Front cowl mounting on a front-engined car places antenna and lead-in very close to electrically noisy parts in the engine compartment. The same antenna mounted on a rear fender will give better reception.

BODY MOUNTS

These require drilling holes but make a very strong installation. Some require larger holes than you can drill but there are ways around the problem, discussed later. You need access from the inside of the vehicle so you can put on inside screws to install most body mounts.

SPECIAL MOUNTS

Campers, RV's and autos with no room to spare in garages need fold-over mounts. Some automatically fold down when the antenna strikes an obstruction—very handy! If you replace a conventional antenna with a fold-over mount, check the SWR.

Motorcycles, snowmobiles, fiberglass vehicles and boats require antennas that operate without a metal groundplane. Such antennas come with a variety of mounting arrangements. They come with more coax than you'll ever need on a motorcycle or snowmobile. You can cut to proper length unless the antenna instruction forbids shortening the coax. Make sure the coax is securely fastened all along the run from antenna to transceiver. Loose connections develop from dangling wires on such vehicles. That degrades performance. What's worse, dangling wires may snap obstacles or riders.

QUICK DISCONNECTS

For ease of removal, install a quick disconnect between the antenna and the shock spring or mount. They fit practically every antenna with a 3/8 in. by 24 thread. With the gadget installed, all you do is press and twist to snap the antenna out. Some have a smooth surface, some have a knurled or hex-shaped grip. The grip type is easier to operate should the device freeze in place with ice or dirt. You can usually do the job while wearing gloves.

The exposed opening in the antenna base when you have removed the antenna invites rust and corrosion. Spray the base socket with CB Climate Barrier.

RETRACTABLE ANTENNAS

You raise or lower this kind of antenna with a dashboard control or wire it so it goes up when your CB is turned on and retracts when

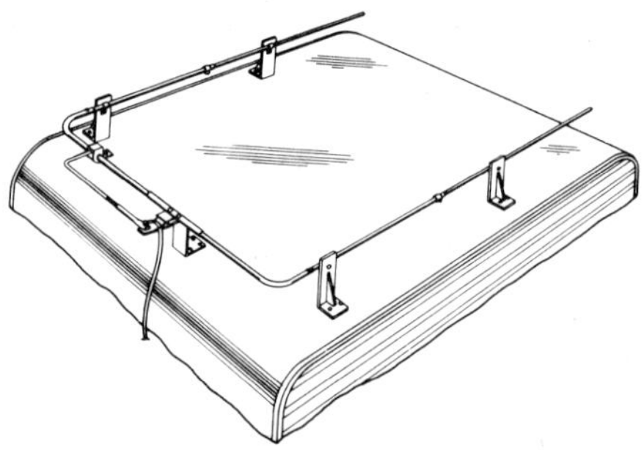

Unique antenna for RV's by Bill Owen, Inc. offers a low profile about 8 inches above any metal-roofed vehicle. Manufacturer claims it has 8.2 dB gain over isotropic and is vertically polarized. Uses coaxial switch to transfer antenna from CB to TV or FM sets.

This center-loaded motor-driven retractible by Tenna needs a 12 V DC line to a switch on or near your CB rig. Keep the telescoping elements clean; look for wear at joints due to wind bobbling. Never key the transmitter unless antenna is fully extended.

you turn it off. Telescoping motorized antennas work well if you keep them clean and coat them lightly once a month with a greaseless lubricant such as LPS 25 or CRC Sta-Loose.

Some retractables may not fully extend when dirty or tired. If so, the antenna's full length is not working as an rf radiator and the transmitter is not properly loaded. This can burn out the power transistors in your set. For that reason, get a motorized unit with a built-in switch that turns on the CB only when the antenna is extended and has reached full length, or one that connects your transceiver safely to a dummy load whenever the antenna is less than full length.

DISGUISE ANTENNAS

These look almost like ordinary car radio antennas—a payoff to the common citizen from the nation's fight against crime. The whip is non-telescoping.

If you have a standard AM/FM broadcast antenna already installed on your car, you can use it with an accessory coupler for CB/AM/FM. These standard antennas offer the same hazard to your transmitter as the retractable type. If you telescope one while going through the automatic car wash and then forget to extend it to the correct length, the next smell of burning transistors will be yours.

BEST BETS

In case you are still coin-flipping, here are my choices in several categories.

Best Bet For Removable Auto Antenna—A spring-clamp rain-gutter mount with the longest base-loaded quarter-wave antenna it can take without a height problem. With the coax fed through the trunk lid, and the antenna stowed in the trunk when not in use, I'm left with only one hazard—forgetting to remove the antenna before garaging the car.

Best Bet For Permanent Auto Antenna—Choose a trunk-groove antenna—the longest that will fit on a flip-over mount for instant stowing in the trunk.

Best Bet For Trucks And RV's—Use mirror-mounted co-phased antennas set way out on the brackets for maximum performance.

TOOLS YOU'LL NEED FOR MOBILE ANTENNA INSTALLATION

For Mounts Without Holes in the Vehicle—Magnetic base, trunk lid, rain gutter, most bumper, temporary mirror mounts, replacement cowl mounts:

- Screwdrivers – flat-bladed and cross or Phillips head
- Allen or Bristol wrenches if not supplied with antenna
- Open-end wrenches—assorted or adjustable
- Wire cutters
- Wire stripper
- Long-nose pliers
- Pliers
- Emery cloth or sandpaper
- Cable ties of tough plastic or
- Plastic electrical tape—cold and weather-resistant
- Fishtape, snake-wire or expendable wire coathanger

For Mounts With Holes Add:

- Electric drill with slow-speed capability
- Hole punches as required
- Center punch
- Hammer
- Good steel drills—1/4 in., 3/8 in., 1/2 in.
- Rubber grommets or bushings for sealing a hole around a wire
- Non-corrosive silicone sealer—See Chapter 14.
- Rat-tail file

For Solder-type Connectors Add:

- Soldering iron
- Rosin-core solder
- File or steel wool to keep soldering-iron tip clean.

For No Loss of Parts Add: One cigar box, cookie tin or equivalent container. Don't laugh. First you lose insulating washers, pins, special set screws, wrenches. Then you lose your composure.

For Improved Grounding Add:

- Braided copper grounding straps.

DOING IT

Most antennas come in plastic shrink-wrapped packages. That means you won't know if you have all the pieces until you get home, unless you open it in the store and check it out as the salesman looks on. Some antennas come with a parts list; others just give you assembly instructions showing labeled parts. Either way, check the parts before you begin. Put the easy-to-lose parts in a cigar box and close the lid tight so it won't spill when you or your volunteer helpers kick it under the car. Put the instruction sheet where it won't blow away—taping it to a board or box is smart. Work in your garage

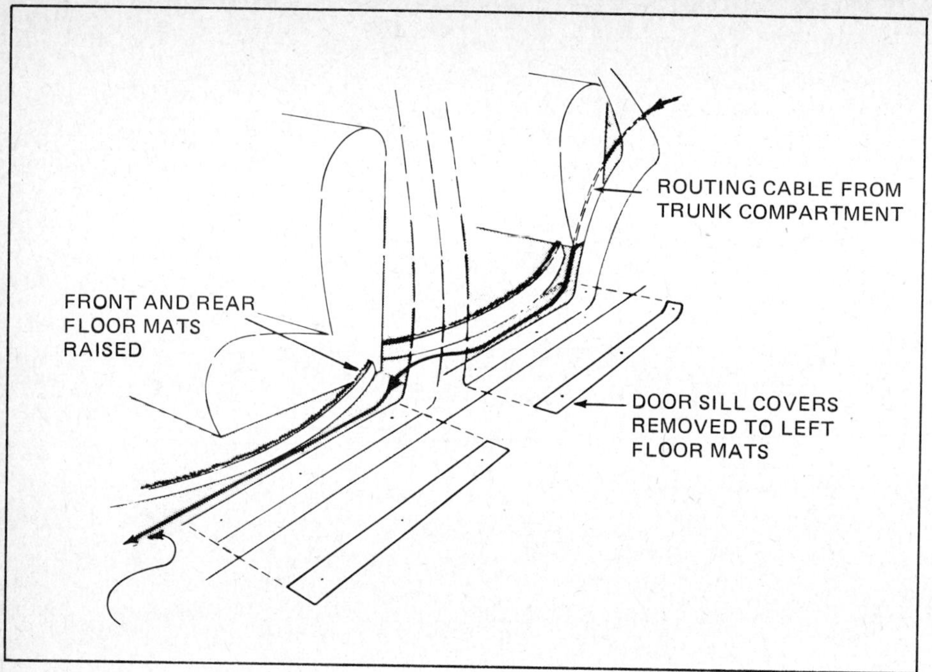

Routing from trunk through passenger compartment goes under door sills so cable is protected from wear and pinching.

or on a paved surface. If you really want to punish yourself, ignore all these little precautions and try it on gravel or grass.

Next, plan your cable run from transceiver to antenna by laying the cable along the general route. If you're dealing with co-phased antennas, start at the transceiver end. The common lead from the transceiver branches at a "Y" to feed two antennas. With a long enough common lead—some are 16 ft.—you'll have only one snaking chore for trunk or rear bumper-mounted twins.

You'll probably have more coax than you know what to do with. Do *not* cut it now. In fact, certain antennas require that you do not cut it at all. If the cable is not to be cut, no matter when, then figure on coiling the excess cable in the tunk or some out-of-the-way place in a truck or RV, such as behind or under the seat.

Does Neatness Count?— In routing and stowing coaxial cable you are up against a critical mechanical consideration. Coax is one conductor inside another, separated by insulating dielectric material. If you make too tight a bend, the center conductor will gradually creep through the dielectric material and short circuit with the outer conductor. The minimum bending radius for the relatively thick RG-8A/U is 10 in. For the more flexible RG-58/U, don't bend any tighter than a 5-in. radius.

Treat coaxial cable like a wet noodle. Push or pull it gently. If the connector is not already attached to the cable, leave it off until you have fished the coax through various parts of the vehicle. If the connector is on the cable, tape it and a few inches of cable firmly to the fishtape or stiff wire you will use to poke or draw the cable. It's better to leave a few inches of fishtape or wire sticking out in front of the connector if you are going to push through a tight opening. Wrap some colored plastic tape on the tip.

Here's How To Go Through the Trunk—The easy way, if you're lucky, is to attach the antenna end of the coax to a stiff push wire. Place it on the rear seat, left or right of center. Shove the wire between the seat cushion and the bottom of the seat back. If possible post a helper to watch for the wire entering the trunk.

Leave a couple of feet of slack at the transceiver in case you ever want to change the mounting position. Route the rest along the kick panel and trim on whichever side you choose. Then push the cable toward that side so it runs alongside the seat cushion and around its back.

An alternative method is to remove the rear seat cushion. It's more of a struggle than you may think. Don't rush into this until you have determined what holds it in place. You may find the seat bolted in place at the floor near each end. From the front of the seat, you should be able to reach underneath and remove the mounting bolts. Then lift the seat up and out.

Older cars and many imports use two retainer clips or hooks to hold the rear seat cushion in place. Locate these by putting your head at floor level in front of the seat and looking underneath. When located, push in on the lower edge of the seat front to release the seat from the retainer and then lift the seat up. Do this on each side and you should be able to take the seat cushion out of the car. Some retainers release with a down, back and then up sequence of movements.

Then fish the coax cable through from either direction, remembering which end belongs at the antenna. Lay the cable around the side where it cannot be pinched when you replace the seat. Better leave the seat out until you've completed the cable run.

If you can't get the cable into the trunk by fishing under the seat back, you'll have to remove that too. First, remove the seat cushion. Then lift the seat back upward out of its retainer clips. That exposes the bulkhead between seat back and trunk.

Drill a half-inch hole—larger if necessary—low on the bulkhead so the seat back won't kink or pinch the coax when you reinstall it. Inspect the back and consider how it fits in place when deciding on the hole location for the coax.

Look in the trunk, too. Some cars have part of the pollution-control system right near the seat back. Drill from the trunk into the passenger compartment so you're sure where the hole is going. Vacuum out the chips so they don't get into your hands. File burrs out of the hole. Insert the thinnest ring of a rubber grommet or bushing through the hole to keep the cable from being cut or chafed.

Concealing the Cable—This is easy and merely takes patience. Don't look for a short cut by running it under the floor mat. It will get worn and pinched by passengers and other heavy objects. Remove trim and run the cable as shown in the accompanying drawing. Remove the front kick panel to route the cable from floor level up to the dash.

Now run the cable to the transceiver or quick-disconnect mount. Make the connection and then use cable ties to keep the portion between transceiver and kick panel from drooping beneath the dashboard. Coil—*do not stuff*—the couple of feet of slack coax behind the front kick panel, observing the minimum bending radius. Will the loop shift enough to make a rattle? If so, secure the cable with clips or tape it securely to keep it quiet. Maybe a piece of rag can be tied onto the cable as a silencer. Refasten all kick panels to enclose the coax. Where necessary, run it under the carpet edge. Replace carpet and trim. Double check before replacing the back seat to be sure the cable will not be pinched.

Now bring the end or ends of the cable out to the antenna mount location or locations and determine how much cable you have left over. If the manufacturer says don't cut the cable, coil it neatly in the trunk where it won't get clobbered by suitcases, tires, bowling balls or fertilizer sacks. Tape the coil so it will maintain the proper bending radius. Use cable ties to keep the rest of the cable lead or co-phase harness where it will emerge from the trunk through the rain-channel gasket. Be sure it will not get snagged when something is put in the trunk or pinched by a descending trunk hinge. Make sure it won't rattle against a panel.

If you can cut the cable, first consider whether the slack is all that bothersome. It won't make much difference in transmission loss, so don't bother about 6 ft. or so, which is about all you'll have left in most installations.

The final two steps are to connect the cable to the mount—if necessary—and attach the mount and its antenna to the vehicle. If you do it this way, skip the next paragraphs on making holes but read the following installation notes.

Making Holes—Many mounts require holes in the vehicle, either for the mount itself or for the coax. An electric drill can do more harm than good to the vehicle and to you if you don't follow simple precautions and procedures.

Be sure there are no wires, glass or obstructions where you plan to drill.

Use the antenna manufacturer's template to mark hole locations accurately.

Center punch each hole to keep the drill from slipping.

A new or freshly-sharpened drill bit is more than worth the cost or trouble.

If the hole you need is larger than the drill you have, and the difference isn't very much, use a round file and patience to enlarge the opening. Or, drill a lot of small holes in a circle and file away until you end up with a large hole. A small grinder and abrasive bits will enlarge and shape holes in a hurry.

When drilling large holes in body metal, the drill usually leaves a tang or projection on the inside of the hole just as the tip of the drill passes through. This projection hangs up in the flutes of the drill and causes the drill itself to auger into the hole, sometimes wresting the drill motor from your grasp. Be very careful just as the drill penetrates the metal. Reduce drilling force; hold on tight; be prepared to stay in charge of the operation. After drilling, file away sharp edges and burrs.

Coming Through a Door—For mirror mounts and some body mounts you will end up bringing the coax through the outer panel of a door, along the inside into the door post through two holes which line up, and then to the set.

In trucks and vans access panels may be screwed to the inside of the door. If they are in the right location, they may give sufficient access to install mounting-bolt nuts on the inside of the door and run cable where you want it.

You may have to remove the inside door panel. The following procedure will usually do it. If the top of the door panel is caught under the trim strip at the bottom of the window, remove the trim strip. Look all around the edge of the door panel for screws holding it to the door. Remove any you find. Remove elbow rests by taking out screws. Push in on the panel near the window winder and door handle. Look at the shaft exposed by pushing in the panel and you should see circlips or pins holding the handle on the shaft. Remove the handle by taking off whatever holds it. Don't lose the small parts. With all that done, the inside door panel should lift off.

You'll be drilling one or two holes per door—one on the outside near the mirror, and the second in the hinged end of the door if there is not already an access hole. Be sure to put that second hole opposite an access hole in the door post if there is one. If there isn't, you'll have to drill one there, too. The holes must be paired so the cable slips back and forth without pinching when the door swings. Lots of vehicles have spare access holes for all kinds of additional cabling through door and door post. If all the access holes are already used you may be tempted to run the cable through holes already occupied by wires. Test first to see if there is enough room for free-sliding passage.

For outside holes, be careful about installing the grommets because they are weather seals and protection against sharp edges. If lugs or terminals are already connected to the cable, they may be too wide to pass through the grommet. You can remove and replace them later or put on fresh terminals—soldered for best electrical performance. Or, gently curl them over with a pair of long-nose pliers to reduce their width until you have run the cable through all the grommets. Then carefully flatten them out again.

Working a grommet into a hole is sometimes frustrating. Grommets look like rubber donuts but act like stale bagels. Cover a dull screwdriver blade with tape to avoid scratching the paint and force the thin side of the grommet into the hole by poking progressively along the outside. When you think you have it all in, seat it with a twist before giving the cable a slight tug.

Hold cable from mirror bracket to the hole in the outside of the door firmly in place with cable ties when you complete the installation. That means the cable should not slip in and out of that grommet as is necessary from the inside door hole to door post hole. Arrange the cable from the last tie-point on the mirror bracket to the grommet in the outside hole in a droopy U. Make sure the bottom of the U is below the hole so water can't run down the cable into the door. Finally, pack some silicone sealer around the cable in the grommet so splashing water or wind-driven rain can't get in.

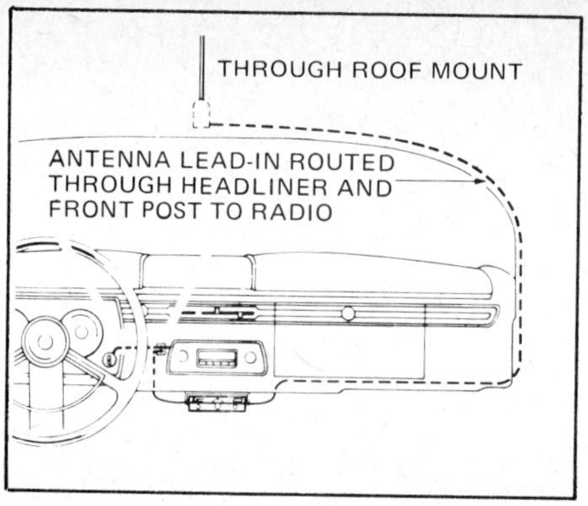

This drawing from E.F. Johnson Co. outlines the roof-mount problem with coax. You can do it, but it's difficult.

This roof-mount installs entirely from the outside. Drill hole, position mount, tighten screws (arrow) which pull claws against roof metal.

Through the Roof—If you must drill through the roof, here's how to avoid drilling through the headliner, too. You can unfasten the headliner, which is tricky, and refasten it, which is trickier yet, even with the special headlining and upholstering tool. I suggest you *avoid* the approach.

If you insist on a roof mount, there are some by Turner and Hy-Gain requiring no access from underneath. These have little togglebolts or claws that fit down through the hole; when you tighten a screw from above the claws expand to grip the hole. Caution: Don't use on a fiberglass roof unless you *do* get access from below to put in a reinforcing plate or large flat washer.

Then get a drill stop that fits the drill bit you'll be using. Adjust this for the thickness of the roof metal so it will just penetrate the insulation between roof and headliner. You can get a fair idea of the metal thickness by examining trunk or hood, even though the metal is usually folded at the edges. Put some tape over the place where you'll drill so the drill stop will not mar the paint. The antenna mount will cover any scratches but some day you may want to remove the mount, insert a hole plug and sell the vehicle for big money. Remember the drill bit is tapered to a point; the full width of the bit will have to go all the way through the metal to give a proper size hole.

If your cable runs through the roof, push it along behind the headliner to one of several points where it can be brought down and routed to the transceiver. You can guide it with your fingers pressing against it through the headliner unless you have unusually thick insulation in the roof. On most autos with a center post between front and rear doors you may be able to remove some trim and bring the cable down behind it to floor level and behind a kick panel to the dashboard, as described in through-the-trunk routing. I say "may" because of the variety of trim fittings used where there is a center post, and because hardtops don't have a post.

Every vehicle has removable trim around the windshield. It's easy to take off by removing the few screws. Trucks and vans usually have a shell around the crossbeam overhead. It's a neat place to run cable for co-phased antennas, especially on mirror mounts. You can come down through the door posts and out through an access hole. Don't forget to leave slack so the cables will not be pulled taut when doors open and close.

CONNECT COAX TO ANTENNA MOUNT

You can get the cable connection reversed by not following the exact assembly sequence. You risk burning out your transmitter if you don't check this before turning on the set. The center conductor of the coax cable goes to the radiating element, or to the center pin of a connector that plugs into the antenna base. If there are terminals on the coax cable at the antenna end, the terminal on the center conductor goes to the antenna radiating element. The other terminal, attached to the braided shield of the cable, goes to the metal of the bracket. Always remember the braided part side of the coax goes to vehicle ground.

You can short circuit your antenna to ground if the coax center conductor touches a grounded point—like clamps around a mirror bracket. Watch the position of the terminal and wire when installing it so it doesn't contact a grounded part of the mount. You can take an extra precaution by insulating the back end of the terminal where the wire is connected. I use plastic shrink tubing for this purpose. Slip it in place over the connection, then heat it gently with a hair drier set on hot, or a hot soldering iron held *near* the tubing. Let it shrink until snug. Shrink tubing is available in electronic parts stores; be sure to get the right size.

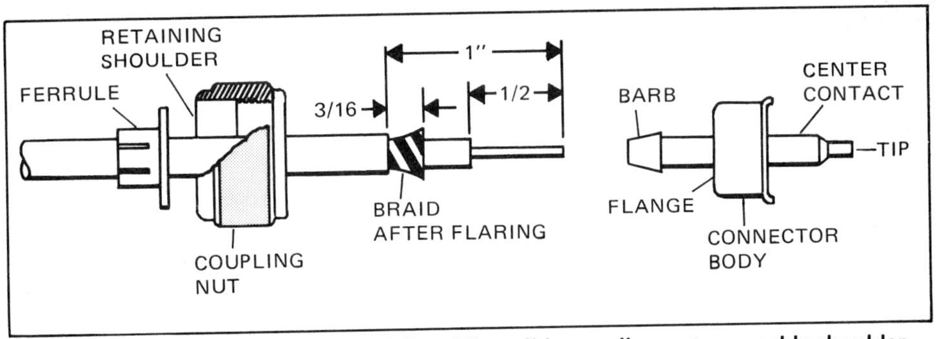

Slide ferrule over cable slotted end first. Then slide coupling nut over cable shoulder end first. Then strip the cable to these dimensions. Be careful not to nick the center conductor. Flare the braid slightly by pushing gently outward in a circular motion. Don't twist insulation or conductor.

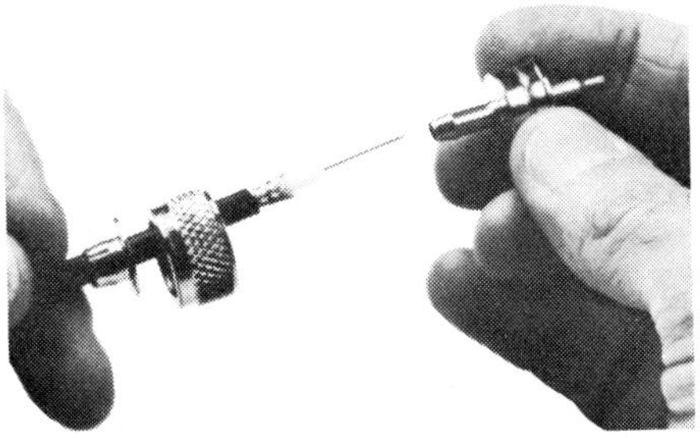

Slide connector body over the center conductor.

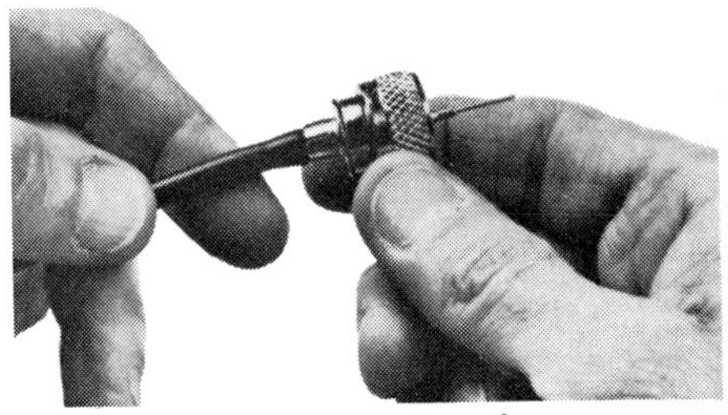

Slide the coupling nut over the connector body. Hold cable in one hand and use other hand to push ferrule over barb until braid is captured between ferrule and connector-body flange.

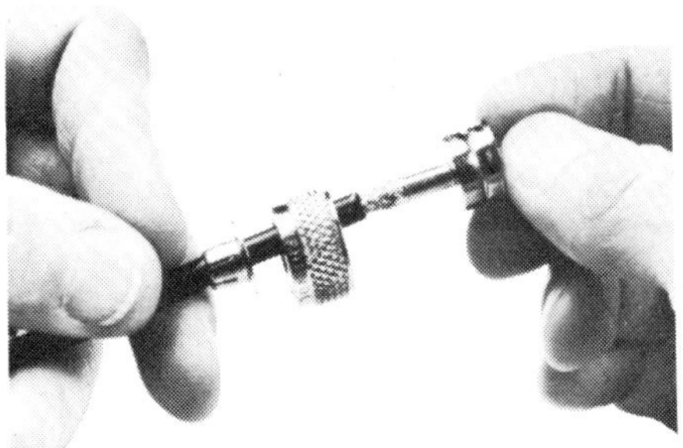

Push barbed barrel of connector body so it fits over center conductor but under flared braid.

Squeeze-crimp the *tip* of the center contact.

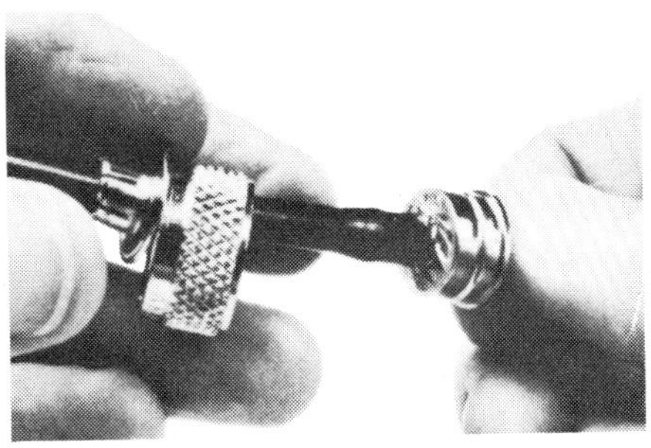

When body assembly is properly positioned, the braid fans out against rear flange or body.

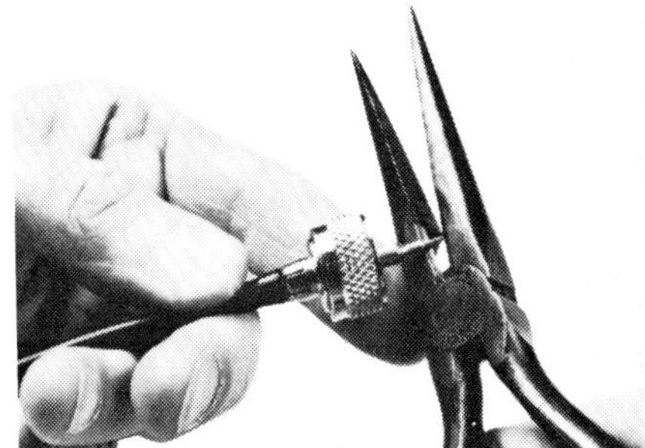

Trim the center conductor flush with the front tip of the center contact.

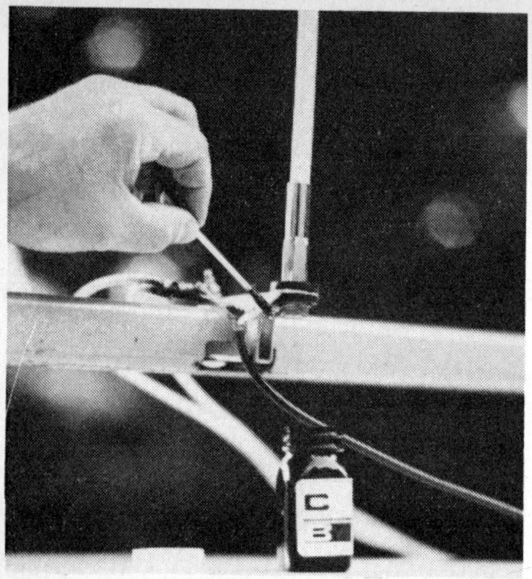

Some antennas have exposed terminals rather than coaxial connectors. Protect the terminals and cable braid from moisture and corrosion with CB Climate Barrier; then seal with Boatlife Marine liquid tape.

When making antenna-base connections, be sure to fan out the braided shielding so it makes good contact with the grounding nut that fits over it. The *insulation* of the center conductor must protrude through the grounding nut, or you will have a short circuit.

Avoid sharp bends, not only to the whole cable, but also of the center conductor where it emerges from the cable. When you tighten the bolt or screw that holds the center-conductor terminal, the terminal can turn with the screw and put a strain in the wire. Hold the terminal end so the wire is straight *while you tighten* the connection.

I once thought the only good coax connectors were the type you install with a soldering iron. After using the newer type that doesn't require soldering, I recommend them for anybody not well-experienced in soldering coax fittings.

MECHANICAL PRECAUTIONS

Tighten threaded antennas and nuts *gently*. They are usually brass threads which may get stripped. Use the right size wrench. Some antenna nuts have threads on the outside, too, to attach a base loading coil. These round nuts have small flats on opposite sides; hold them *only* on the flats when tightening so you don't damage the threads.

CHECK STANDING WAVE RATIO (SWR)

After installing the antenna and hooking everything together, check SWR.

Expensive CB sets have a built-in SWR meter which can be switched into the antenna circuit for a reading. Follow the manufacturer's instructions for calibration before you take readings on any SWR meter.

If you don't have an SWR meter, accessory units are available at CB dealers and electronic parts stores from about $15 up to about $80. When you buy the meter, be sure you get an adapter cable to connect the meter into the antenna output of your CB. Then the antenna cable is plugged into the SWR meter so the output power of the CB passes through the SWR meter on the way to the antenna. Keep this adapter cable as short as possible—preferably less than 6 in.

For an SWR test, tune the set to its center channel—10 or 11 on 23-channel sets, 20 on 40-channel sets. When the channel is quiet, briefly key the transmitter and observe the SWR meter reading. If it's more than 2, tune the antenna—but first be sure you are getting an accurate reading.

Where the coax cable between transmitter and antenna happens to be 12-ft. long, an SWR may read inaccurately for complicated technical reasons. If you have a 12-ft. cable, lengthen it temporarily by adding a 6-ft. section with a male and female connector. Take another SWR reading. If it is worse, the SWR problem was being concealed by the 12-ft. length of cable. Leave the 6-ft. section in while tuning the antenna, then remove it.

Before tuning, recheck connections at antenna and set. If anything is loose, tighten and check SWR again.

TUNING YOUR ANTENNA

Theoretically an antenna can be tuned to only one wavelength and therefore to only one channel. However if it works well on one channel, it will normally work nearly as well on the adjacent channels. On channels which are farther away, antenna performance will be still worse.

Some antennas are advertised as broad band. That means they work well over a broad band of frequencies and therefore give good performance on channels 1 and 40 when you have tuned the antenna to channel 20. Antennas which are physically longer normally have better broad-band properties—a full quarter-wave antenna is best but hard to mount on a vehicle as discussed earlier.

If you are using your old 23-channel antenna with your new 40-channel set and the tuning procedures discussed here don't get good performance over all 40 channels, suspect the antenna. Some antenna manufacturers claim their 23-channel products are so broad banded they'll work just as well on 40 channels. Don't believe it if the antenna is *much* shorter than 102 inches of fiberglass or 108 inches of metal.

HOW TO DO IT

First, take SWR readings at the highest and lowest channels your set can receive, such as 1 and 40. If the upper channel has a higher SWR than the lower, your antenna needs shortening. If the lower has a higher reading, the antenna needs lengthening.

Some antennas can be tuned by sliding the whip in or out of its base to change length. Some have a short section at the tip that can be raised or lowered. Either way, you'll find a set screw you loosen to make the adjustment.

Change antenna length about 1/8 in. each time, and recheck SWR at the high and low channels. One should go up as the other goes down. When both are the same, the antenna is tuned for best match at the center channel and the SWR is as low as it's going to get with that particular arrangement of

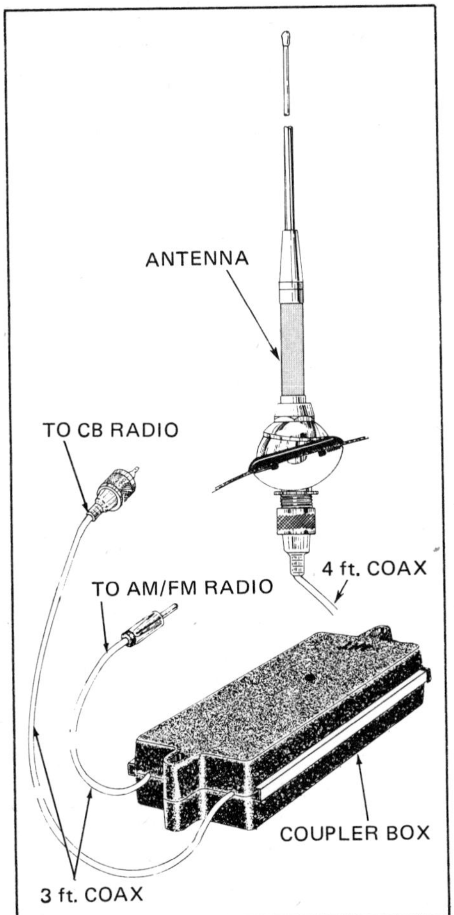

A disguise antenna looks *almost* like an ordinary radio antenna on the outside. The whip is one-piece; the loading coil appears to be part of the mounting hardware. The coupler box goes inside the vehicle, feeds your CB *and* your standard broadcast receiver. This Hy-Gain antenna gives superior FM reception.

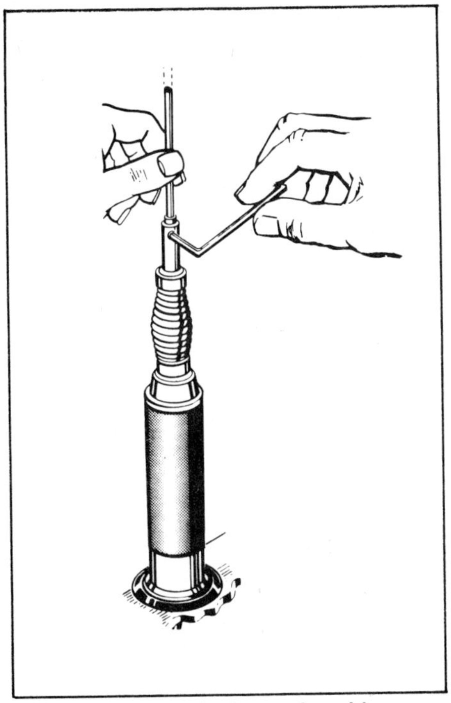
The set screw at the base of a whip may require an Allen wrench or a small screw-driver. To tune the antenna, move the whip in or out only about 1/8 inch at a time, then recheck SWR.

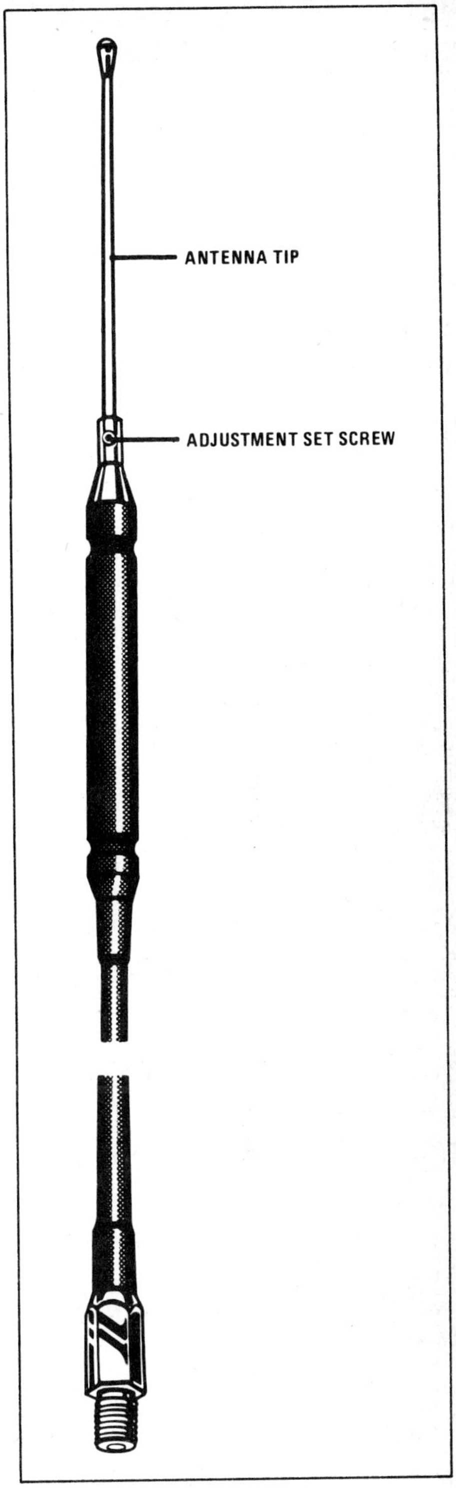

If it doesn't tune at the base see if it tunes at the tip, like this GC Grabber antenna.

hardware.

You may run out of length-adjustment. You may have to shorten a metal whip by removing 1/4 in. at a time. At the bottom, make two opposing notches with a fine steel file, then carefully snip off the end.

If the whip is too short, you can't lengthen it physically but you can use accessory antenna matching devices that plug in between transmitter and antenna. These also solve the problem when you can't get the center-channel SWR lower than 1.5:1 which is the highest figure you should accept.

Some authorities suggest opening up loading coils and snipping off portions; unwinding the spiral wire buried in fiberglass stalks and similar measures. I don't because these tricks are beyond the capability of most non-professionals and you end up destroying your antenna. If you can't get a low SWR by the methods described here, buy an antenna matcher. See Chapter 19.

When finished, record your final readings on high, center and low channels for reference when you test again, which should be every 4 to 6 weeks.

CHAPTER 13

HOW TO SELECT AND INSTALL BASE-STATION ANTENNAS

This chapter deals with mounting methods, problems, and what you need to do to make an installation that is legal, safe and mechanically strong enough to stay put. Base-station antennas are not toys. Some big beam antennas look like science fiction designs or something to communicate with space stations.

As I write this, Hurricane Belle has just come ashore a few miles north of my house. People not far away were evacuated. I battened down while the hurricane exhausted itself. I had Hy-Gain colinear antennas on my roof when the blow began and are still there.

As in the mobile antenna chapter just preceding, the emphasis here is on what the instructions sheets don't tell you.

FEATURES OF OMNI ANTENNAS

Most base station omnis are made of aluminum although a few are fiberglass. Seamless aircraft-grade aluminum alloy elements and radials are especially strong. Most structural strength is needed in the lower part of the antenna and the inside ends of the radials.

Toward the outer ends of the radiator and radials, smaller tubing can be used. Indeed, it is desirable to have smaller tubing because it is lighter and offers less wind resistance. All of this is logical mechanical engineering, but it takes smart design to pull it off. Check the antenna to be sure there is good electrical contact where different-sized tubing joins.

If you live near the ocean, or around industrial pollution, you will probably be better off with a plated antenna which resists corrosion from salt and chemicals.
Iridite plating makes aluminum look gold. Your neighbors will think you are rick. Buy one anyway if you're not in a clean-air belt; for ten dollars more you can add years of top performance to your base station.

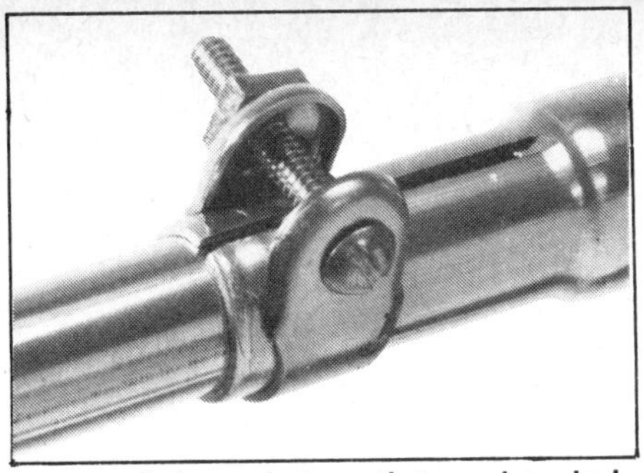

Long radials need strength to resist wind and snow-loading. The outer tips of my Hy-Gain test-base antennas are smaller diameter for less wind resistance. The larger tubing fits over the smaller element and is slotted for a compression clamp which gives good electrical and mechanical contact.

BEAM CONSTRUCTION FEATURES TO CHECK

As in my advice on omnis, look for the strength and weathering qualities of seamless aircraft-grade aluminum-alloy tubing. Booms carry a lot of weight, so they should be 2-in. or 3-in. diameter heavy-wall tubes. Hardware to hold a beam together consists of clamps because it is better *not* to drill holes in the antenna elements to bolt them in position. These clamps are made of rugged aluminum, secured with steel hardware which should be plated or preferably made of stainless steel.

FEED POINTS AND MATCHING

Beam antennas such as the Yagi and quad have two feed points on the driven element, neither of which is grounded. Such an antenna is said to be *balanced* and it must be fed with a balanced transmission line with neither side grounded. Coax is the best transmission line but normally the outer shield is grounded at the transmitter. Solve this problem with a *balun* transformer which you can buy at the same place you get the antenna. A balun transformer, sometimes called a balun coil, is mounted on the mast near the antenna feed point.

With beam antennas, matching devices are often needed to present the correct impedance of 50 ohms to the transmission line. There are several types in use, usually adjustable in one way or another. Check the instructions with the antenna and be sure the antenna matcher is supplied if required—or else purchase what the instructions call for.

If a beam antenna uses a *gamma match,* you can feed it with an unbalanced coax line without using a balun.

If the antenna uses a *beta match,* it requires a balanced line and therefore a balun must be used when you feed it with coax.

A SERMON ON THE MOUNT

Here I really mean to preach care and caution about installing base-station antennas. One reason is, FCC legalese can be misinterpreted easily and often is by eager CBers and carelessly written radio catalogs.

A second reason is safety. An improperly protected antenna invites lightning into your house which can burn out your transceiver and your radio shack faster than you can call 10-70 to report the fire.

Some base-station antenna supporting structures are so large even when shipped knocked down (disassembled) that you spend almost as much in shipping costs as for the mount.

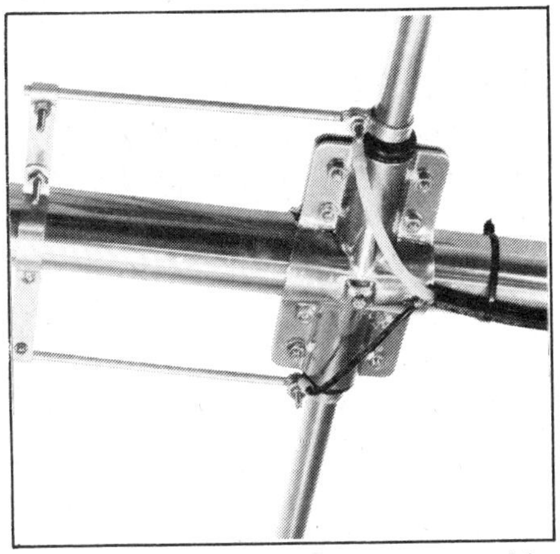

If your beam antenna has a *beta matching device* like this, it should be fed with a balanced transmission line. Run the coax cable to a balun transformer on the mast nearby. Insulate the coax braid as shown because it is not at electrical ground after the signal has come through the balun.

According to the FCC, there are two ways to mount an antenna: On something that is already there, or on a tower that you erect *from ground up.*

In the first case, something that is already there is usually called a man-made structure but it includes trees and tall rocks. The way out of this dilemma is in the official language of Part 95: *A man-made structure is any construction other than a tower, mast or pole.*

On a man-made structure, tree or natural formation on which it is mounted, you can use an antenna of any kind provided its highest tip is *not more than 20 ft. above the highest part of the structure.* That means you can put an omni antenna or a beam antenna or any other kind of a CB antenna on the roof of your house.

You can't increase the height of a structure solely for the purpose of getting your antenna higher. If you try it, the add-on to your house is not counted as part of your house—it's counted as part of your antenna and, oops, you got it too high in respect to the original structure you started out with.

If you start from ground up and erect a tower, mast or pole, the highest tip of the antenna can be 60 ft. above the ground, but you can use *only an omni*—no beams.

Either way, you can't separate the antenna fuctions and put a receive-only antenna higher than the law allows just because it isn't radiating anything.

If You Are Near an Airport—A special rule applies to antennas near a *runway* of an airport. It applies to any antenna but it is more likely to affect the 60-footers. The highest part of your antenna must not exceed one foot in height above the established airport elevation for each 100 ft. of horizontal distance from the nearest airport runway.

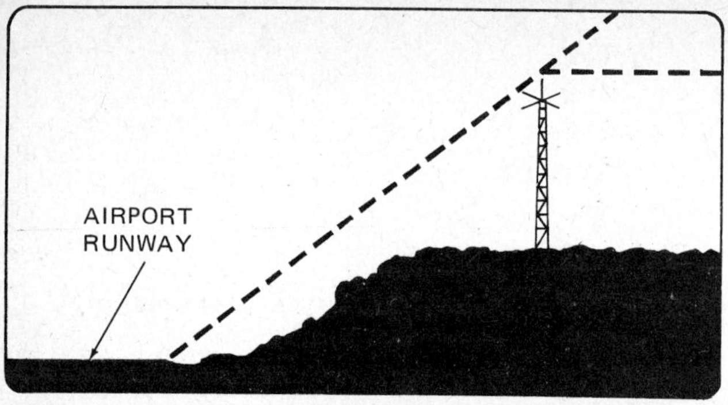

An antenna near an airport must comply with FCC and FAA restrictions: 1 foot of antenna height for each 100 feet of horizontal distance from the runway. Compare antenna tip elevation above sea level with runway elevation above sea level.

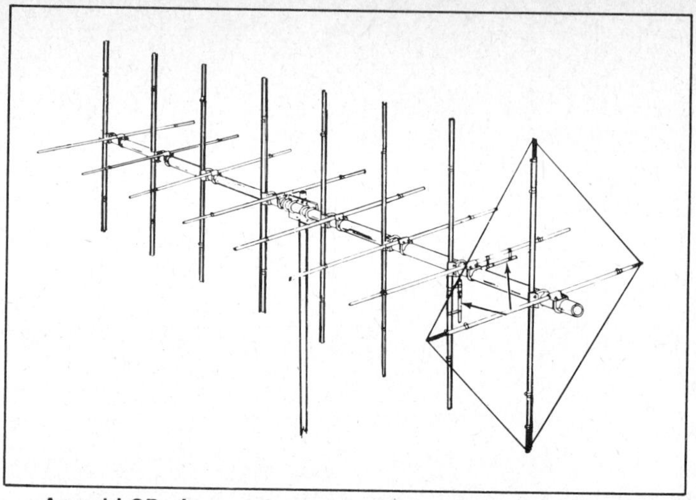

An avid CBer's eyes glaze when he day-dreams about owning a big beam. This Wilson Electronics Super Laser 500 uses a 40-foot boom with dual polarization. It's tricky to install but worth it if you want 18 dB gain. The antenna weighs 70 pounds. *Gamma matching devices* **(arrows) allow feed with unblanced line.**

Zoning Boards and Ordinances–Before spending big bucks and a lot of time installing either a tower in your backyard or a beam antenna on your roof, check your local laws or zoning board about what is legal. They don't care if the FCC says you can put up a 60-ft. tower–their own laws may not allow it.

Going Piggyback–If you already have an antenna supporting structure with an antenna for a radio station licensed for some other communication service, you can put a CB antenna on the same structure. But you can't get any sneaky advantage out of doing it. The tip of a CB omni can't be more than 60 ft. above ground and the highest point of a CB beam antenna can't be more than 20 ft. above ground.

ANTENNA HEIGHTS VERSUS DIRECTIONALITY

You can't have both height and directionality because of the rules given earlier, *unless* you can put a beam antenna on top of a tall building. There the 20-ft. rule still applies but 20 ft. above a 10-story building is pretty tall.

If there are intervening hills or tall buildings between your base station and distant points you want to reach, get every foot of height you can by any means possible.

Is your major communication interest base-to-base or base-to-mobile? If it's base-to-base, use a beam antenna unless you must have more height. Example: You don't have a tall building to put the beam on, so it can be only 20 ft. above your one-story garage–say 30 ft. total. This lets the antenna look at a nearby hill rather than look over it so you can talk to a fixed base station on the other side. But using an omni on top of a tower at the full 60 ft. gets your signal over the hill to the desired location.

If you communicate base-to-mobile and you don't know where the mobile units may be, use an omni antenna on a tower.

You can get both omni and directional beam operation from an electrically-scanned antenna but the FCC considers it a directional beam unit because it has that capability. Therefore you have that 20-ft. limit above the sturcture.

GROUND-UP ANTENNA CONSIDERATIONS AND COSTS

A 10-ft. omni requires a 50-ft. tower or mast to reach the 60-ft. ceiling. If you're going to all that trouble and expense for maximum performance, put some of your money into a longer antenna for a lower angle of radiation. I recommend a 40-ft. tower or mast and a 5/8-wave omni.

Approximate Costs: About $130 for a 40-ft. self-supporting tower, not including shipping and taxes. About $40 for a 40-ft. telescoping mast, guy wire and additional fittings. Add approximately $30 to $50 for the antenna.

DIRECTIONAL BEAM ANTENNA CONSIDERATIONS AND COSTS

Steerable beams require some thought about turning radius and weight. Antenna booms can be as long as 40 ft. Let's say you choose a small beam antenna, but still big enough to be significantly better than the best omni. A five-element beam with about 12 dB gain has a 24-ft. boom. Because of the projecting elements it requires a turning radius of about

14 ft. Be sure you can clear trees, poles, chimneys, other masts and guy wires before choosing this kind of radio artillery.

Physical rotation requires a heavy-duty antenna mount for big beam antennas. You'll need an electrically-operated heavy-duty rotator, thrust bearing and special electrical cable for controlling a steerable beam. Don't forget to figure the weight of the antenna rotator assembly into the total weight your mount must support. Guying is advisable for all that weight 20 ft. up.

Approximate Costs: About $25 for a mount that can handle a beam and rotor system, $25 to $75 for the rotator, $40 to over $200 for the steerable beam antenna.

SCANNING ANTENNA CONSIDERATIONS AND COSTS

Scanning antennas put less burden on the mount and require less clearance of surrounding objects.

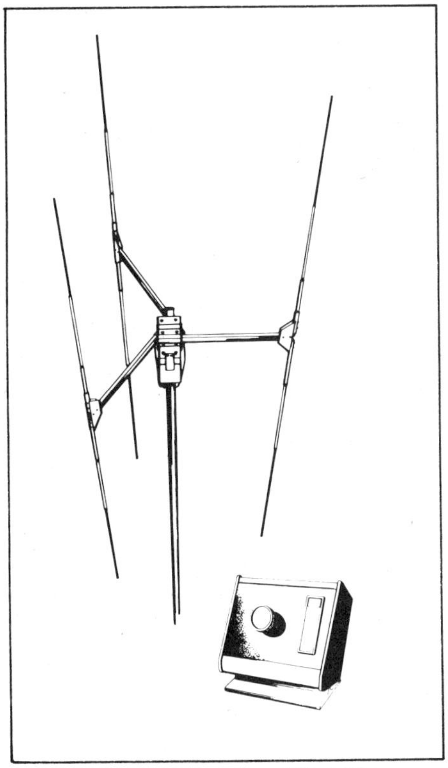

This Antenna Specialists MS 119 is a 3-element electronic scanning antenna. Select omni or electronically steerable beam, using control box at transceiver.

Electrically conductive nearby objects may interfere with the signal and upset the directional scanning pattern.

You have to wait a few seconds for a rotator to swing a steerable beam on target but electronic scanning antennas rotate 360° in milliseconds so there's less chance of missing out on a communication. A special multi-conductor cable runs from a scan control box at your base station transceiver to the relay box on the scanning antenna for switching among its elements. You can use a relatively light antenna mount.

Approximate Costs: About $12 minimum for a heavy-duty aluminum mast, plus about $8 for base, guys and fittings. An unguyed tripod tower costs roughly $30. It is also a much more visible structure. Scanning antennas aren't cheap: About $130 to nearly $260.

BEST BETS

I'd pick a scanning antenna for all-around use even though you get maximum performance with the biggest steerable beams. Here's my reasoning: The least expensive steerable beam installation will cost about $100 and bring occasional electrical or mechanical troubles with the rotator. It will have about the same forward gain and front-to-back ratio of a scanning antenna in the directional mode. For about $50 more for the least expensive scanning antenna installation, I have both an omni *and* a directional antenna plus faster pick-up of other stations.

ROOF MOUNTING BASE-STATION ANTENNAS

From the ground it looks easy. When you get on the roof it seems harder than you could have imagined. After you're finished, you wonder why you thought it might be difficult. The secret is advance planning and allowing lots of time for actual installation.

Advance planning does not mean simply eye-balling the possible mounting sites from below. You have to get up on the roof to inspect and measure *before choosing* the best combination of antenna and mount. Don't buy anything without planting both feet on the potential antenna site and rotating yourself thoughtfully through 360°.

Read the following sections on mounts even before you go up on the roof so you'll know what to look for and measure when you're aloft. I'll assume you have decided on the kind of antenna you want–omni or beam. And if it's a beam you have decided about polarization–vertical, horizontal or dual. You've thought about the amount of gain you want and have the dimensions of a few antennas that fit your needs. Now you have to find out if they'll fit your roof.

Check Out The Antenna Site– You'll want to be more sure-footed than on the ground. Wear workshoes with non-skid treaded soles, firmly laced. High shoes give more ankle support, which you'll appreciate on a sloping roof. Use a ladder that extends several feet above the point where you'll step off onto the roof, so you'll have something to grab when coming down again.

Make sure the ladder is firmly on the ground and its base is far enough out from the building so it remains stable when you put your weight on it. Don't use a cheapie ladder; repair work to your body can be expensive. Lightweight aluminum ladders tend to blow over in a wind, so lash such a ladder in position or you may be stranded up there. It's best to have a partner, to hold the ladder and aid in measuring and the trickier work of erecting, positioning and guying.

If you use an extension ladder, check it over carefully before trusting it. Be sure you know how the locking device works to hold the extended part so it doesn't slip when your weight is on it. Then be sure you have set the locking device properly.

Flat rungs are more comfortable to your feet than round rungs.

The following discussion of roof

mounts also considers the difficulties of working from a precarious perch and tells how to simplify the task.

Clearances and Heights—Dimensions you haven't thought about complicate siting omnis as well as big beams. Some omnidirectional antennas have vertical radiators nearly 23-ft. high. Clearly you can't mount these on the peak of a sloped roof without exceeding the legal limit, unless there is some higher point *of* the structure like a good-sized chimney. Measure chimney height above the roof peak. If it is going to make up for the length of a 23-ft. vertical radiator, you're OK on the roof peak. Be sure to allow an additional 6 in. minimum for a stub mast and mounting base. Keep the coax feed connection at the bottom of the antenna clear of the roof. If you live in an area where there are deep snows, allow enough clearance so snow won't short out the antenna connection.

Then there is the problem of groundplanes. If you're below the roof peak, clearance for the radials can be a problem. Consider every part of the antenna and how you are going to mount it.

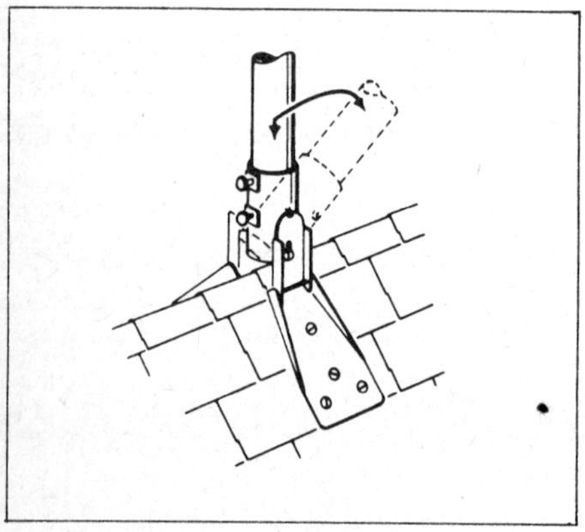

An adjustable mounting base fits on a peak, slope or flat roof. The swivel is handy, but don't try to "walk up" a tall mast and heavy antenna. Guy any mast with a base mount.

I already mentioned the turning radius of big beams. Look for clearance problems not only at the boom height, but also at roof height if your antenna is vertically polarized. Remember half of the elements extend down from the boom. If you're mounting below the peak of a sloped roof, allow for tip clearance of the elements as the beam rotates.

MOUNTING BASES

Every problem can be solved by some kind or mounting base, even if you have to manufacture it yourself. Some on the market are so flexible in their application they can be used to solve several different problems. Bases are less noticeable on your roof than small tripods and larger roof towers. They bear the full weight of the mast and antenna on flat roofs, peaks and slopes. The catalogs are full of such mounts, but the best type is adjustable to adapt, saddle-like, to a roof peak of any angle or fit flat against a sloping or horizontal roof. On a slant, the mast socket should swivel to a vertical position. Some bases have a double swivel so you can orient horizontally as well as vertically—a lot easier than trying to turn a mast in the socket.

Hot-dip galvanized or cast aluminum gives better protection against rust and streaking. Be sure to examine the spacing on the mounting holes. Not all bases are designed so the screw holes line up with rafters for firm anchoring.

Careful installation will defeat the invisible forces of wind and vibration at work to destroy your investment in a good base-station antenna. I have heard of perfectly guyed masts that simply rest astride a roof peak and are said not to creep or walk. Bravo! Nevertheless, your antenna is safer if you secure the mount firmly to rafters when you can.

Rafter Location Method #1—Go into the attic or crawl space and measure rafter location from the underside of the roof. Choose a reference point that projects upward *through* the roof such as a vent pipe or chimney. When you're topside again, measure carefully from the reference point to the nearest rafter. Was that measurement from the reference point to *the edge* of the nearest rafter, or to *the center*? If you're not certain, it can send you back into the attic to double-check. Always measure to the centerpoint and avoid the problem.

Method #2—This works only with flexible composition shingles, not slate or wood shingles. From topside, lift up a shingle and drive a long nail under the shingle so it goes through the roof as a reference point where you intend to locate the antenna mount. Then go into the attic or crawl space and locate the nail as a reference for measurement to the center of the rafters you will use for support.

In cold weather, shingles get brittle and may crack when bent. Soften by covering it with thick cloth, then pour boiling water over it. Or use a torch but don't set the roof on fire.

Dip or spray your locating nail in bright paint and let it dry thoroughly before driving it through the roof. Paint clinging to your locator nail will confirm its identity.

Method #3—In certain kinds of construction, rafters may be seen under the edge of the roof. Hooray!

Method #4—On some roofs with composition roofing, you can lift up the edge of the roofing carefully and see the rows of nails used to fasten the wood roof sheathing to the rafters. To do this, you must lift both composition shingles and the layers of roofing paper underneath the shingles. If it's a cold day, heat the area as described earlier so the composition material doesn't break or crack. If you lift a large area, consider tacking it back in place or use roof-patching material—the black stuff—to glue it back in place. Otherwise the next strong wind may show you the rows of nails all over again.

MOUNTING BRACKETS

Don't be discouraged by a limited variety of these items in stores. Most stock the standard, fast-moving merchandise, but there is a mount for every problem.

Ask to see a manufacturer's full catalog and get the store to order what you need. Certain kinds can give you problems. Here's what to check:

Chimney Mounting Brackets—These are simple and popular but in some localities, they are prohibited by zoning laws. Make sure before installing. Some chimneys look sturdy and may be strong enough for TV antennas and omnis or beams on a short mast. On the other hand, they may not be strong enough and it's hard to tell, Bricks are held together by mortar. Look for cracks or loose mortar; check to see if the mortar is solid or soft and crumbling.

Chimney mounts are relatively easy to install when you can stand on the roof on all sides of the chimney. Things get difficult when the chimney runs up the side of the structure.

Banding straps support the mounting brackets that hold the antenna mast at one corner of the chimney. Make sure you pick a corner you can get to easily when you want to install the mast; it's better to work from the roof than from an outside corner where you have to stand on a ladder. Measure the circumference of the chimney *before* buying the mount kit; straps vary in length from 10 to 18 ft.

You can get chimney mounts in galvanized steel, zinc-plated steel, cast aluminum with steel straps, and all stainless steel. Straps are the weakest part of a chimney mount and a penny saved is sometimes an antenna overturned.

Get stainless steel for maximum strength and corrosion resistance. Don't just stop at stainless-steel banding straps; get *all* stainless steel—mount *and* hardware. One of my antennas has been on a stainless-steel chimney mount for 19 years in rugged Northeast weather, with no problems.

It's easier to use a snap-in mast-holder bracket because you may not have enough hands to hold the mast, insert a U-bolt and fit the nuts. This is twice as hard in any wind above 10 MPH. A snap-in holder is riveted at only one point, so you are trading some bracket strength for convenience. I recommend the strongest bracket you can find, and an agile helper.

Not all strapping materials are the same; most tend to stay coiled like a spring when you want them to flatten out so you can lasso a fat chimney. They get especially ornery when kept tightly rolled in bubble packs on store display racks. I've discovered South River and GC brands behave better.

One problem with strapping is noticeable only when you've run out of hands or your fingers have gone numb: initially securing the banding loop. Normally this is done by cinching the strap through a buckle-like arrangement. When working with only one hand, it's easier to use a clip, then pull the banding tight and fasten with a set-screw. Not all clips are the same; look for a notch that holds the strap so it won't slip out before you get to the set-screw.

All the thoughtfully designed mounting hardware shown in the drawings of this chapter is made by South River Metal Products. I have been using their mounts and fittings for many years because they are reliable and easy to install.

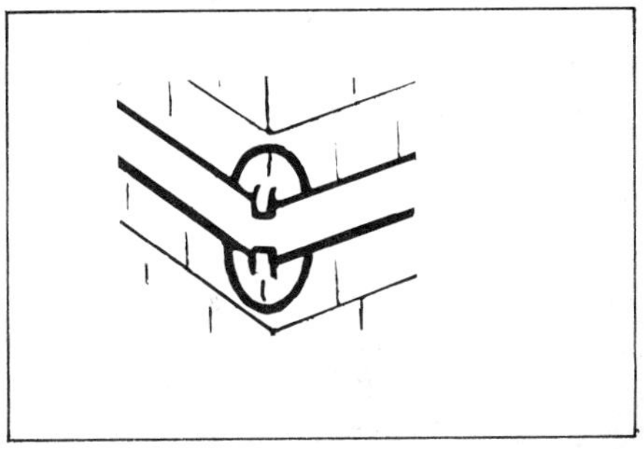

Corner mounts improve appearance and durability of a chimney installation.

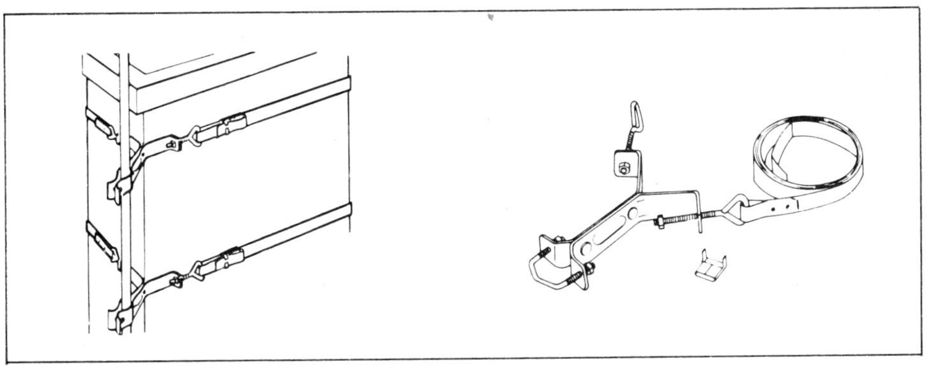

Chimney mounts use snap-in brackets or U-bolts to hold the mast. Get two more hands from a helper and use the stronger U-bolt bracket.

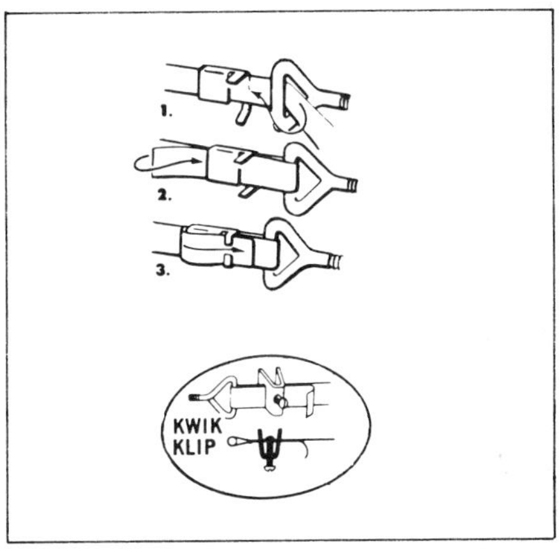

After passing the stainless banding strap around the chimney, secure it with either of these devices.

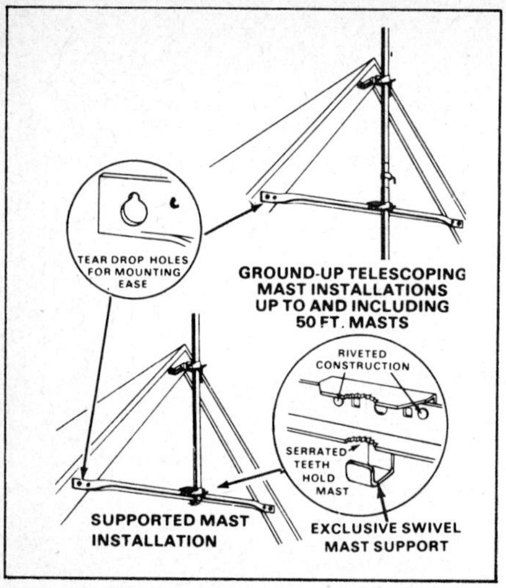

This eave-mount is a versatile type of wall bracket. The features I find helpful are shown enlarged.

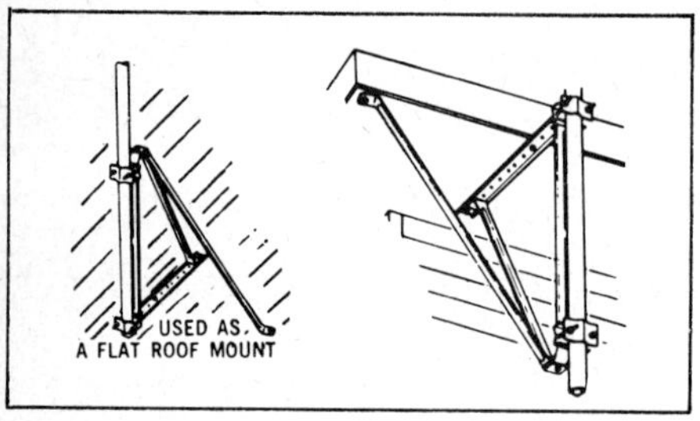

This mount is useful under an eave or on a flat roof.

By far the easiest, fastest strapping method for banding a mast bracket to a chimney is a ratchet mount. You just loop the banding straps around the chimney and tighten up the ratchet with a wrench.

I'll put up with a little additional work to use all stainless steel. The ratchet mounts I've seen use stainless steel only for the banding straps, so I don't use them.

One chimney mounting accessory should be on your shopping list—corner guards. I suggest them because they protect both chimney and banding straps from abrasion. They shouldn't rust but don't have to be made of stainless steel. Most are aluminum, which is OK.

Wall Brackets—Come in various sizes to give clearance of 2-3/4 in. to 36 in. from the wall. Check the eaves and other overhangs before choosing a wall bracket or you'll have a terrible time. Be sure to get supporting legs on brackets which are longer than 6 in. Some shorter brackets are made of steel bar stock. I prefer stainless steel to avoid rusty drips on the wall.

Roof Overhang Brackets and Corner Brackets—These fasten to angled surfaces. When checking out the antenna site, determine if such a bracket should have a swivel so you can bring the mast to vertical.

BEST BET

No matter what anybody says, use a heavy-duty roof mount or brackets the *first* time. This advice applies not only to your own installation, but also to one made by a professional installer. Offer to supply the mount, if necessary, to get the strongest you can find. Cost is only a couple of dollars more than economy versions. Moreover, you can use masts of larger diameter, now or when you change to a different antenna.

WORST BET
VENT PIPE MOUNTS

Don't use them. They *look* strong. They may even be strong. But even with a properly guyed mast you run the risk of flexing and damaging pipe or joints down inside some inaccessible place.

WISE GUYING

To keep your antenna safe from toppling in high winds, plan on guying any mast which extends 10 ft. or more above firm support. Most bases with masts need guying.

A catalog describing a tower supplied with a 10-ft. mast says 3 ft. of the mast will be supported *within* the tower. Because the tower is a firm support, you needn't guy provided you use a strong mast—16-gauge steel or 0.125-gauge aluminum alloy.

You can even telescope a 1-1/4-inch O.D. mast inside a 1-1/2-inch O.D. mast for greater strength, assuming wall thickness is 0.125 inch. And you needn't guy the radiating element of a large omni unless you buy a poorly-designed super cheapo. I've seen it done out of necessity with insulated clamps on cheapie antennas, but good manufacturers assure wind survival at 100 MPH.

The accompanying table gives the length of each guy wire for various mast guying points. Multiply by the number of guys before buying the wire.

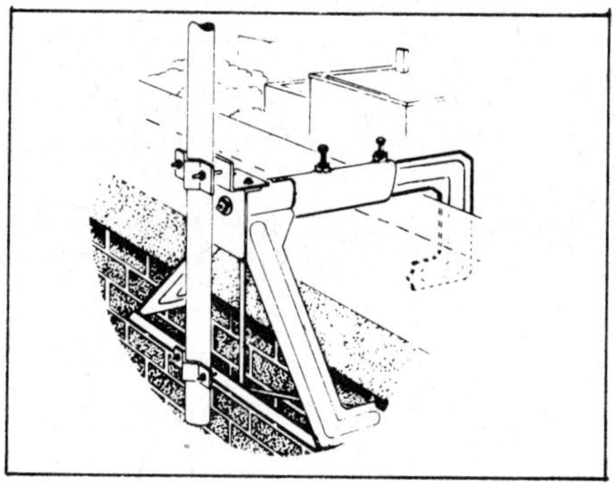

It's no fun to work leaning over the parapet on commercial buildings and apartment houses. This parapet mount lets you work safely, standing on the roof.

How To Buy Guy Wire—Guy wire comes in 50-ft. interconnected coils. Don't accept separate coils or you may be wasting wire. Not all guy wire is the same.

Guy wire is twisted from several individual strands of wire of one gauge or another. The lower the gauge number of the strands, the heavier and stronger they are. Guy wire also gets stronger with each additional strand. It also gets fatter, so look for tightly twisted wire for minimum wind resistance. Steel is stronger than aluminum.

I favor heavy-duty aluminum wire made by the Alclad process to assure full strength. Don't be conned by the counterman at the parts store into accepting wire that looks just as good.

Your choices usually are galvanized steel wire, aluminum wire, vinyl-clad steel and rope. For a do-it-once installation, use aluminum. If you live in an area without much rain or corrosive fumes in

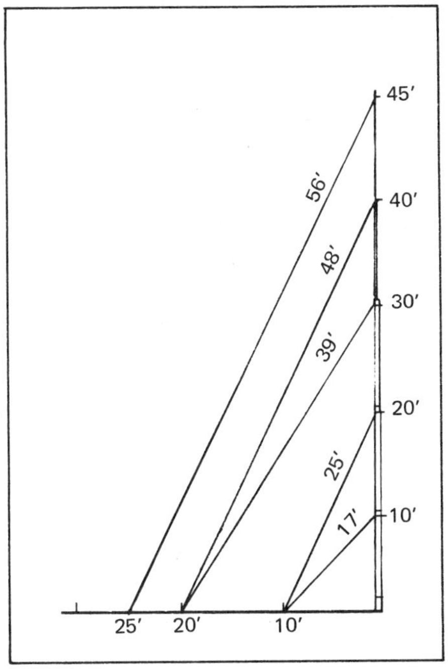

Guy any mast that is 10-foot tall or higher. Estimate the length of each guy line from this chart which shows different heights and distances along the ground from mast center.

the atmosphere, galvanized steel will last a long time.

Plastic-coated steel wire looks like a good deal but I don't recommend it. Very few types of plastic last very long in direct sunlight and when the plastic cracks there's nothing left to protect the inner steel wires from fast rust-out. Plastic-clad cables are difficult to tie-off or clamp and the final result may not be secure over a long period because the plastic may creep or change its shape.

You may have read or heard about guying with rope. It does eliminate rust and solve the TV interference problem discussed in the next paragraph but I don't recommend using rope. It has a limited life.

GUY WIRES AS RF RADIATORS

The problem of TV interference (TVI) due to CB sets has been increasing as the number of CB operators increases. There have been a lot of complaints.

This caused the FCC to tighten the rules for spurious emissions from CB transmitters along with the other rule changes announced on July 27, 1976. The main victim of TVI is channel 2 because it is almost exactly double the frequency of the CB band. A CB transmitter radiates not only the CB carrier frequency of about 27 MHz but also spurious frequencies which are multiples of 27 MHz, called harmonics.

The tiny bit of 54 MHz signal allowed by FCC rules to be emitted by a CB transmitter can cause TVI. The situation gets worse if antenna guy wires help pump out the 54 MHz harmonic.

This is commonly referred to as the problem of guy wires reradiating TVI. The cure is to break up the continuous lengths of guy wires into small segments, using insulators between the segments. Minimize TVI from your base-station antenna by using insulating strain insulators in all guy wires so no continuous length of wire is longer than 3 ft.

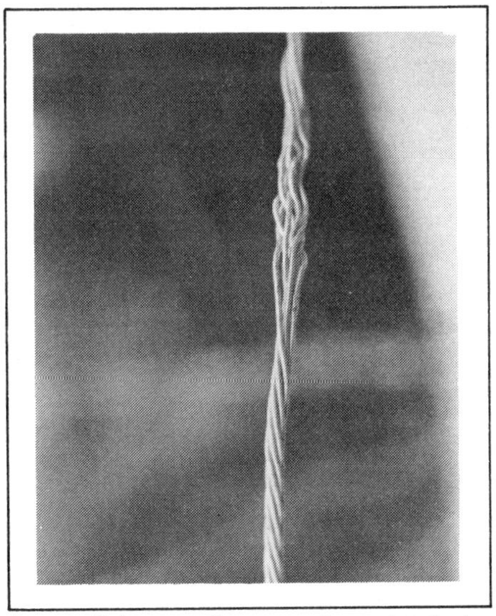

Be careful not to let guy wire twist during installation. This kink took only 60 days to become separated strands.

Reduce TVI caused by re-radiating guy wires. Install strain insulators every three feet.

ROOF TOWERS FOR BASE-STATION ANTENNAS

Roof towers are self-supporting and sturdy. Their legs are equivalent to guy wires so no guying is necessary for the tower itself. If the mast held by the tower projects more than 10 ft. you should guy the mast. The appearance of a tower is a matter of taste; you may be happier with it on a commercial building than on a house. But for heavy antennas—particularly with rotators—a tower is unsurpassed.

One dimension to check very carefully is the spacing between the tripod-shaped tower and the nearest element of a vertically polarized beam. Because the elements are vertical, you may need a taller mast to provide clearance for the lower end of this element. In most cases a rotator may add the additional height needed—it depends on the type of rotator and the angle of the tripod tower.

More Towering Wisdom – Semi-knocked-down models can be assembled in about 15 minutes. Plan to do this on the roof if you can because towers are devilishly clumsy to carry up or haul up.

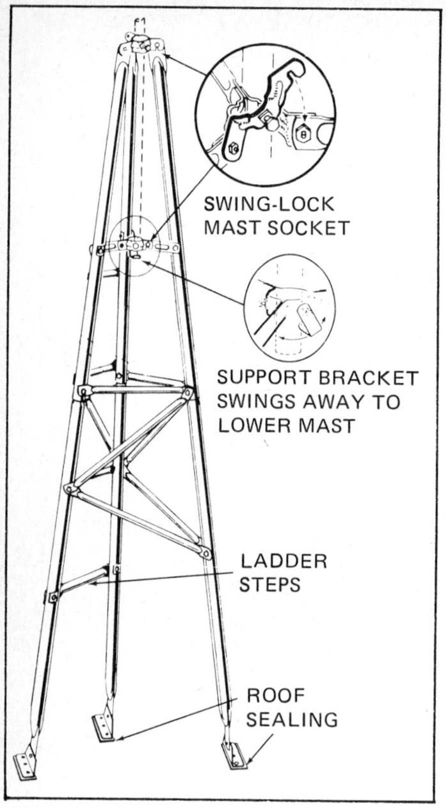

When you order a roof tower, be sure the legs will fit your rafter spacing. Some construction is not the usual 16 inches on centers. This tower has latches to let you slip in the mast from the side—which is much easier than lifting a mast and heavy antenna over the top and dropping it down into the tower.

You really can't tip up an assembled 10 or 15-ft. tower from horizontal to vertical. Final assembly should be done in the vertical position.

Wind blowing through a tower causes vibration. It can loosen bolted joints. A good tower will come with vibration-proof stopnuts. Don't substitute any other kind if you drop them and they roll off the roof. Get some more of the same type.

You can deform tower tripod legs by improper tightening. Always use the preformed washers provided by the manufacturer. Make no substitutes if they roll off the roof. Get some more of the right washers.

Be sure to use pitch patches—usually packed with individually boxed towers—to make a watertight seal where the tower legs screw to the roof. Ask for these when you buy the tower.

Secure the Mount to Rafters—Same as any other roof mount, get strength and rigidity by driving the mounting screws into rafters.

Mounting Roof Towers Without Locating Rafters First—Be sure you can work from the underside of the roof before deciding on this method. After fastening the tower feet to the roof with lag screws, go inside the attic or crawl space and mark the location of the protruding screws. Now go topside and back off the screws so they just show on the underside. A helper with a loud voice is handy.

Cut a length of 2" x 6" or larger board to fit snugly between the rafters. Use more than one if needed. Place it flat side up against the underside of the roof where the tower-feet screws are located. Secure the board by nailing through the rafters into the ends of the board. This makes a solid platform, tied into the rafters. Then you can screw the antenna-mounting screws into the board. Or you may use 1/4-in. or 5/16-in. bolts through holes drilled in the board.

TELESCOPING MASTS FOR GROUND-UP INSTALLATIONS

When a ground-up structure makes a *significant* increase in

It is extremely difficult to lift a mast and antenna over the top of a tower and lower it into the socket. Look for features like a mast socket and braces that open so you can place the mast in from the side. Even so, towers higher than 7 ft. pose difficulties because the mast socket may be beyond reach without a ladder. Leaning a ladder against a tower on a sloping roof is *extremely dangerous!* Buy a tower with welded rungs *meant* to be used for climbing; don't buy any old tower and expect to climb the cross-braces.

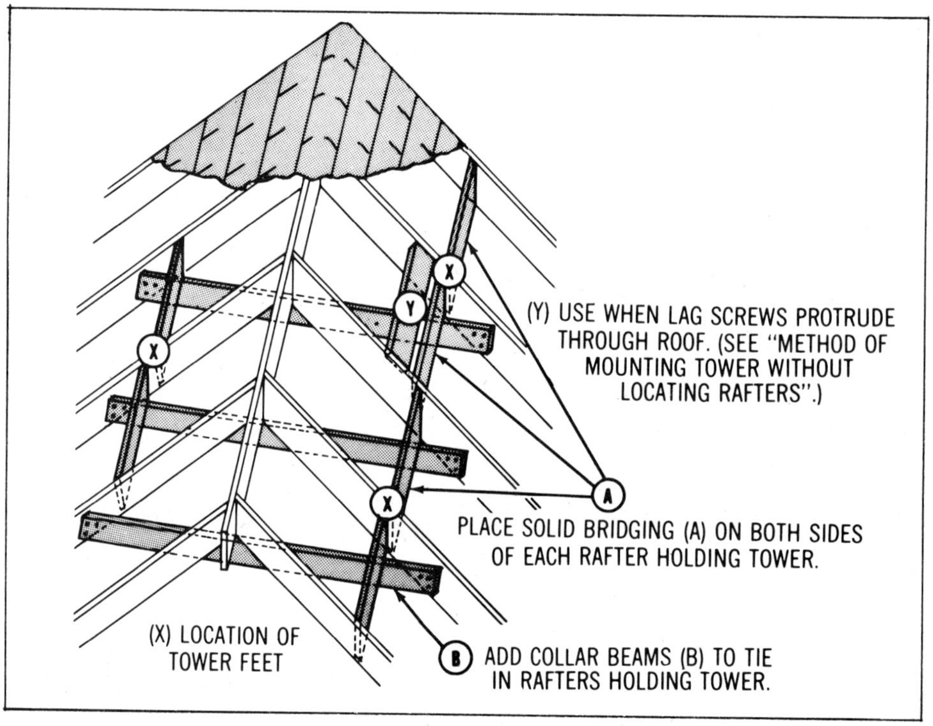

You won't believe what wind forces can do when you have a big beam mounted on a roof tower. Here's how to reinforce the rafters where the tower is fastened.

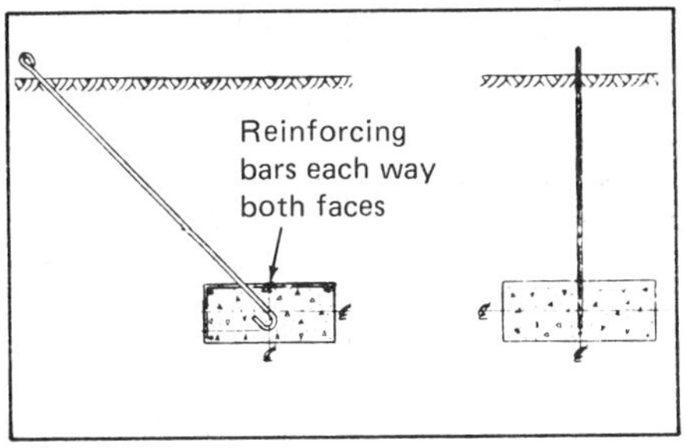

To anchor a guy wire in dirt, bury a "dead man." This drawing shows poured concrete. Depending on the type of soil and strength needed, you can bury concrete blocks or creosote-treated logs and attach guy wires.

height over a roof-mounted antenna, a telescoping mast is easy, neat, inexpensive and unobtrusive.

When I say easy, I mean relatively easy with an omni and harder with a big steerable beam. Things get more difficult as you go higher because you don't tip up a telescoping mast. You push it up a section at a time. Let's say you are putting up a dual polarized beam that weighs 40 lbs. You are using a heavy-duty rotator and thrust bearing which adds about another 10 lbs. You'll be pushing up 4 sections of a 50-ft. mast—about 27 lbs.—and trailing a few hundred feet of guy wire—say, another 8 lbs.

That's about 85 lbs.—and I said you *push* it up. The aggregate weight explains why you can't tip it or walk it up from horizontal to vertical without rigging some clever tackle. Don't be misled by advertising references to walk-up pivoting mounts for telescoping antennas. Those catalog copywriters erect only ivory towers. Don't buy crank-up telescoping masts unless they are delivered to the site by a local dealer. The sections are long and costly to ship.

I said telescoping masts are easy, and they are. But they're not simple. You need a lot of room for guy wires. Most ground-up installations are adjacent to a building. Wall brackets on the building can support a telescoping mast as far as the roof, but above that you need guy wires.

Three-point guying may give two anchor points on the roof. But for the third anchor point you'll have to anchor to a pole or to the ground. Figure the guy anchoring points carefully to be sure they don't fall on your neighbor's property.

Another problem is attaching a vertically polarized beam to the mast. The lower halves of each element extend down about 9 ft. from the boom. These elements must clear the roof so they don't get bent. That means that you'll have to reach at least 9 ft. above the roof to attach the boom to the mast. It's easy with a stepladder on a flat roof, tricky on a slope.

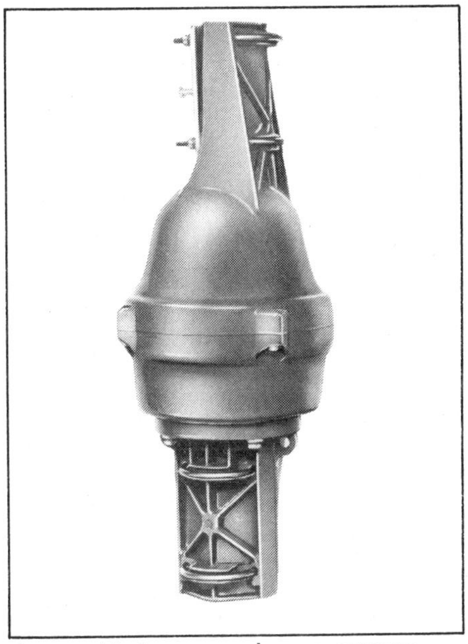

A steerable beam requires an antenna rotator. This rugged in-line Ham II Rotor by Cornell Dubilier has an electronically controlled brake that positively locks the rotor in any of 96 positions.

ANTENNA ROTATORS

If you intend to point your super beam antenna in more than one direction, you need a rotator. This is installed atop a mast or tower and turns a stub mast to which your beam is attached. The control unit is located near your base transceiver. More about controls in a moment.

There are two rotator arrangements. An *in-line* drive system is on the same axis as the antenna mast. The drive unit appears to connect two masts end-to-end. This arrangement evenly distributes the weight of the antenna and wind-loading forces, but heavy antennas or strong winds may require the stronger support of a thrust-bearing. Most rotator manufacturers supply thrust-bearings for heavy-duty service.

Offset rotators clamp to the side of a mast top with a stub mast rising from the drive unit to support the antenna. Weight distribution is not as even as with in-line drive and the bottom mast is subject to more stress. Offset rotators are usually lighter in weight and smaller than in-line drives. The additional weight and sail area of either type stresses the main mast so I advise guying either type. Guy-wire lugs are built into the drive-unit housing.

If you live in an area where icing is common, get a heavy-duty rotator. Starting out with a strong motor could mean it will be capable of breaking ice years later.

Rotator Controls—A *manual* control turns the antenna in the selected direction until you release the control switch. You don't know where it is pointed—unless you stick your head out the window.

A *semi-automatic* control overcomes this problem with an indicator showing the direction of the beam. If you keep a consistent communication schedule with certain fixed base stations, you can mark the display with their relative bearing. You hold the control switch closed until the desired bearing (direction) is indicated.

An *automatic* control is a rotating pointer. Rotate it to the bearing you want and the antenna turns until it reaches that position, then stops.

Not all automatic controls are alike. Electromechanical controls can get out of synchronism so the beam is not pointed where the indicator says it is. Antenna pointing is in fixed increments or steps of 4 to 6 degrees so you can't point the antenna *exactly* where you want it.

Fully electronic controls cost more, stay synchronized, and point in about 2-degree steps, so you can be more sure of homing on another station—especially one with a narrow beam of its own.

You'll need 4- to 8-conductor wire to link the control unit at the base transceiver to the drive unit, depending on the model you buy. Use dual standoffs for the control cable and coaxial cable run up the mast. A T-shaped cable standoff is the neatest, easiest to use. Be sure to allow a couple of feet of slack in the coax between the stub mast atop the rotator and the main mast so the antenna can freely rotate through 360°. It won't turn more than a full circle or a little more. At limit stops, it reverses direction.

LIGHTNING PROTECTION

It's really simple to achieve good lightning protection as shown in the step-by-step photos of an antenna installation. A lightning arrestor is connected in the coax cable—preferably at the antenna end—so you don't bring high voltages indoors before bypassing them to ground. *Electrical grade* #8 aluminum wire or heavier is connected to a terminal on the outside of the lightning arrester. Run this down the antenna and building wall through standoff insulators to a copper pipe driven into the ground or a cold-water pipe that enters the ground. Follow your local electrical code.

Lightning arresters really don't arrest anything. They provide a spark-gap so high voltages are discharged to ground. Sometimes they're called static-discharge devices. They do bleed off static electricity accumulating on your antenna and give less noisy reception. Don't expect an arrester to save your antenna from a direct

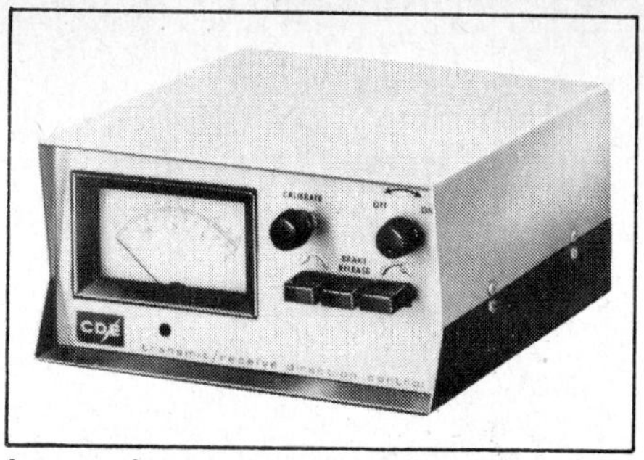

Locate the rotator control box next to your transceiver up to 150-feet away. This Cornell Dubilier control requires 8-conductor cable up to the rotator on the mast. Depress BRAKE RELEASE button and either direction selector.

hit, but a sturdy unit, not a cheapie, will bypass a heavy discharge to ground.

Even so, high-voltage electricity will still continue down the coax into your set. When an electrical storm approaches, disconnect the antenna lead from the transceiver. Be sure to tag the *front* of the set with a prominent warning: DO NOT OPERATE. ANTENNA DISCONNECTED. Disconnect the power cord from the wall socket or battery and attach another tag to it with the same warning. Somebody may come along later and not know you've disconnected the antenna. Without that people-arrester tag they may turn on the rig, key the transmitter and blow out the final stage.

THINGS YOU NEED

After you're sure the antenna of your choice can be erected, make up a shopping list for mount, fittings, hardware and tools. You may already have the tools you need. Every brand of mount and antenna is different but here are the most likely tools and things you'll need:

- Marking pen (to code antenna parts)
- Level
- Hammer
- Roofing nails
- Waterproof sealing compound
- Duct tape
- Open-end or socket wrenches in assorted sizes for turning nuts, bolts, lag screws
- Socket set with ratchet for quick assembly of bolted joints.
- Heavy duty wire-cutting pliers
- Pliers
- Long-nose pliers
- Heavy rope

If you have to anchor in cement or masonry you'll need:

- Electric drill
- Long, outdoor power cable extension (three-wire)
- Masonry bits
- Expansion sleeves for setting lag screws in masonry.

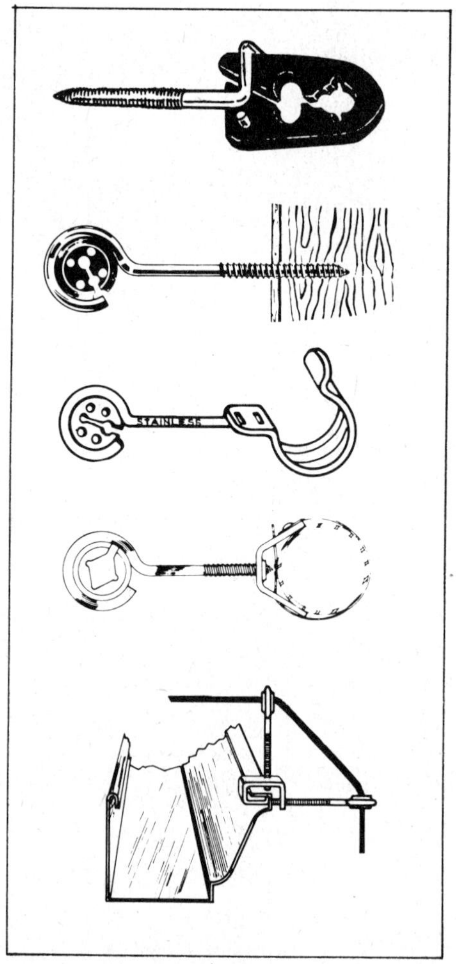

Here are stand-off insulators to route cable down a mast, across the roof, over gutters and down walls.

It's easier to select mounting hardware if you know what to look for. Guy-wire cable clamps secure a loop and protect it from wear by a thimble inside the loop. Turnbuckles can be used to tighten individual guys and are necessary in each guy connected to a common anchoring point to allow separate tension adjustments.

And if you're guying a mast you'll need:

- Guy wire
- Eye bolts or screweyes
- Screw hooks
- Guy-wire thimbles
- Guy-wire strain insulators
- Guy-wire cable clamps
- Wire tighteners or turnbuckles or ratchet guy-wire anchors (best)
- Floating guy-wire rings and supports

No matter what type of antenna mount you're putting up, you'll need:

- Stand-off insulators for coax cable that will fasten to –
  - Mast
  - Walls
  - Roof
  - Possibly rain-gutters
- Insulator lead-in tubes for walls

ALLOW ENOUGH TIME

Antenna installations don't take long compared to the life you expect from them. But they always take longer than you think. You can't stop work before guy wires are secured, so don't erect a mast or antenna unless there is enough daylight left to guy it. *Never* install when there is an electrical storm in the vicinity.

Wear a jacket with lots of pockets for tools and parts. On a sloping roof, most things you put down just keep rolling along. If you take a tool box aloft, nestle it in the crook of a chimney so it won't slide away.

ASSEMBLING THE ANTENNA

There is no universal procedure. Instructions are different for every model. Though you'll be following the manufacturer's assembly manual for your own antenna, let me give you some general tips I learned the hard way.

Open the antenna carton on a clean floor, not on grass, gravel or dirt. You want to see parts and hardware you'll surely drop. Put a rock in the box to keep the wind from causing a spill. Anchor the instruction sheets under a rock whenever you put them down.

You'll see lots of similar looking tubes. They may be slightly different in length or diameter. Identify each part, checking against the list or diagram in the instruction sheet. Put the corresponding number on the part with the marking pen. You may have to measure each piece to be sure. Step-by-step instructions are easier to follow with numbered parts and there is less chance of making a mistake.

Follow the order of assembly recommended by the manufacturer.

If your antenna has a little top hat of static precipitators, put these on last, just before you're ready to raise the antenna and clamp it to the mount. Otherwise you'll bend the soft wire elements by letting them take some of the weight of the antenna when it is horizontal during assembly.

Install coax, check and adjust SWR and you are *10-8*.

WHAT A POOR SWR DOES TO THE POWER

No transmitter, cable and antenna are perfectly matched at all frequencies, so a standing wave ratio of 1:1 is virtually impossible. Here's what happens to your full 4 legal watts at various SWR's.

SWR	ACTUAL POWER INTO ANTENNA (WATTS)	COMMENT
1:1	4	Virtually impossible
1.05:1	3.997	Unlikely
1.1:1	3.991	Achievable on base antennas and some mobiles
1.2:1	3.960	Achievable on base antennas and some mobiles
1.5:1	3.840	Likely on mobile antennas with no other vehicles nearby
2:1	3.520	You can improve this if you work at it.
2.5:1	3.280	You're losing 12% of the power
3:1	3.000	Losing 25% of the power and reflecting dangerous amounts of power back toward transmitter.

BASE-STATION ANTENNA TIPS

IF YOU USE GUY WIRES

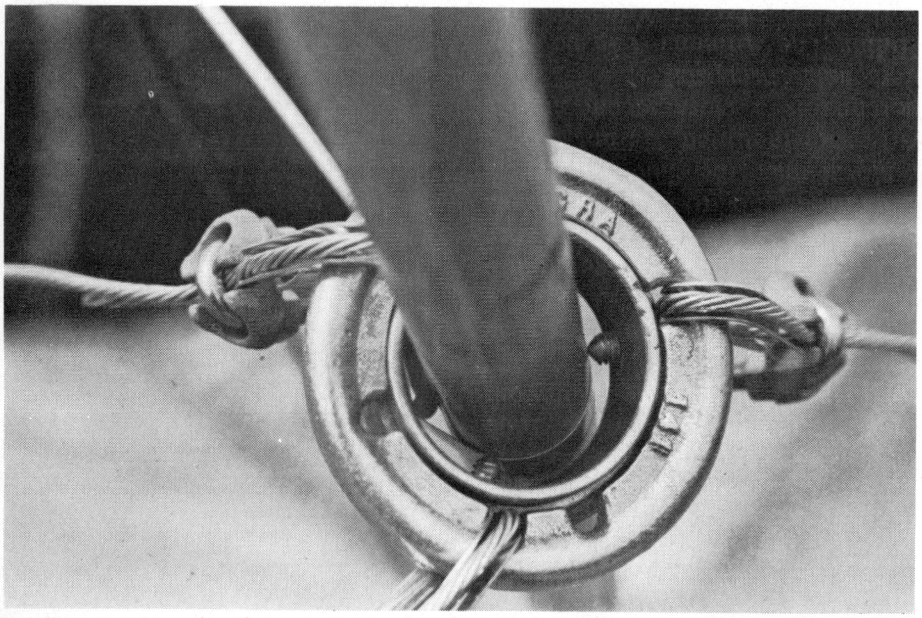

To allow turning a mast, connect guy wires to a ring that floats on a fixed collar. Be sure to tighten the collar setscrews firmly and evenly so wind-induced cable vibration doesn't loosen the screws. If the collar slips, your mast can crash.

I prefer ratchet guy-wire tighteners to turnbuckles especially in cold weather. You don't have to fumble with guy-wire clamps and thimbles and you can tighten the ratchet with a wrench. Tension each guy evenly, leaving only enough slack so it won't overtighten and snap in cold weather.

WHEN YOU ASSEMBLE A GROUNDPLANE

Even if you see how the pieces fit, follow the exact assembly sequence given in the manufacturer's instructions. Once you clamp certain radials, you won't be able to get a wrench on the remaining clamps.

Protect slender radial tips from bending by attaching them with the mast already mounted. If you're working near the edge of a roof, leave the mast loosely fastened so you can rotate the groundplane like a turnstile to get at all the radials.

USING A MOUNTING BASE

Use lag screws with hex heads so you can fasten the mount with a *socket* wrench. For safety, use the flanged type of lag-screw head so your wrench doesn't slip and spin you off the roof or ladder.

Always use 16-gauge steel tubing or 0.125-gauge aircraft aluminum alloy tubing for your mast, even for a stub mounting. Otherwise, this can happen.

Because round inserts in stand-off insulators sometimes pull out, I find swivel-lock insulators best for securing coax cable.

FOR LIGHTNING PROTECTION

Most long ground-rods come in two sections. Tapping one into the sleeve of the other is trickier than you would think. Put the pointed end on a board; stand on a ladder; use a helper.

Install a lightning arrester at the antenna in line with the coax cable. Attach heavy electrical-grade aluminum ground wire to the arrester terminal.

Bring the lightning-arrester ground wire through stand-off insulators to the ground rod driven at least 4 feet into the earth, or to a cold-water pipe, according to your local electrical code. Be sure to tighten the ground clamp or strap securely to get a good electrical contact.

CHAPTER 14

PORTABLE AND MARINE ANTENNAS

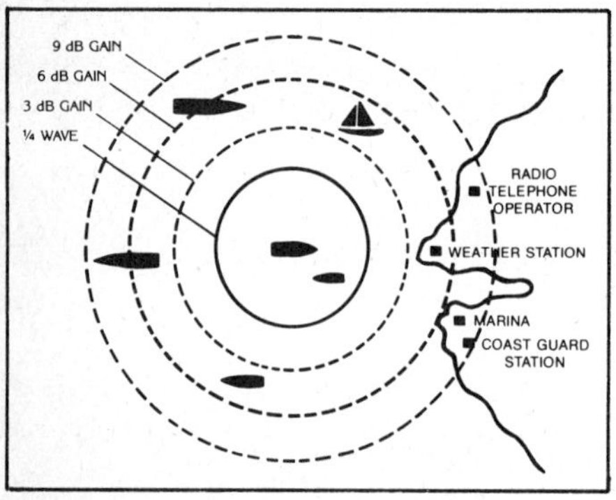

When you're on the water you learn to appreciate what a good antenna does to push back that horizon. CB at sea reaches out farther with longer antennas that don't waste sky waves on sea gulls. The Coast Guard wants you to have VHF for even greater reliability at ranges out to 20 miles. This drawing from a Hy-Gain publication shows how antenna gain helps you talk farther. Typical CB gain-antennas are 3 dB—Marine VHF antennas have more gain.

On large craft, the CB transceiver may be a permanent installation connected to a marine CB antenna. On small craft, including rowboats, canoes and outboards, a good 4-watt portable wrapped in plastic protective covering works well for short ranges. Some boaters even bring portables aboard cabin cruisers and hook up to a permanently installed marine CB antenna.

MARINE CB ANTENNAS

Don't be lulled into a false sense of security by having CB aboard your boat. It is no substitute for more dependable marine VHF which has greater range and the advantage of a Coast Guard monitoring on VHF channel 16. You'll be safest with both systems on board. But if you depend only on CB, you'll want to get the best possible performance from your relatively low-powered transceiver and limited antenna height.

Here are some problems you're up against. You have the same FCC height restrictions as on land—20 ft. above the highest part of the man-made structure. Sea level is discouragingly low and the horizon can be very near for small craft, so count on CB operating ranges of from 3 miles to as much as 35 miles to a station with a 60-ft. tower on a high shoreline. Unless your boat has a metal hull, you'll need an antenna that operates without need of a groundplane. Marine antenna mounts are more complex because you need to lower the antenna to pass under bridges, tow the boat on a trailer or stow the antenna out of the way of other operations. The wet, corrosive marine environment is an enemy of good radio performance. Installations require great care to assure long-lived electrical continuity.

Larger craft can easily accommodate half-wave whips designed to work without a groundplane. There is a variety of mounting options on powered craft but sailboats pose unexpected antenna mounting problems. Small craft using quarter-wave whips require some form of groundplane. All vessels need lightning protection for antennas.

HOW TO PICK A MARINE CB ANTENNA

You want maximum efficiency of radiation. A full half-wave or even 5/8-wave omnidirectional antenna wastes little radiation into the sky-wave and has 3 dB gain over standard quarter-wave whips.

Why not a beam antenna? Some

Even small craft have their ears on when they put out from shore. One of the two half-wave marine antennas on this runabout is for CB. Boaters use the name of their vessel as a personal handle.

vessels could accommodate a small beam, but any rocking motion points your radio signals at sky or water.

The selection of marine CB antennas is not as broad as for mobile use. Full-size half-wave antennas get pretty long—about 18 ft.—and require some form of mast support. Center-loaded half-wave whips come as small as 7 ft.—a fine height for runabouts. Most free-standing half-wave whips average 9 ft. in height. A combination CB/VHF antenna—really two antennas in one unit—has height of 20 ft.

As with all CB antennas, the best mounting location is the highest permissible.

Now think of the antenna length so you can anticipate problems when it is in the horizontal or laydown position for bridge passage or stowing.

Almost all marine antennas are made of fiberglass to protect against corrosion. Stainless-steel antennas require more care than fiberglass. A coating of plastic or marine paint is sufficient provided you are careful not to get paint on the insulators or the base—some paints are conductive at radio frequencies. Corrosion is more of a risk wherever parts of an antenna are joined, so examine the fittings and seals in two-section antennas. Fittings should be stainless steel, monel, bronze, copper or brass. Avoid any antenna using self-tapping screws; vibration may take it apart on every voyage.

Finally consider the mount available with the antenna. Some antennas are supplied with mounts; others come with bases to fit a marine mount. For a sloping deck, be sure the mount has leveling plates. A cabin or bulkhead mount should swivel to allow vertical positioning. A laydown mount folds over without difficulty when wing nuts or a ratchet is loosened. A lift and lay mount requires lifting the antenna out of the mount base in order to lower it. Consider where you will run the antenna cable into the cabin. Some mounts have a hollow center bolt directly below the antenna for this purpose.

Mount materials require you to do some thoughtful shopping, too. Your antenna mount will be doused by even more spray and green water than the antenna. Cycolac and Lexan, two high-impact plastics, are favored by many marine-mount manufacturers. Both are strong and corrosion-proof. If you choose a metal mount or one with metal fittings, your best bet is stainless steel. Monel, bronze, copper and brass are next in that order.

After you've done all that checking and scheming, use the longest antenna you can get by with, made of non-corrosive materials with a practical mount. Keep in mind that you have to live with the antenna in two different ways. One is the radio antenna that you hope will get you out of trouble someday.

Antenna height restrictions apply equally to marine and land CB—20 feet to antenna tip above the highest point on the structure. On a boat that means the superstructure so most owners simply put a whip atop a cabin or bridge. You *can* mount an antenna on a mast for convenience but the tip still must not measure more than 20 feet above the superstructure. This 10-foot freestanding antenna is on, well, a twin-masted structure—a ladder above the cabin. For low-bridge passage, a crew member climbs up to unship the whip from the lift-and-lay mount.

The other is an on-board nuisance that may constantly interfere with something you want to do.

WHERE TO MOUNT YOUR MARINE CB ANTENNA

You wouldn't think *sailboats* with masts and so much rigging, would pose the greatest mounting problem. Sailboats have a low superstructure and you want every foot of height you can legally use for your antenna. You don't want the antenna to interfere with sail operation. And you certainly don't want metal masts or metallic rigging in close proximity to detune or shield your antenna. Should heavy weather damage the masts or rigging, you don't want your antenna fouled by whatever is broken or hanging loose just when you need your radio most.

I fear you are stuck with mounting on a yardarm or a few feet out from a mast on a small platform unless you figure out something better. This location may be annoyingly inaccessible if you are underway and utterly useless if you are dismasted. There is lots of swaying up there due to sea motion, so choose a light, flexible antenna that won't break off.

Cabin cruisers allow a choice of deck or bulkhead mounting. Put your antenna as high as possible to maximize CB operating range, but be sure it is accessible for antenna laydown. If you don't use a free-standing antenna, be sure there is a rail or bulkhead for a support bracket. You'll also have to consider accessibility to the support unlatching mechanism for antenna laydown. If you don't plan ahead for laydown, you may discover your mistakes when you are under way and ready to negotiate your first bridge passage.

Very *small craft* leave you little choice of mounting position. If you are using a half-wave antenna of manageable height, say, 7 ft., you've saved yourself some work for you won't need a groundplane. Any deck or rail mounting location will do. If you use a quarter-wave whip, try to locate it on the centerline. You'll need to run radial wires in a horizontal plane on both sides to provide a groundplane.

HOW TO INSTALL MARINE CB ANTENNAS

Sailboat installation is easier if you don't have to work in mid-air. Unstep the mast and begin your installation from top to bottom while everything is conveniently in the horizontal position. Attach the antenna mount to the yardarm or special platform, using stainless-steel through-bolts and locknuts. With luck your antenna will come with enough cable already attached. If the cable is not attached, clean all metal surfaces before fastening it to the lugs designated by the manufacturer. Solder the connection if possible; if the connection is by means of a terminal slipped over a screw, be sure to use a lock-washer. Now immediately coat the connections with non-corrosive marine silicone sealant and tape the cable in place so the sealer won't stretch thin due to movement of the cable during the time it requires to set. Setting times are usually on the label.

When the sealer is set but still sticky, double-wrap the connection with cold-weather-resistant electrical tape. The connection will now

If you don't have a freestanding antenna it will need an above-deck support like this. Be sure you can reach the support to unlatch it for rapid laydown when you need to.

WHICH SILICONE SEALER FOR CB?

There are all kinds of space-age silicone sealers on the market. Don't ask the hardware clerk which one is good for CB; he'll think you mean Caulking Bathtubs. You're going to be sealing electrical connections and caulking cable entries. The curing agent in some silicone sealers can have harmful corrosive effects on metal.

Some curing agents have an acid base which reacts to metal. Amine curing agents have a non-corrosive ammonia base.

It would be simple if I could advise you to read the label on any tube of silicone sealer. But even the sealer I recommend for CB doesn't name its curing agent on the tube, though the package says the cure process is non-corrosive. The manufacturer tells me it's amine.

So head for a marine supply store, probably the only place you can get Boatlife Marine Silicone Sealer which is white, and Boatlife Marine Liquid Tape which is black. Except for the color and consistency of the goop, these products are essentially the same. Use them to seal antenna mounts, cable entries and antenna connections.

A plastic or metal cable strap or clip like this Perko marine fitting holds coax cable inside a hollow mast. Attach them to the cable every two feet or so and screw them to sliders which run on a track down the inside of the mast.

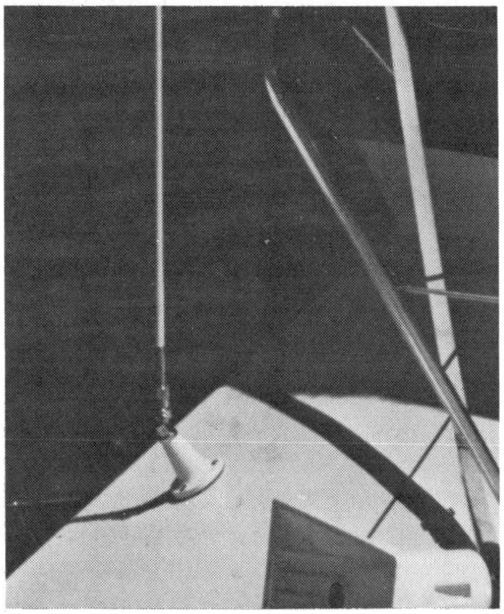

Your boat will rock—and so will the one you're talking to. Nevertheless you want a vertically polarized antenna so be sure to get a swivel base for mounting on the sloping surfaces found on most craft.

be waterproof, condensation-proof and so frustratingly tamper-proof you'll have a terrible time getting it apart again yourself.

If you have a hollow mast, you may wish to run the cable down inside. Don't assume the cable is safe just because it's inside. Hollow masts may also contain halyards, electrical wiring and instrument cables. Your CB coax can be chafed by rubbing against these lines. Here's how to protect it.

Inside modern hollow masts you'll find a metal track running down one side. The track holds traveling sliders. Attach cable clamps or clips to the coax with a spacing of about 2 ft. Attach each cable clamp to a slider, using a metal screw furnished with the slider. Then feed the cable into the mast, drawing it through with a fish-wire, and insert each slider into the track. This is best done with the mast horizontal. Make sure that halyards and other lines inside the mast are taut and out of the way while doing this, so you don't entangle the fish-wire or cable. Then step (erect) the mast.

If the mast doesn't have a track or groove for sliders, encircle the cable with discs of rigid plastic foam slit to the center so you can insert the cable. Use these every 2 ft. or so, holding each in place with a plastic cable tie. As you pull the cable through the mast, feed these foam spacers into the center. center.

Clearly, hollow masts are a lot of work. Some masts are foam-filled; terrible for installing CB cable unless there are conduits inside and messenger lines for drawing cable through. You'll find it easy to run your coax through such conduit but the exits will have to be existing ones at the masthead and at the foot of the mast. If the masthead is well above your antenna position, lead the cable down the mast again through rigid conduit attached to the *outside.* The short external run to the antenna mount will do no harm. At the bottom of the mast you have two choices. A mast stepped on deck may have a through-deck pipe or fitting too small to accommodate coax connectors or even heavy RG-8A/U cable. Your CB cable must compete with other lines already passing through the pipe. You can terminate the cable in the mast with a connector that

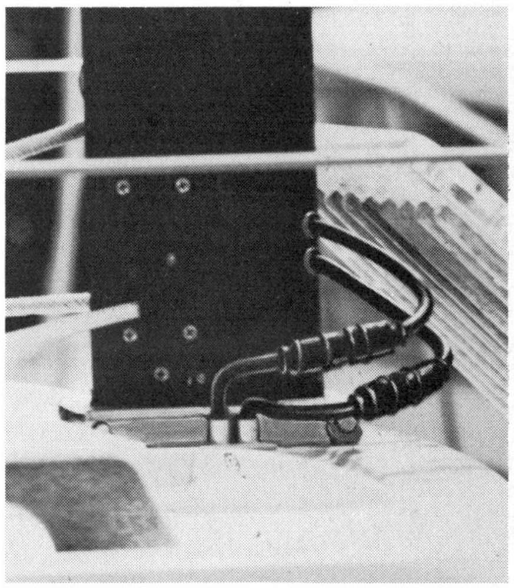

Sometimes there's not enough room at the base of a hollow mast for halyards and rf transmission lines. These coaxial cable exits for a mast stepped on deck are neatly sealed by rubber grommets.

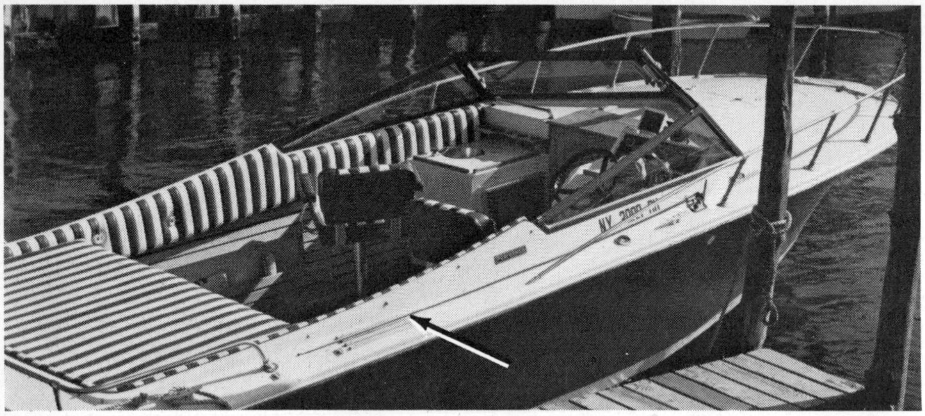

Never kick a good antenna when it's down, especially fiberglass. Plan your mounting location for laydown so the antenna won't be damaged by other on-board activities. The skipper of *L'il Ana* stows this whip in the grooves of the step plate before he lets any landlubbers clamber over the starboard gunwale.

plugs into a second cable of thinner RG-58 C/U. This will probably pass down through the pipe provided you haven't put a connector on the lower end of the cable yet. Be sure to waterproof the cable connection inside the mast. If you ever have to get at the connection because it has become corroded, it requires unstepping the mast again.

With a hollow mast, you can lead the cable through an exit in the side. If the mast is stepped on deck, seal the exit hole around the cable with a grommet surrounding the cable, then seal the grommet with silicone. The cable may then be passed below deck through a watertight fitting. A mast stepped on the keel is a bit easier; just make the side exit hole below deck.

Leading the cable along the outside of a mast is relatively simple. Even a sloppy job won't make any performance difference for a while but the cable will take more of a beating if it is slack and flaps around. Use rigid plastic conduit attached to the mast.

If you use a combination CB/VHF antenna, you can't put it on a metal mast. These antennas have two coax leads but the one for VHF cannot be run close to a metal mast. Find another location on a vessel with a metal mast.

If your boat has a lightning rod or another antenna higher than your CB whip there is a cone of protection over your CB antenna; all you'll need is a sturdy in-line lightning arrester near the transceiver to bleed off static discharges. Lightning protection is discussed later in this chapter.

Cabin cruisers and fishing vessels offer a variety of antenna-mounting positions. Don't settle on any one of them until you have positioned the antenna in the laydown position to be sure it can be stowed without damage. In cold climates, fiberglass gets brittle and may crack if bumped or kicked.

If you are going to use a base mount rather than a bulkhead mount, be sure it will not allow water seepage through the mounting holes. Vibration is a major problem on every powerboat; sloppy mounting holes are enlarged by vibration, loosening your antenna and making openings for seepage.

Now, before you fasten the mount, apply liberal amounts of silicone sealer to the undersurface of the mounting base to keep water from seeping beneath. For a neat appearance where sealer will ooze out, choose an appropriate color—black or white. As you insert each mounting bolt, coat the bottom and sides of the head with sealer so there will be a thorough seal when the bolt is snugged down tight. You may even put a *bit* more sealer on the top of the bolt so it is completely enclosed. Don't use too much or you'll have to carve it off with a knife if you later want to wrench on the bolt. Work fast with sealer because it dries quickly on the outside. Don't attach the antenna or otherwise stress the mounting until the sealer is thoroughly dry throughout. Read the instructions for drying time—not merely setting time.

Unless you use a mount with a hollow center bolt, the coax cable must be fed below via a water-tight cable outlet. An alternative is a water-tight grommet. Use sealer on the cable entry *after* the coax is secured in place so the seal won't be disturbed by any pull on the cable during drying.

If your cable entry is through a bulkhead, make a drip loop below the entry so water can't run along the coax into the hole. If you don't use a water-tight cable outlet, slide a grommet on the cable with the round side out so it will be on the exterior. Carefully work the back of the grommet into the hole with a dull screwdriver blade so you don't cut the rubber. After the cable is secured on both sides, use sealer on cable and grommet.

The best way to run transmission line through a deck or bulkhead is with this Perko watertight cable outlet. It is superior to an ordinary grommet stuffed in a hole because the outer collar is screwed down to compress the rubber ring.

Water runs downhill, right? Even with the best hardware, it's smart to make a drip loop below the entry. I prefer more droop to the loop because uphill can become downhill when a vessel is rolling hard.

HOW TO WATERPROOF CABLE CONNECTIONS

Coax is waterproof wherever the outer jacket is unbroken, but the ends are thirsty. Braided-copper shielding is like a wick; water will travel along it underneath the jacket. Water—especially salt water —and copper braid don't mix. Even with tinned braiding, corrosion is only a matter of time. Ugly green corrosion develops where you can't see it, weakening the rf signal traveling along the transmission line. Here's how to prevent water seepage into your coax:

Use coax connectors instead of attaching center wire and braid to screw terminals. Even if the braid doesn't get wet, high humidity on any body of water makes the braid gather condensation whenever the temperature drops.

Use a continuous length of cable from antenna to transceiver where possible, rather than using one or more short lengths plugged together.

Some antennas come with the coax already attached to the antenna and some have the plug already installed on the end that connects to your rig.

Cut off the plug so you can drill smaller holes to pass the cable through on the way from antenna to set. Then reinstall the plug on the end, as described in Chapter 12.

Obviously, if you can make the cable run without drilling any holes, you don't need to cut off the coax plug.

Use CB Climate Barrier on the plug tip and locking ring threads so the connector won't corrode and stick. Then use Boatlife Marine Liquid Tape or Marine Silicone Sealer to complete waterproofing the coax connections. Never use a rubber boot—condensation will form underneath it. Insert the plug into the socket and tighten the locking ring. Coat the entire outside of plug and socket plus an inch or two of cable. Let it set until tacky.

Neatly double-wrap the plug, socket, and a few inches of cable with heavy-duty plastic *electrical* tape—not duct tape. I found the toughest grip in this type of tape made by Nashua. Select a tape specifically marked *cold and weather resistant/flame retardant.* Some tapes with this marking are called *cable wrap.* There are many grades on the market. Most of them tend to unstick after a week or two. Don't hurry; wrap each turn correctly the first time because good tape clings so strongly you'll have a hard time undoing it without pulling with pliers.

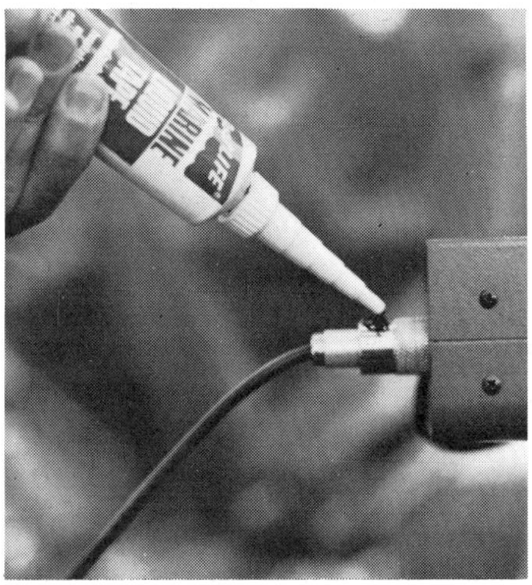

After a little more squeezing this entire connector will be covered with waterproofing that will form a skin in 5 minutes and cure in 24 hours. Wrap it with cold and weather-resistant plastic tape when the silicone sealer becomes tacky.

Other Cable Precautions—Stay out of bilges even it means a slightly longer cable run. Make *gentle* bends in the coax. Run the cable along control pipes or fasten it along carpet edges with plastic clips, *never* staples. Leave slack at the transceiver so you can move it with the cable still connected—a water-proofed connection is hard to undo. Add about 10 inches to the desired slack length before cutting the cable to attach the connector. If you have trouble cutting and trimming the cable, you'll have extra length.

LIGHTNING PROTECTION

If your CB antenna is the highest point aboard the vessel, it is actually a lightning rod. When properly grounded it can protect your transceiver from static discharges—reducing electrical noise, too—and will provide a cone of protection over the vessel. You're not absolutely safe from lightning, just *safer.*

Most small lightning arresters ordinarily sold for base stations on land are little more than static discharge devices. They bleed off accumulated static and nearby lightning. Lightning discharges a couple of miles away can damage your transceiver. Marine lightning arresters are designed for heavy duty; some are claimed to safely bypass 10 or more direct hits on an antenna. Don't expect a CB antenna to survive even one hit, but a good arrester may still save you, your transceiver and boat.

The antenna downlead connects to a socket on the lightning arrester. The cable continues from a socket at the other end of the arrester. The arrester must be connected to marine ground. Use electrical grade ductile aluminum or copper wire no smaller than #8. You're safer with #4 wire even though this size is harder to obtain.

If your boat has a metal hull, make a solid bolted electrical

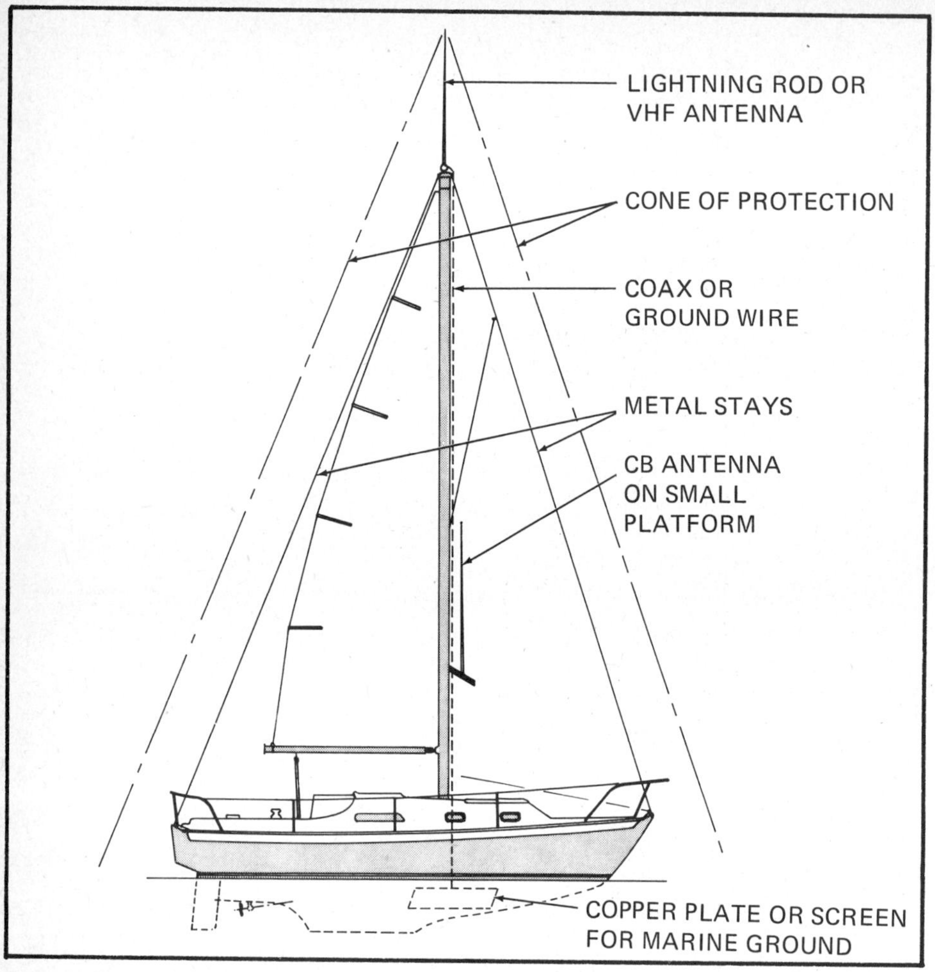

Even with small sailboats like this 28-footer you can't put your CB antenna more than half-way up a 40-foot mast; the tip is limited to 20 feet above the superstructure. But you do have the advantage of the antenna being within the cone of protection provided by a mast-top lightning rod or VHF antenna and metal stays all connected to a good marine ground.

A lightning arrester has a spark gap on the inside to divert the discharge to ground. Aboard a ship or boat, a metal plate below the waterline is connected to the case of this Hy-Gain marine lightning arrester. The sockets on top and bottom are for coax connections to antenna and receiver.

connection between ground wire and hull. Solder or crimp a lug on the end of the ground wire, with a hole large enough to fit over a bolt somewhere on the hull. Take the nut off the bolt, clean all surfaces of terminal lug, washers, nut and the metal of the hull surrounding the bolt. Capture the terminal lug securely under the nut, then seal everything with non-corrosive marine silicone sealer, painting an area extending 2 or 3 inches around the bolt.

If your boat does not have a metal hull, the ground wire from the lightning arrester must be bolted or soldered to a marine ground made of copper screening or sheeting fastened *beneath* the waterline on the *outside* of the hull. Do not connect to the engine and drive system even though it makes contact with the water; it is not a satisfactory marine ground. *Antenna groundplanes are not the same as marine ground.*

Caution: Though you may have done everything possible to provide lightning protection, it is still safer to disconnect your antenna from the transceiver during lightning storms as follows:

Disconnect the antenna at the set or at the downlead side which continues from the lightning arrester. The arrester should still protect the boat. Don't be discouraged by the permanence of your waterproofed antenna connection; the effect of lightning can be even more permanent.

This owner quit before the task was done and simply taped his fiberglass whip to the metal railing for support. Wonder how far he can talk?

Hang a *warning* sign on the transceiver power switch: ANTENNA DISCONNECTED. DO NOT OPERATE. You don't want to go back on the air after the storm and key your transmitter without a proper antenna load or you'll burn out your rf power transistors.

Keep everybody on board clear of the antenna down lead and arrester during the storm. Lightning is notoriously capricious.

After the storm, reconnect the antenna lead to the transceiver or arrester. Waterproof with silicone and wrap with tape as you did in the initial installation.

GROUNDPLANES

Quarter-wave whips require a groundplane to bring antenna impedance to a 50-ohm match with your CB transceiver and transmission line. Radials to make the groundplane should also be a quarter wave in length–9 ft. for the 27 mHz CB band. On a sail boat it may be possible to run radials down the rigging. On a small craft you may tape insulated wire to the deck or hull starting in a horizontal plane fanning out from the base of the antenna where they are connected to the braid of the coaxial cable. Move the radials around while watching an SWR meter as the transmitter is keyed. Tune for lowest SWR no matter how crooked or unsymmetrical the radials become or how much they droop. Be sure to waterproof the coax braiding where the radials are connected. Take care so the radials do not trip your passengers or crew.

Forget about connecting your transceiver to marine ground. It's a waste of effort at CB frequencies because you won't be able to make the ground lead short enough–it should be no more than a couple of feet.

PORTABLE ANTENNAS

Transceivers you carry have no groundplane to work against. Usually they come with whip antennas operating as loaded half-wave or quarter-wave dipoles. They radiate at right angles to the long direction of the antenna. Some are fixed length and others–even center-fed antennas–telescope for compactness when carrying the unit. When you extend a telescoping antenna, you may find it unwieldy in cramped quarters, but it must be the correct length for a proper match to the transceiver.

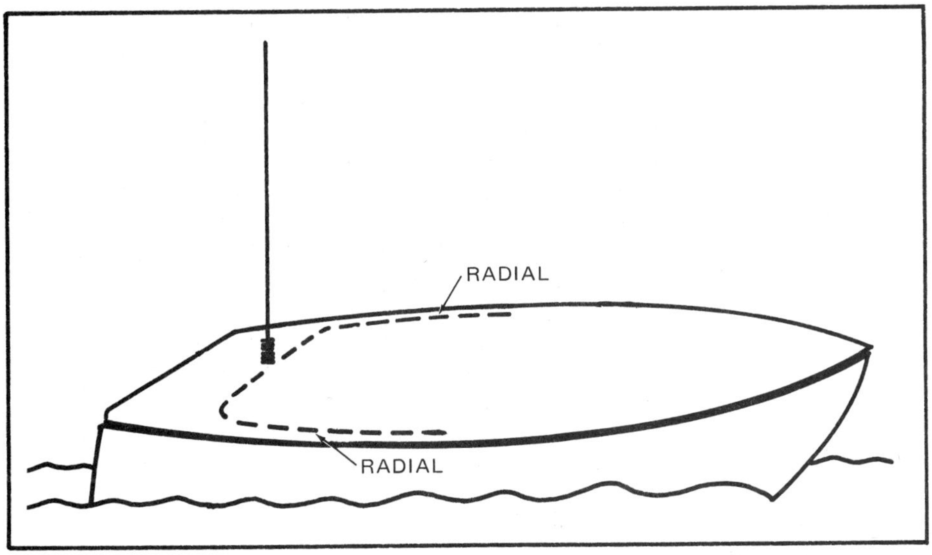

If you use a quarter-wave whip on your runabout you'll need some form of groundplane to provide an efficient 50-ohm match to your transceiver. Simplest is a counterpoise made of at least two insulated wires one-quarter-wave long (9 feet) connected to the ground side of the antenna and strung horizontally over or just under the superstructure. You'll have to fiddle with their position while watching a meter for lowest SWR; then tape them in place.

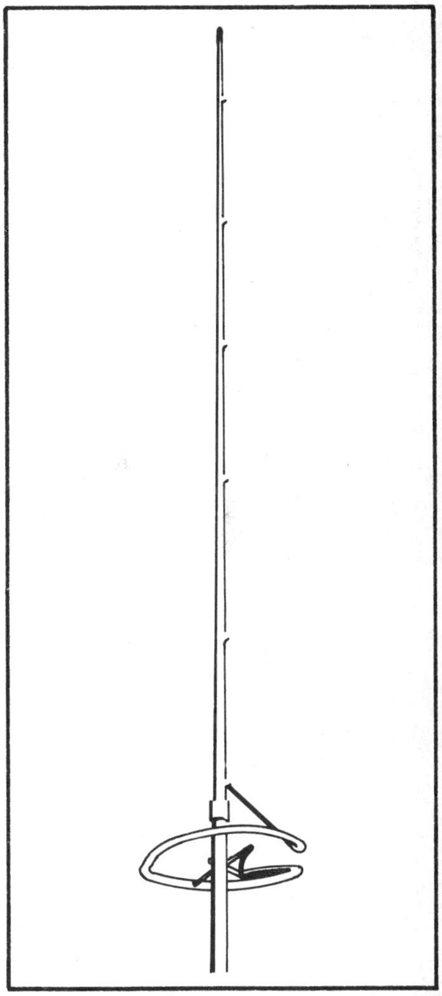

Even a 5/8-wave vertical radiator 21 feet tall can be used as a portable antenna. The individual sections of Wilson Electronics VI telescope into about 3 feet. Total weight is 5 pounds. Sanctimonious CBers who back-pack this antenna through the woods can be seen wearing the loading loop as a halo.

The telescoping whip on this Realistic TRC-200 hand-held rig is center-loaded. You become part of the antenna system by grasping "Range-Boost" panels (arrow) on the sides of the transceiver to capacitively couple your body into the antenna system. The carrying case for this model has holes in the sides so you can touch the Range-Boost panels.

You can substitute a shortened fixed-length whip with more loading coil. This solves the physical problem of an antenna getting in the way but gives decreased operating range. It does make the transceiver more portable.

Don't despair. You can get top performance from a portable transceiver by hooking it up to a large portable antenna—or even a permanent antenna. For example, a light-weight two-section fiberglass antenna intended for base station use can be stowed in a camper or carried to a campsite and quickly erected. Half-wave dipoles of this type don't need radials for a groundplane. You can do the same with a telescoping 5/8-wave groundplane antenna without radials. A 21-ft. antenna telescoped into 3 ft. weighs only 5 pounds, no more than twice the weight of a good 4-watt portable transceiver.

And nothing says you can't use a large portable antenna with a mobile or base station rig in your RV. Mount the antenna on the vehicle roof temporarily—don't exceed 20 ft. above the roof—or perch it on a hill near your camping spot. If you put the antenna in a tree, your antenna tip can't be more than 20 ft. above the tree top. So run for the giant redwoods.

You can use a portable hand-held with a mobile antenna to get greater range. Most portables don't have noise control circuits, so operate with your engine off.

CHAPTER 15

HOW TO AVOID RIP-OFFS

Everyone wants a CB radio. CB Bandits want yours.

Snatching CB's has become more popular than tape-deck pilfering. Your set is a target for thieves whether you live in the city or the suburbs. In Lansing, Michigan, population 135,000, 149 CB's were stolen in a single month. Manufacturers estimate 40% of their sales are replacements for ripped-off rigs. The average time-to-theft of a new CB is about 3 weeks.

BEWARE OF CB CON ARTISTS

You are expecting only burglars and masked bandits? Don't be fooled by the slick operator who steals your set while you watch him do it.

The fake FCC agent con game has been played very successfully. Someone posing as an FCC field investigator demands to see your license. If you can't show one he insists on confiscating your set right on the spot.

This con is aimed at CBers *running barefoot*—operating illegally—and is impressively intimidating. I hope you have your license, though it doesn't mean the end of the con game even if you do. The "investigator" may then complain that your station identification is not properly displayed for his inspection. It has to be affixed to the transceiver or control head, or at least be reasonably near. For the bogus agent, this is a ploy to keep you on the defensive. He'll then ask for proof of ownership of the set, knowing that you probably don't carry it with you. At any one of these three stages, the fake Fed may try to confiscate your set. *Don't let him take it.*

An actual investigator for the FCC carries an identification card with a green seal and his photo. He also carries a badge. These days it's easy to fake such things. Don't waste time guessing about their authenticity because FCC agents do not have the power to confiscate equipment.

They may inspect a station and its records, but seizure requires a court order. Even so, an imaginative con artist might show up with a partner posing as a US Marshall waving a counterfeit court order. Don't be intimidated. In actual practice, the FCC first *mails* a Notice of Violation.

It might be a real FCC agent. If you have doubts, write down his full name and badge number. Ask him for the telephone number of the FCC Field Engineering Office from which he operates. Then look up the number yourself.

Prowling a parking lot, here's what a CB bandit sees. Antennas stick out like sore thumbs, so he reasons those cars must have CB's somewhere. If you're going to a movie, he'll know how much time he has to break in and steal your set.

Gone! Some CB snatchers smash and grab to rip-off a rig. Poor KYC 6649 thought his set was safe on an Army post like Fort Sill. He was away from his car for only an hour. Remove temptation—remove your rig when you leave your vehicle.

FCC Field Office and request confirmation of the agent's identity. A genuine agent will wait for you to make this check. A crook will try to bully you, or simply take off.

BURGLARS AND BANDITS–IT DOESN'T PAY TO ADVERTISE

Chances are you've been advertising to thieves and don't know it. Here's a list of common things inviting theft of your set.

Mobile Sets—Thieves prowl parking lots looking for CB antennas and other evidence of CB-equipped vehicles. You are most vulnerable in parking areas of movies and motels, sporting events, factories with regular shifts, schools and colleges. In these places, you'll be parked for a definite period of time. A smart thief knows this and counts on stealing several sets. Sometimes a thief works with an attendant of a commercial parking lot or garage and shares the take. The average stolen mobile rig brings $50 to the thief.

Transceiver mounts, though emty, suggest your rig might still be in the trunk. Prying it open won't take much time. "How long will you be?" asks the parking-lot attendant. Gulp!

Call signs displayed on windows, bumpers or license frames announce you're legal—prey! All stickers and decals say the same thing to a crook: "Come on!" A competent thief can remove your set and say "We gone." in less than a minute.

Marine CB Sets—Antennas are one tip-off. It's hard to hide a quarter-wave or half-wave monster, or make one look like a fishing pole.

Broacasting your boat's name and local berth is like a radio commercial for a free CB.

Base Stations—Deluxe base stations bring out the more discriminating thieves, although house burglary is a more formidable task than entering a vehicle.

Antennas give you away every time. A knowledgeable thief looks for expensive beams or high-gain omnis, reasoning you must have an above-average transceiver inside.

Central registries with lists of handles and CB units owned by registrants *could* become catalogs for steal-to-order gangs.

Broadcasting details of your equipment gives the crook a chance to appraise the loot in advance.

HOW TO KEEP YOUR SET A SECRET

An antenna on your car is such a giveaway that you should remove it, hide it or disguise it when leaving the car unattended.

Removing The Antenna—It's easy

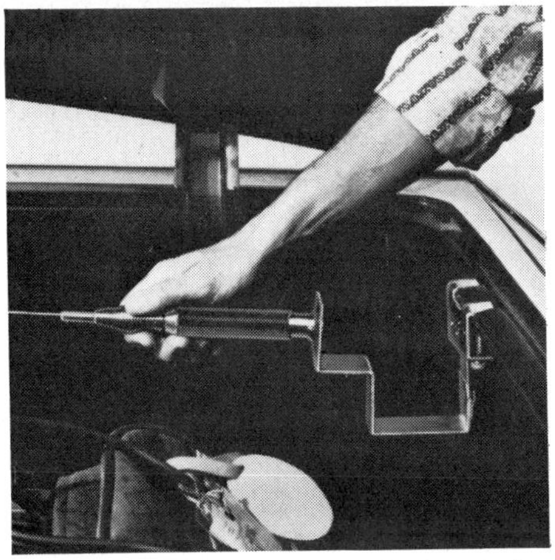

Here's a way to make a trunk-mounted antenna disappear. A trunk-*groove* antenna held by a Tuk-A-Way mount flips under the trunk lid.

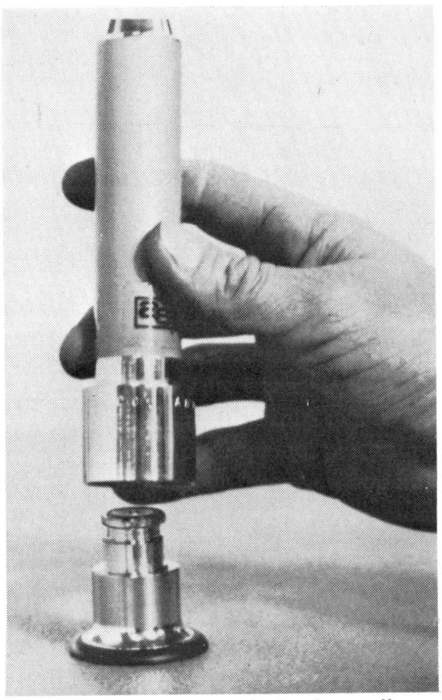

This quick-disconnect device is a collar that inserts between the antenna loading coil and the mount base. Antenna Specialists M-460 has a combination lock on the collar with 8 possible settings to slow a young whip-snatcher (street gangs use them as weapons). If you remove your antenna, the mount base tattles about your CB, so remove the rig too.

to do with a quick-disconnect device featured on some mobile antennas or available as an accessory inserted between the antenna base and whip. But thieves aren't so naive; an empty mounting base still says there may be a valuable transceiver in the vehicle. You must remove all *evidence* of an antenna.

Magnetic mounts and spring-clip gutter or mirror mounts are the fastest *detachable* antennas I know of.

Trunk-lid antenna removal requires a small tool—hex wrench, flat-bladed or cross-head screwdriver. You won't want to bother in rain, snow or freezing cold. A thief won't care—wearing foul-weather gear he'll home in on your antenna.

Most trunk-groove antennas need tools for removal, too. Look for a slide-mounted trunk-groove mounting bracket; it's sold as an accessory and needs no tools for antenna removal once the slide is installed. A slide-mounted trunk-groove antenna is a close second in my detachable sweepstakes.

Hiding The Antenna—Can be done without detaching it. You can make it disappear by retracting into the vehicle body or by tipping it horizontally into the trunk.

Retractable antennas disappear by telescoping into a cowl or fender. Some made to look like ordinary radio antennas telescope manually. Don't forget to *fully extend* such an antenna, before keying your transmitter.

Fold-down brackets for trunk-groove mounts are less convenient than retractable antennas. You have to get out of the car to hide the antenna. But it's cheaper and there is less to go wrong with a fold-down installation. Once your antenna is securely installed on the bracket you simply open the trunk, fold down the antenna, close the trunk.

All fold-down brackets give a good, reliable, permanent ground connection. The tuck-away idea is

Hide the antenna and receiver before you get to the place where you park. Thieves watch parking lots of factories, schools, sporting arenas, cinemas—anywhere they know you'll be gone for a predictable length of time so they can work unmolested.

An attache case is a good way to carry your set around without being obvious. Handy for travelers and car renters, this case holds a Johnson Messenger 123A transceiver and an Antenna Specialists magnetic-mount whip.

great if the antenna is short enough to go into your trunk. Measure your trunk-lid opening and the additional width inside on one side *only*. Trunk groove antennas run from 27 to 54 inches in overall length but most can flex to tuck a reasonable length into the trunk beyond the lid opening. Keep a rag handy to wipe the antenna before stowing or you'll soil whatever else you carry in the trunk.

Disguise antennas are fairly new to CB. For years, Smokeys have had them on unmarked cars—it hardly makes sense to put a label on a plain wrapper. A disguise antenna looks just like an ordinary auto-radio antenna. A good one should perform as well as a loaded whip of equal length. Manufacturers frequently offer these as tri-band antennas—AM broadcast band/FM broadcast band/Citizens Band. You get a coupler for all necessary lead-ins. You may have to tune for lowest SWR on CB; if there's no adjustment on the antenna, insert an accessory matching device between your transceiver and the antenna.

As with the retractable antennas, the loading-coil position can make a difference. Some antennas perform best with the loading-coil above the cowl. You can also use your standard auto-radio antenna for your CB transceiver. When you buy the coupler for this, look for one that loads and tunes the antenna.

Caution: Be sure your disguise antenna is fully extended before keying your transmitter or you may get excessive SWR.

REMOVING YOUR TRANSCEIVER

The best protection against theft is to remove your transceiver from a vehicle or boat and carry it with you. Next best—you hope—is to hide the transceiver from view by stowing it.

I tried out each of the sets discussed in the performance guide of Chapter 8 before turning them over to our engineers for field testing. I used them for a day or so in two different cars. I make it a practice never to leave a set in the car; so I carried or stowed a lot of sets. That's how I came to appreciate the really compact units. When shopping for a new unit, don't overlook size and weight.

The Case For and Against Cases—If you are going to carry a set with you, you are walking around with a big dollar sign under your arm. Disguise can be a virtue; put the set in an attache case or typewriter case. You could put it in a case made for a camera or tape recorder but these say you are carrying something of value. Pads of polyurethane foam will keep the transceiver from sliding around. This approach is also ideal for people who fly to a destination, then rent a car. Once you're hooked on wearing ears, you hate to leave your set behind.

Truckers who switch vehicles a lot often use undisguised cases with a whole set of entertainment and communication gear installed. The controls are on panels set into the sides of the case, for convenient operation on the road.

The case draws power from a cord that fits a cigarette lighter socket on the dash. There may also be a polarity inverter for truckers who change rigs frequently and must be prepared to hook up to positive or negative ground systems.

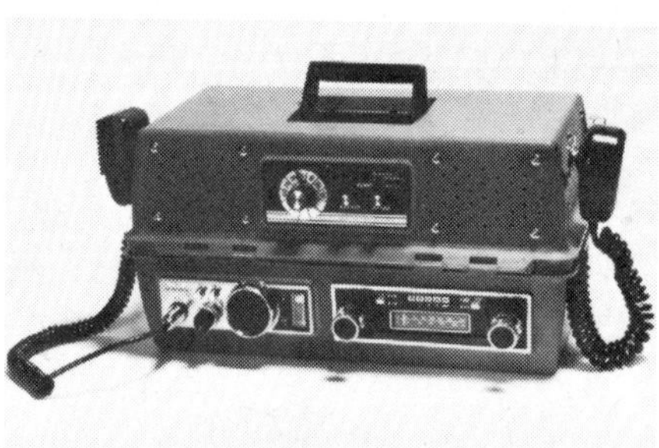

Your boat at a marina or mooring isn't safe from CB buccaneers. Take this communication package ashore with you. This fitted Trans-Comm case contains an XTAL CB, Regency marine VHF radiotelephone with weather channel, and an AM/FM/stereo tape unit plus speakers.

Enterprising CB set makers package their own sets, or offer custom cases. You can get empty cases from CB accessory makers.

I advise you to carry any of these cases with equipment around the store for a while before you buy it. The concept is attractive but the actual product could be too heavy or bulky to suit your taste, unless you mean to lock it away in a vehicle or boat. Consider how far you'll be carrying your rig and what *else* you'll have to lug. Can you put a small antenna in the case? If you fly and drive, will the case fit under a commercial aircraft seat?

Quick-Disconnect Mounts—With one pull you disconnect the power supply, antenna and set. Unfortunately for purists, the antenna connection is not electrically shielded. With some mounts, there may be a slight loss of rf power, and an increase in electrical noise in the received signal. Some mounts disconnect everything but the antenna lead; you then unscrew the antenna plug locking ring and pull it out of the socket on the set.

Most slide-out mounts consist of a pair of metal plates with matching electrical contacts. Your CB set attaches to one plate; the other plate is fastened under the dash, on the floor or to a transmission-hump mount or saddle. Read the product label carefully. There are mounts which work in only one or two of these positions. You may want a universal model with extra mounting plates so you can pop the same CB into other vehicles, boat, a base station adapter or the family airplane.

Look for heavy-duty wiping-action contacts; you want all connections to stay tight despite bumps and vibration. One variation I like is the C-D brand mount which uses pairs of heavy-duty magnets instead of spring-type wiping contacts. The magnets also take the place of a mechanical slide arrangement to mate the plate on the set to the mount; just

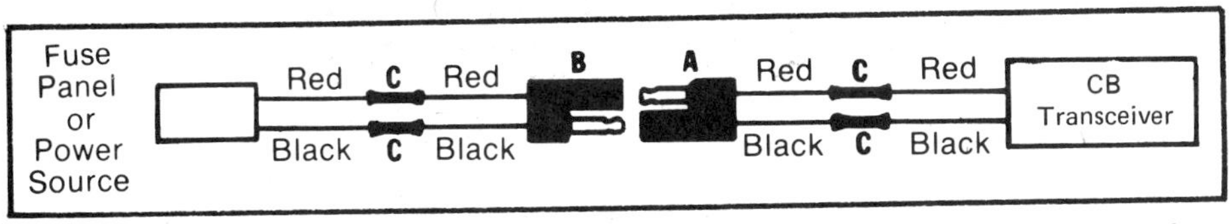

When your power cable is permanently wired to the set and you leave your CB in the vehicle, a crook will simply cut the wires to take your rig. Install this simple quick-disconnect device for power leads and you'll be able to remove your set in seconds. If you use this type of connector, Part B *always* connects to the power *source* exactly as shown so you don't expose "hot" terminal upon disconnecting.

The thief will use a crowbar to defeat a locking mount and not worry about what it does to your car.

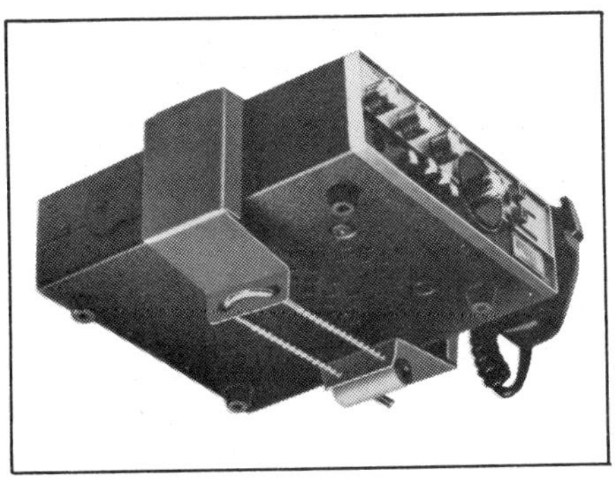

Locking mounts may discourage the amateur thief or a pro who doesn't want to make noise or attract attention while ripping up your dash.

position your set so the contacts engage and, "thunk!"

I advise picking a mount with lots of spare contacts. You may not want to use the pair for the antenna if you are concerned about a small rf loss. You'll need a minimum of one pair for power, plus a pair each for external speaker, public address and burglar alarms discussed later in this section.

Non-locking mounts are the easiest to rip-off—gone in the half minute you may take to visit the gas-station restroom. *Lock your car if you leave it while getting a fill-up at the pump.* Locking mounts slow down a thief but most slide-mount locks are too flimsy.

All not-so-quick mounting systems should have cable slack behind the transceiver so you can remove the set to get at the connections. It's better than reaching behind the dash and working by feel. Hide that slack when you have the set out, or a patient crook with lots of time may pry open your trunk to look for the set the dangling cables tell him you own.

Locking mounts slow things up. Be sure the thief can't simply remove the set and leave the locking mount in place.

The simplest deterrent is to make it difficult to remove the screws or knobs holding the set to the mounting bracket. One manufacturer of highly prized rigs supplies trick screws whose heads can be turned only by a special wrench that comes with the set. *Don't* leave that wrench in the glove box. Expect full-time CB snatchers to have a wrench exactly like it—*the manufacturer sells extra wrenches for $1 apiece!*

You can guard ordinary mounts by replacing the pair of screws or knobs with locking screws of the type with cylindrical heads. The heads spin freely and the screws can't be engaged for removal without a key in the cylinder. Not all locking screws fit all transceivers. Read the label carefully or try before you buy.

Hasps, padlocks and locking frames which enclose the set give you a nice warm feeling about security. But such measures merely slow up the weaklings and discourage crooks too poor to spend $2 for a tungsten-carbide hacksaw blade.

Access to bracket mounting screws can make locking mounts vulnerable. Be sure the underdash mount you buy has a cover plate over the screws when the set is in place. Floor mounts and saddle mounts bolted over the transmission hump should have special bolts to prevent removal from above. No thief wants to crawl under your car; it's too difficult to get up and run. On the other hand, the nuts can easily be removed when your vehicle is on a lift or over a grease pit. Many auto service and repair shops post prominent signs disclaiming responsibility for CB sets and tape decks left in a vehicle to be worked on.

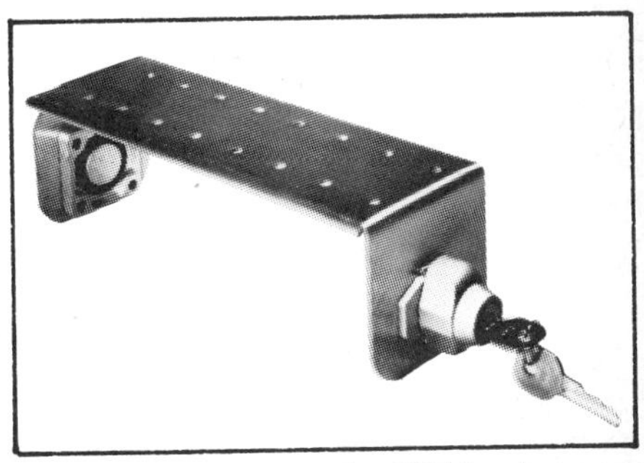

This Universal Mount by SST has key-locks at each end. You can get the mounting bracket in stainless steel for marine CB installations. The more bolts you put through the bracket, the harder it will be to pry.

ALARMS AND ALERTS

Alarms can be a deterrent to a thief or a warning to you. Audible alarms turn an amateur crook into Chicken Little; it sounds like the sky is falling and he runs. Lots of amateurs are employed by professional CB rip-off rings because teenagers without records get off easy when they are caught. But an alarm's effectiveness depends on the neighborhood. In a quiet area where any extraordinary sound would attract attention, only a brazen master of the 22-second CB snatch stays around to finish the job. In a city full of sirens, horns and roaring traffic, an alarm blends with the scenery.

The least-expensive alarm system sets off your vehicle's horn or a siren. It can be turned on by removal of a ground connection on your CB or by a switch operated by removal of the CB set. A thief familiar with this simple alarm set-up can jumper a connection to ground or tape the alarm switch.

Better systems use a drop-out relay which can't be reset by restoring a connection. Once the alarm is triggered by removing the antenna, opening a door or disconnecting the set, the horn or siren sounds continuously. In some places, sirens are illegal on a private automobile even if intended only for a burglar alarm.

Don't be encouraged. Some crooks take no chances; they open the hood and disconnect the battery before starting work. If your hood is also wired to trigger the alarm, he can disconnect the horn in a second. Who will pay attention to one honk of a horn or one sigh of a siren in today's noise pollution?

Ah, but your hood is locked! Good thinking. But why is that fellow sliding under the front end of your vehicle with a pair of loppers? Sorry. You took too long to answer. He has already cut a battery cable. What you need is an alarm system run by a separate battery enclosed in a steel shell—

For about a dollar you can make it harder to burglarize your vehicle. In most cases smashing windows is too obvious; thieves prefer to fish up a door-lock knob. Replace the regular knobs on your doors with these cone-shaped knobs that can't be lifted with a loop of wire worked past the window. Then never lock yourself out.

called *jewelers' specials*—which may cost more than your CB rig.

If you still want an audible alarm, follow these precautions:

Protect the entire vehicle against intrusion, not just the transceiver and antenna. You want the alarm to sound when a door is opened; it's too late when the CB set is already cut away. Put switches on doors, trunk, hood *and* transceiver.

Hide the arming switch. It sets the alarm or turns it off. If you put this switch inside your car, you need an arming-delay circuit to give you a few seconds to get out. When you return, be prepared for the alarm to sound for the few seconds needed to get to the arming switch. A thief who doesn't know where the arming switch is may not hang around long enough to find out. Don't make it easy for him by putting it in the glove box; that's too obvious. Next to the ignition switch is also naive. If you put the switch under the hood, wire it to the voltage regulator where it is less noticeable because other wires also connect there. Dirty up the switch wiring for camouflage.

Use a hidden battery to power your alarm system. Hide a heavy-duty dry cell under the rear deck or behind a wheel well. A 6-V or 12-V dry cell won't be rapidly drained by an electronic whooper. It's easy to hide the 6-inch horn that makes the noise. Most thieves won't wait to disable a second battery they can't find, or an unnerving whooper they can't get at quickly.

Boat-owners can use pressure-sensitive door mats under the cabin rug to trigger an alarm system. This supplements switches on hatches, doors and CB sets.

Silent Alarms—These alert you at a remote location. If you put this kind of alarm in a vehicle or boat, you must be within the approximately 1200-ft. range of the typical system's tiny radio transmitter. It sends out a coded signal to a receiver you carry, or keep nearby, or leave with somebody who will call the law.

The concealed transmitter sends signals using an existing antenna on your vehicle or boat whenever somebody tampers with anything triggering a switch. Then the receiver emits a continuous beep, alerting you to a thief at work. You should call the police. You might also interrupt the crook by asking him for his call letters. First, get his attention by tapping him with a tire iron. If he answers, you didn't tap hard enough.

Look for these features on a silent alarm system:

The transmitter should have its own batteries so it won't be disabled should the crook cut off the main electrical system.

The transmitter code should be sufficiently complex to minimize false alarms at the receiver caused by spurious radio signals. You should be able to set up the code in the transmitter and receiver so it will be different from any other alarm units of the same type nearby.

The receiver should be triggered

Who can sell a CB with somebody else's name engraved across the front and side? A "fence" wouldn't give $5 for hot merchandise like this.

by the transmitter and continue to emit an alerting signal even if the transmitter is discovered and disabled.

SIMPLE DETERRENTS

After all that technology, simple deterrents may seem refreshingly welcome. Don't take the easy way out. The two I am about to describe *never* can be as effective as removing your CB from access.

Change the door-lock knobs in your vehicle. For less than a dollar at a discount auto supply you can get tapered knobs that can't be lifted by a wire loop slipped past the window. Nor can such knobs be lifted by people with arthritic fingers; so they're not for everybody. Don't get child-proof knobs designed to be lifted by the tip of a key; the tip of a wire coat hanger will lift them, too.

Operation Identification is a nationwide program to engrave identifying marks on personal possessions. Locally you may find the program sponsored by your police department, a civic organization, or an enterprising insurance agent or hardware store.

For your CB set you'll have to buy or borrow an electric engraving tool that marks on metal. This tool costs $2 to $12. Unless you're going on an engraving spree, try to borrow one from your local Operation Identification sponsor and get some stickers too.

A prominently displayed "Operation Identification" sticker is the main deterrent wherever the program is strong. Most thieves take its message seriously: "Don't steal this CB because an engraved identification number will make it impossible to sell." Few fences are willing middlemen for such marked equipment, nor do they want to be bothered with sets where an identifying number has obviously been obliterated.

A sticker is also an advertisement. Some thieves won't be stealing your set for a fast buck; they simply want a nice CB of their own so they don't care if it has your identification number on it.

What should you engrave for identification? A popular practice is to use your Social Security Number. *Don't do it.* If your stolen set is recovered, it won't be returned to you unless you can prove rightful ownership. A properly suspicious policeman will want to confirm your Social Security Number, but he'll have trouble getting it from the Federal government. So it's better to engrave your driver's license number; cross-checking can be immediate through procedures normally available to the police. In addition, engrave your initials in some secret spot as further identification you can point to should engraved numbers be obliterated. This requires opening the set, but it's worth it.

Retain your sales slip and warranty registration stub; both should carry the serial number of the set you purchased. This may not be the best identification but it helps. If the seller of the set will not give you a sales receipt, or will not record the serial number on one, *don't buy the set.* If it turns out to be hot merchandise and it can be proven you knew it was stolen, you may be convicted as an accessory to the crime.

There are a number of commercial counterparts of Operation Identification. For a yearly fee you are issued a number to put on your set. If the set is stolen, you report the theft and your number is earmarked. Sometimes dealers and repair shops are notified of it by hot sheets. Anyone may query most registries by calling a toll-free number.

INSURANCE

The CB crime wave is reflected in insurance rates and coverage. As recently as 1975, a CB radio was included under comprehensive policy terms—your transceiver, when securely fastened in the vehicle, is part of the covered equipment. If your present policy still reads that way, you're lucky.

Some insurance companies have instructed their agents to use and charge for separate riders specifically covering CB sets.

If you remove your CB from its mount in a vehicle or boat, it may be covered under your homeowner's policy. Check the situation in your state; ask your agent what *your* policy covers. Buy additional coverage if you think you need it.

> **IMPORTANT NOTICE**
>
> Loss or damage to sound reproducing equipment including Citizens Band Radios is excluded from this policy. You may purchase coverage at an additional cost. Contact your agent for further information.

The insurance agent giveth and the fine print taketh away. When you get a renewal policy in the mail, don't file it without looking over every page. You may find a sticker taking away coverage of your CB. For more money you get the coverage again.

CHAPTER 16

HOW TO STAY LEGAL AND KEEP CB A NICE PLACE

Being legal is easy. Get a CB station license. It's cheaper than a $10,000 fine and a year in jail for operating a radio station without authorization. If you're running barefoot, get legal with the instant Temporary Permit plus license application described in Chapter 2.

Staying legal is a bit harder. To operate within the law requires you to pay close attention to proper use of channels, correct communication procedures and equipment technical requirements. You can be in violation for infractions of FCC Rules and Regulations and for operating your CB station for any purpose contrary to Federal, state or local law.

If penalties like loss of license, heavy fines and jail aren't good enough reasons for staying legal, you may be sentenced to a worse punishment: Suffering the illegal operations of CB air-hogs, self-appointed channelmasters, creeps who step on your signal and overmodulating maniacs bleeding into adjacent channels. At best, you'll miss important Smokey reports and may start collecting gift certificates from bears doing their duty. At the worst, you may not get through in a genuine emergency.

With 750,000 newly licensed CBers each month, you can play the odds on getting caught in violation of almost any regulation. But other CBers operating in violation may be depriving you of your rights. Staying legal becomes important to every CBer if millions of stations are to co-exist. CB is becoming a tight little island where good manners and playing by the rules are a necessity if we are to survive. Your island covers 5 to 10 miles in all directions from your antenna. If 40 people nearby all want to talk at once, all channels will be busy. It's against the law to intentionally interfere with the communications of another station.

A CB license authorizes use of specified channels when they are available. It's not a guarantee that any channel will be available. Lawyers are pretty shrewd—they put a paragraph in the license application that says you agree not to sue if you can't get access to a channel. The FCC makes it specific that *all* CBers are created equal: ". . . the frequencies available to stations in this service are shared without distinction between all licensees . . ."

Sharing applies to everybody. Equally. No channel is reserved for the exclusive use of any type of operator. A four-wheeler equals an 18-wheeler. A private citizen equals a Smokey or any other agent of a Federal, state or local government operating a Citizens Band station. Monitoring teams cannot play king of the band. There are no home channels for base-station operators. Self-appointed channelmasters cannot direct your message traffic. Sidebanders cannot chase you to another channel. That's the law. And it is not changed by the popular practices of mobile operators using channel 19 and 10, or mariners using 13 or 3, or sidebanders gathering on 16, or flakes congregating on 8, or any use of any channel by anybody!

Within that framework of equal rights for all, there are some priorities you already know about. Any emergency message on any channel takes precedence. Until January 1, 1977 channel 11 is reserved for calling and an agreement to switch to another channel for conversation. And, of course, channel 9 is strictly for emergencies as defined back in Chapter 4.

CB's Most Popular Sins

Using Channel 9 when it isn't an emergency	—	Help keep this channel clear. Someday you may need to use it.
Hogging a channel	—	Your time limit is five minutes followed by one minute off the air for any exchange of communications *between* stations. For units of a single station, use the minimum practicable time.
Gossiping on a business station	—	When a business gets a license, it's only for business. Do personal communications with your own call sign and your own rig.
Failing to use your call sign	—	Transmit each letter and number separately and distinctly, at beginning and end of each communication between stations. Every ten minutes or oftener among units of the same station communicating over a long period, and at beginning and end.
Entertainment	—	No music, whistling, singing, sound effects, raspberries or other wierdo behavior. Use plain spoken language only.
Commercials	—	You can't sell or advertise anything.
Technical violations	—	Stay on frequency, with legal power fed to a legal antenna and don't overmodulate. Don't sneak onto unauthorized frequencies or bands.
Transmitting without a license	—	Inexcusable now that you can issue yourself a temporary permit.
Communicating with an unlicensed station	—	If he won't give his call sign, assume he is illegal. If we don't talk to the illegal operators, CB land will be a better place to work and play.

CHAPTER 17

HOW TO UPGRADE YOUR SYSTEM FOR BETTER PERFORMANCE

Sooner or later, you will want better performance from your CB. For some, the urge comes very soon after buying a CB rig. A mobile operator starts getting clues that his short communication range is denying him information, interesting exchanges with other travelers or longer notice of a lurking Smokey. It's like traveling in a moving bubble which limits the range of his ears. Very soon he wants a bigger bubble.

Some operators feel the need to perk up their equipment only because its performance has gradually degraded. One day the CBer realizes he could do about as well lip-reading as using his rig, so he decides to improve things.

Some develop a strong hobby interest in the CB equipment itself and the technical aspects of good performance. This leads to improvement of existing installations and purchase of new items as much for the pleasure of ownership as for the pleasures of longer communication range. These folks are good candidates for an early switch to SSB because it is technically more advanced than ordinary double-sideband AM.

Whatever the reason and whatever your budget, when it comes time for you to improve your system, there are several ways to do it. Some won't cost you anything. Some don't change your rig—they change the environment it lives in.

In this chapter you'll find upgrading strategies and lists of things to check or points where improvements can be made. I'll tell you what to do and refer you to other chapters of this book for more details or instructions. In some cases, you'll be turning back to earlier chapters—perhaps for details on antennas. In others, I'll refer you to chapters that follow this.

The main idea in this discussion is to decide on a basic plan or strategy fitting your needs and then carry it out. If you don't have any surplus in the family budget, shopping for new equipment is only a form of self-torture. Dig into your present system and make it work better! If you have a few bucks, spend it where it will do the most good—probably your antenna system.

MAKE THE BEST OF YOUR PRESENT EQUIPMENT

This is practically free. Basically, it's a check-out of your installation for things that have deteriorated or come adrift in the past few months. It's a good idea to perform this check every four months, no matter what kind of equipment you have.

In your enthusiasm to get on the air your original installation may have been done in a hurry, without careful attention to details. Also connections loosen or deteriorate as time passes.

You can bring your system up to first-class working order by cleaning and tightening connections, sometimes replacing cable, possibly relocating an antenna for a better radiation pattern. Here are the

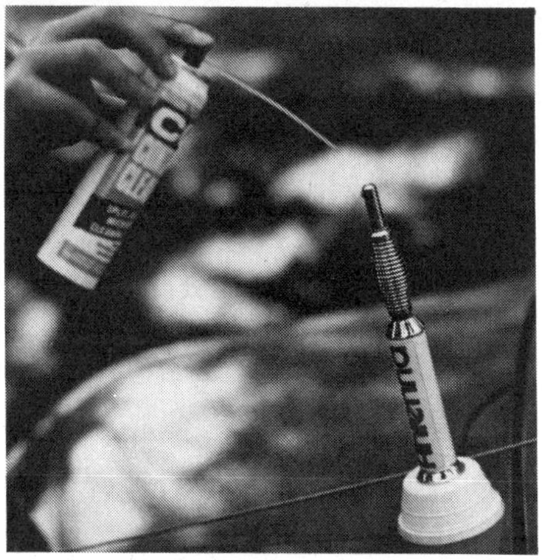

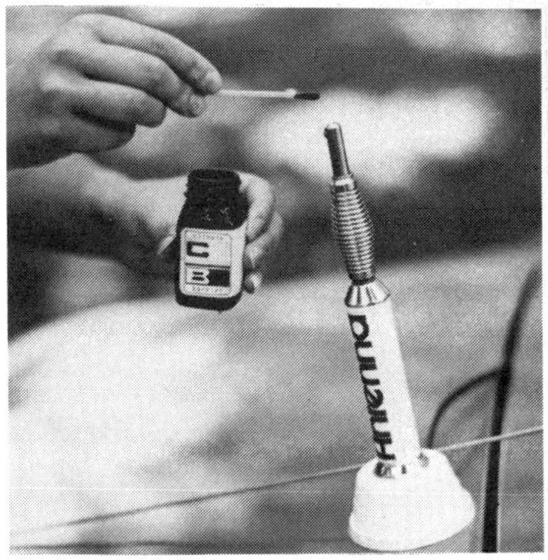

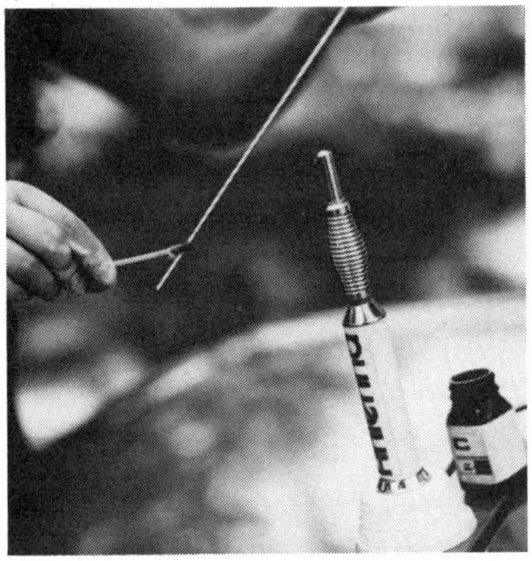

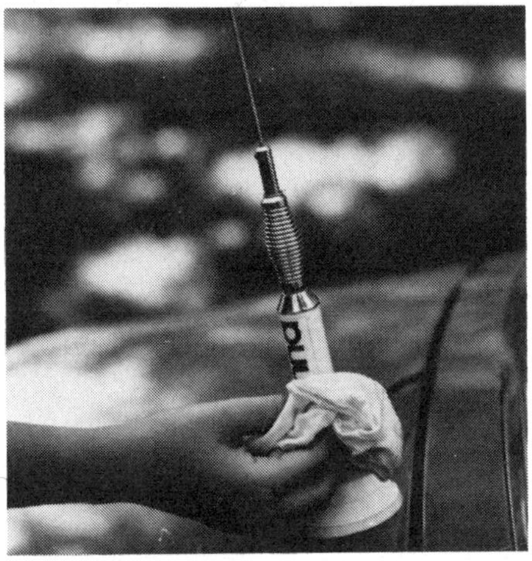

Make the most of your present equipment by restoring its original performance. Give your antenna top priority. Remove the whip from the loading coil and spray with an electrical contact cleaner. Apply CB Climate Barrier to bottom-end of whip and socket of coil. Wipe loading coil and mounting base free of all dirt and grease. Re-tune antenna for lowest SWR.

check points.

Mobile and Marine Antennas—*Look* for signs of oxidation or corrosion where: Whip is seated in loading coil or mounting base. Loading coil or antenna attaches to base. Coaxial cable attaches to antenna mount. Coax connects to transceiver.

Fix by cleaning lightly with medium or fine steel wool. Or spray with an electrical contact cleaner. Don't use any old TV tuner cleaner. I use only a contact cleaner that leaves *no residue.* Read the label, including the fine print for the manufacturer's statement about residue. I prefer CO Contact Cleaner and LPS Instant Contact Cleaner, in that order. Then protect the clean connections against corrosion with a few drops of CB Climate Barrier. Electrical and electronic supply shops should stock these; if you can't find the products, manufacturer's addresses are at the end of this book.

Clean the exterior of the loading coil and any plastic insulating parts with a cloth dampened in mild detergent to remove dirt, salt or grease that may short-circuit your rf signals.

Inspect coaxial cable for green corrosion of braided shielding and yellowing of insulation around center lead. To see the condition of the cable, detach a plug at the end exposed to the outdoors. If the cable is deteriorating, replace it with RG-58C/U. Check the instruction sheet for your particular antenna to be sure cable length is not critical; in a few cases you may have to cut the replacement to exact length.

Check the electrical path from antenna base to chassis ground of a car or truck, using an ohmmeter or simple continuity tester. A meter is better. On gutter and trunk-lid mounts, one screw should bite into bare metal. Mirror, bumper and trunk-lid mounts may be isolated from chassis ground. Install bonding straps as needed—see Chapters 12 and 18.

Retune your antenna for lowest SWR as described in Chapter 12. Short whips will be especially critical. Tune to the center of CB band or to the channel on which you want best performance.

Base Antennas—*Look* for oxidation or corrosion where antenna elements are joined, matching stubs

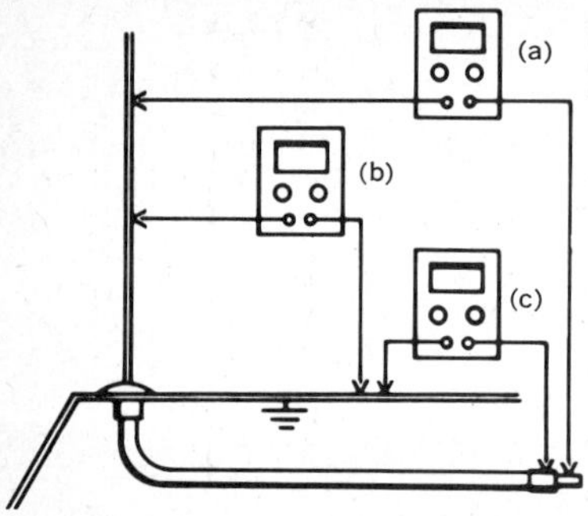

Make these three ohmmeter checks on your mobile antenna: (a) Set meter to lowest scale, touch the prods to whip and center contact of plug. Resistance should be a fraction of an ohm. (b) Set meter to highest scale, touch prods to whip and vehicle ground. Meter should probably stay at high-resistance end of scale. Some antennas with built-in matching transformers will show short-circuit. Check antenna specs or instruction sheet before worrying. (c) Set meter to lowest scale. Touch prods to outside of antenna plug and to vehicle ground. Meter should show zero resistance. Tests (a) and (b) will work on fiberglass antennas by touching *center* conductor on terminal at base of antenna.

are fastened, coaxial cable is connected.

First scribe marks with a nail or screwdriver wherever antenna elements are joined so you'll be able to restore the same length. Then remove each element and polish mating surfaces lightly with steel wool. Coat these mating surfaces with CB Climate Barrier or electrically conductive grease—one or both can be found in electronic parts stores—to protect against oxidation and corrosion. After reassembling, you can keep most moisture out of the joints by spraying them with clear acrylic.

Check coax connectors and condition of cable.

Fix by replacing cable if needed, using RG-8A/U if run is longer than 50 ft. to minimize loss. Install a new plug using a soldered connection or use one of the new plugs that require only crimping the tip with pliers.

Spray plug and socket with non-residue contact cleaner, apply CB Climate Barrier and seat plug *firmly* in socket before retightening locking collar. Then use the modern tricks of the ancient mariners by covering these non-waterproof connectors with sealer as described in Chapter 14 on marine CB antennas. Use a sealer with a non-corrosive curing agent like Boatlife's Marine Silicone or Liquid Tape—available in marine supply stores, marinas and boatyards.

Guy wires may be detuning your antenna or re-radiating. You can improve performance by placing insulators in each guy wire every three feet so there is no continuous length of wire close to CB wavelengths.

Retune your antenna for lowest SWR as described in Chapter 13.

Portable Antennas—*Look* for cracks in fiberglass whips, excessive play between elements in telescoping antennas. Consult the owner's manual for the correct extended length. If the antenna can be disconnected temporarily so an SWR meter can be inserted between the transceiver and the antenna, check SWR with the antenna fully extended. If SWR is 3:1 or higher, replace the antenna.

Power Connections—Mobile and marine connections to a power source are more vulnerable to dirt and corrosion than base-stations.

Check for clean connections. If you've inserted the end of a wire under a screw head or into a clip on a fuse block, there may be oxidation between surfaces. Caution: Disconnect terminals or wire connections from battery before cleaning. Clean terminals or wire ends with contact cleaner, emery cloth or a file, depending on condition and type of surface. Reconnect the CB power leads and protect with CB Climate Barrier or an ignition system spray like Wire-Dri. Open the in-line *fuse holder* which is usually about 10 inches down the plus or red lead from the back of your transceiver. Polish the contacts of the fuse holder with terry cloth, emery cloth or a squirt of non-residue

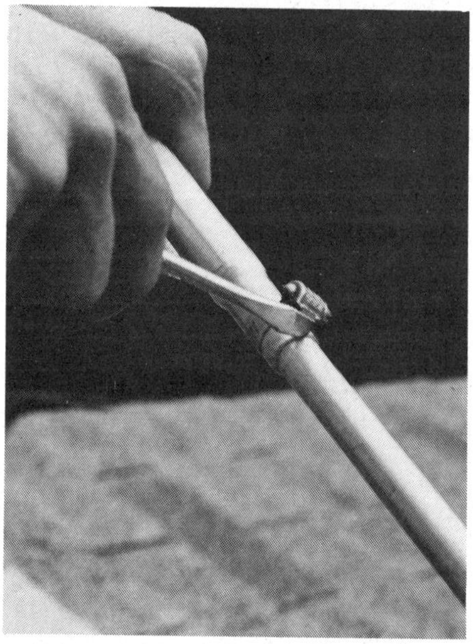

Take apart base station antenna *after* making scribe marks where elements fit together. Remove each element to clean and polish the mating surfaces. Restore to the original length by matching scribe marks. Don't polish off the scribe marks.

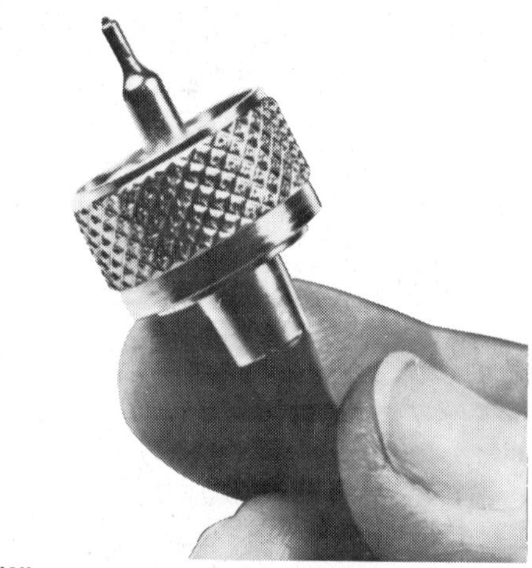

When your old transmission line shows signs of corrosion on braid or yellowing of insulation around center conductor install new coax cable and connectors. If you're not good at soldering, use a crimp-on connector like the Amphenol FCB shown here or Cambridge Products Corporation's field-installable PL-259.

contact cleaner. Caution: In some transceivers, a tiny noise-suppression coil may be inside the fuse holder; open carefully over a piece of white paper or cloth to be sure you don't lose the coil if it should drop out. Reassemble fuse holder, then reconnect power wire to battery.

Check CB set *ground lead* for proper contact to chassis or battery.

Microphone Connections – Coil cords are most sensitive to flexing. If your microphone hanger is in a position where the cord remains stretched when the mike is hanging up, or if the mike is constantly used at the limit of the cord length, you may eventually have problems. Watch the transmitter indicator or meter with the mike switch closed and no modulation while flexing the cord. Intermittent indication points to trouble in the mike cord. Make this test as short as possible so your carrier does not interfere with other communications on the channel. If other stations report your voice is intermittent, the trouble may be in the mike cord. Disconnect from the CB and check for continuity of each lead. Usually there are five color-coded wires; place ohmmeter or continuity tester prods at opposite ends of the same wire. Replace a bad cord *after* making a sketch to show color code and connections.

IMPROVE YOUR ANTENNA SYSTEM

You should get more performance improvement by upgrading your antenna system than any other thing you can do to your CB rig. There are two good strategies: Reach for more height when legally possible. Get more efficient radiation of the rf signal.

Mobile Antennas–Longer radiators are more efficient than shorter ones mounted in the same position and their tuning is less critical.

Direct-contact grounding is an improvement over grounding to the vehicle through a magnetic base.

Loading coil position is critical

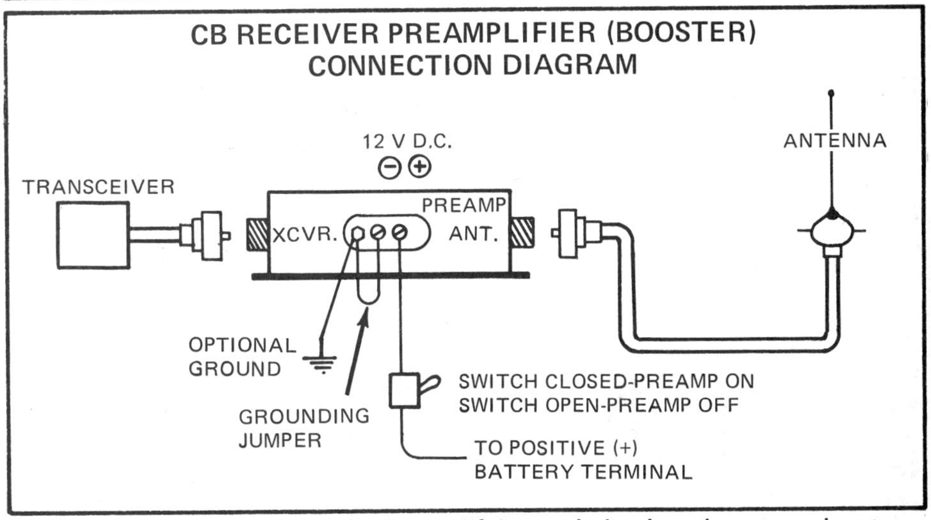

A preamplifier improves reception by amplifying weak signals at the antenna input to your rig. When your rig is receiving this Telco BoosTwenty amplifies the signal. It automatically switches out of the circuit on transmit so rf power output goes directly to the antenna. Use the optional ground connection if you have a magnetic-base antenna. Get an SSB model if you're a sidebander.

If your boat sits low in the water, a taller antenna will increase your range. Runabout owners needn't feel restricted to quarterwave whips and improvised groundplanes. Reach up and out with half-wave whips like bigger boats do.

A beam with selectable horizontal or vertical polarization is a sure way to upgrade your base station. This Cushcraft Superfire Beam has 12.5 dB gain.

in a disguise antenna. When the loading coil is below the sheet metal of the vehicle, the slim wire-whip radiator is physically very close to vehicle ground where it passes up through the insulated hole in the car body. Rain, snow or dust can bridge that small gap between the metal whip and the metal of the car body, causing a short circuit at the radio frequencies used in CB. When the loading coil is *above* the sheet metal, its plastic insulated exterior separates the metal whip from the car body by about two inches so it is less likely to be shorted by dirt or moisture.

Dual antennas improve performance significantly only when mounted about 8 ft. apart.

Relocating the antenna for better radiation pattern improves performance. See Chapter 12 for all the compromises you'll have to make among variables like height, mounting position, radiation pattern and stowing.

A *booster* will improve rf input for receiving; it should switch out of the circuit automatically when transmitter is on. A booster is legal; a linear amplifier is an offense —to the FCC and to your CB buddies. For information on boosters see accessories in Chapter 19.

Marine Antennas—A *longer* radiator means more efficiency.

Increase antenna *height* to the legal limit—of 20 ft. above the highest point on the superstructure —for greater range.

Half-wave and 5/8-wave antennas improve range by a radiation pattern nearer the earth's surface.

A *booster* may improve reception.

Base Antennas—*Groundplanes* will improve performance of straight vertical radiators.

A 5/8-wave antenna will give you the *lowest radiation pattern* of any omnidirectional.

If your omni's tip is at the legal limit of 20 ft. above a structure, check to see if a *ground-up* installation will give you a height advantage over low structures. The limit for a tower-supported antenna is 60 ft. from ground to antenna tip. See Chapter 13.

Any *beam* will give you more gain in a given direction than an omni. The more beam elements, the more gain you can expect.

A booster *may* help, or it may just add noise.

GET MORE MODULATION

You won't gain much in intelligibility of a message by increasing modulation from 85% to 100%. You will improve performance if modulation is below 85% and you can increase it to near 100%. One solution is inexpensive: Improve your microphone technique. The other solutions require you to put your money where your mouth is

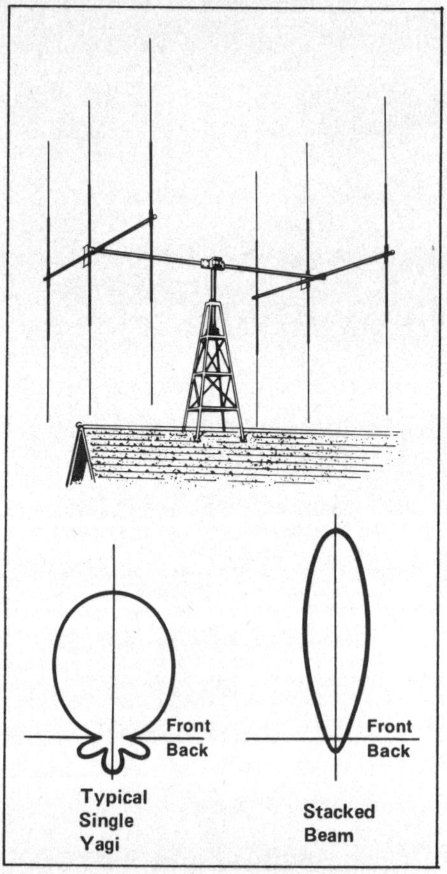

Stacked beams give improved performance over single beams. The more beam elements, the more gain you get. This stacked 3-element beam is Hy-Gain's Super-Duo with 12.7 dB gain concentrated in a narrow forward pattern. Compare the antenna pattern to a single Yagi.

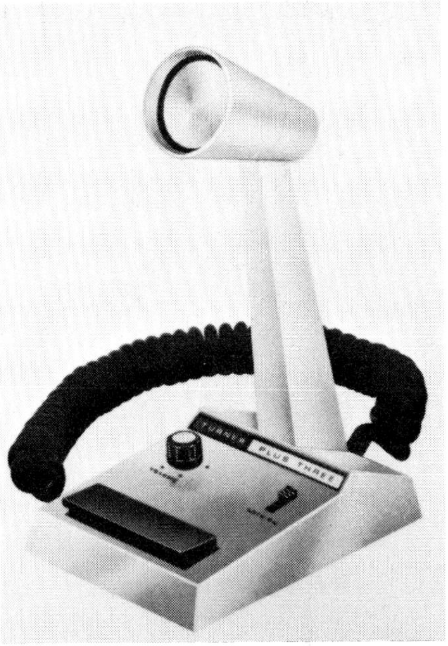

Add a power mike with speech compression for improved modulation. This Turner Plus-3 has battery-powered mike amp in base; the black bar is a press-to-talk switch that can be locked on for convenience.

by buying power mikes or speech compressors.

Mike technique is the way you speak and where you hold the microphone. If you speak *some* words softly, they will not produce a strong signal compared to other words. If you speak *all* words softly, move the mike closer to your lips or learn to sound off! Careful, though—your transceiver may not limit modulation to 100%; overmodulation produces a distorted signal and may even splash your signal into adjacent channels.

If you don't keep a constant distance from the microphone, words spoken farther away will produce less modulation and a weaker signal. Watch how professional entertainers keep the mike at the same distance from their mouths no matter how they move their heads. If you can't put your mouth where the mike is, put your mike where your mouth is—with a boom or cheek mike. These were designed for aircraft pilots and now are popular with 18-wheeler drivers. See accessories in Chapter 19.

Power mikes have a built-in amplifier to increase the input signal to your transmitter for full modulation. If you talk softly and find it a burden to talk louder, a power mike will comfortably step up your weak audio input. If your transceiver has a control marked mike gain, you have a built-in microphone amplifier so you don't need a special power mike. Increase the gain for full modulation.

A *speech compressor* limits strong audio inputs and amplifies weak inputs. The net result is higher average modulation—what some manufacturers like to call *talk power.* Your voice will have improved clarity and your signal will come on strong. A compressor may be part of your transceiver, a separate accessory or part of a power microphone. See Chapters 6 and 19.

SWITCH TO SSB

If you want more range, single sideband is the way to get it. I don't mean just theoretically. In our performance tests, SSB consistently added 50% more distance to reliable two-way communications. SSB pumped stronger signals into shadow areas between hills where AM simply did not produce useable signals.

You'll need a new transceiver for SSB. Choose one with easy fine tuning and good frequency stability—crystals better than ±0.005%, phase-lock loop or both.

REDUCE NOISE AND INTERFERENCE

If your receiving range is limited by noise in the immediate vicinity of your transceiver, you can improve performance by reducing or eliminating local interference. Chapter 18 tells you how to deal with various noise sources around vehicles, boats and bases.

A second way to cope with noise is to get a transceiver with better circuitry. When you've eliminated interference from your own vehicle, boat or building, other noise sources nearby and beyond your control can only be dealt with by the noise circuits in your CB rig.

ADD OPERATING CONVENIENCE

Operating convenience can be part of the CB dream machine you buy—or you can add accessories that accomplish whatever you wish.

If you upgrade by buying a *new transceiver* with lots of complicated circuit features, there will be more things that can go wrong. I'm not knocking complexity, but it reduces reliability. On the other hand, for months our test engineers shuttled a feature-laden Tram D-201 base transceiver among three test sites with never a performance failure. Buy the very finest set you can get.

Adding accessories put complex features where they can be disconnected from the system for repair. Personally I don't like a rat's nest of wires around my rig and prefer built-in features. This route can save you money because many accessories are offered in the emerging CB flea market—often by people scratching for cash to buy a deluxe outfit with built-in goodies.

Here are convenience features I think most desirable. You can add them as accessories.

Tone alert for private calling.

Monitoring or scanning preselected channels and priority interrupt.

Voice-operated transmitter with voice-sensitivity and delay controls.

Phone patch.

For a complete discussion of accessories, see Chapter 19.

CHAPTER 18

HOW TO COPE WITH NOISE AND INTERFERENCE

Your CB operating range is drastically limited by noise and interference.

You can do several things about noise: Control it. Suppress it. Eliminate it.

Controlling noise with receiver noise controls sometimes means slightly reducing the strength of the desired signal to get a lot of noise reduction. You can often control the worst noise sources that affect your receiver—the ones nearby such as in your car or your base station. The purist approach is to suppress all environmental noises around your set so you can operate with minimum electronic noise control in the receiver.

I am not going to be a purist about noise in this chapter. I'll tell you practical things to do about noise in vehicles, boats and base stations and what priorities to put on them. I'll also give you a few tips about interference *you* cause so you don't run afoul of FCC regulations about spurious emissions from your transmitter.

NOISE THAT ANNOYS AND WHERE IT COMES FROM

Noise is any sound you don't want to hear. If it weren't there you'd hear your good buddy loud and clear. Around base stations most noise is a rushing sound, punctuated by buzzes and occasional popping. The rushing sound is atmospheric noise. Punctuations are likely to be static and nearby appliances and motors. Mobile stations are affected by noise from vehicle engines and electrically-operated accessories. Power boats have the same noise problems, usually worse because the engine is not shielded by a metal enclosure.

AUTOMOTIVE NOISE SOURCES

When a car rolls out of the factory it is equipped with devices to reduce engine-generated rf interference so you can listen to AM or FM radio. But ordinary radios are designed to receive strong signals from stations transmitting a thousand watts or more. The noise-proofing isn't enough for a CB receiver designed to be very sensitive.

The farther away a station is, the weaker its received signal. But the rf-noise sources on your vehicle are very close and often stronger than distant CB signals. A commonly-used engineering term to describe the relative strengths of signal and noise is *signal-to-noise ratio.* A high ratio means there is lots of signal compared to the noise and intelligibility is good.

Radiated and Conducted—Noise gets into your CB receiver in two ways. *Radiated* interference comes from sources that act like miniature broadcasting stations. Anything that makes sparks, for example, makes a burst of rf noise with each spark and you hear it as a pop in your receiver. *Conducted* interference travels along the wiring

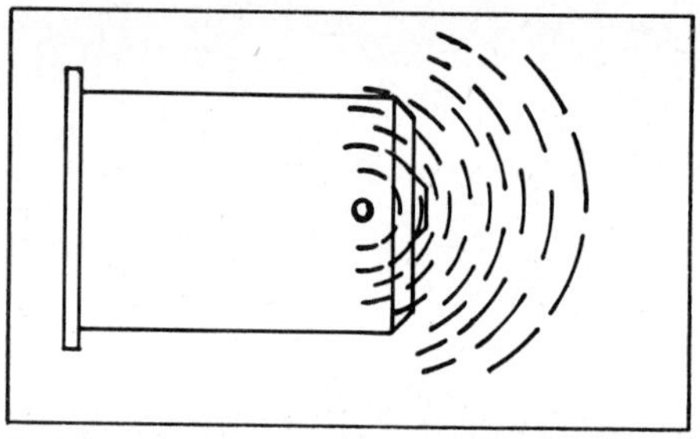

Radiated noise comes from home appliances and parts of a vehicle that act like small broadcast transmitters. You pick up this kind of interference with the antenna of your CB set.

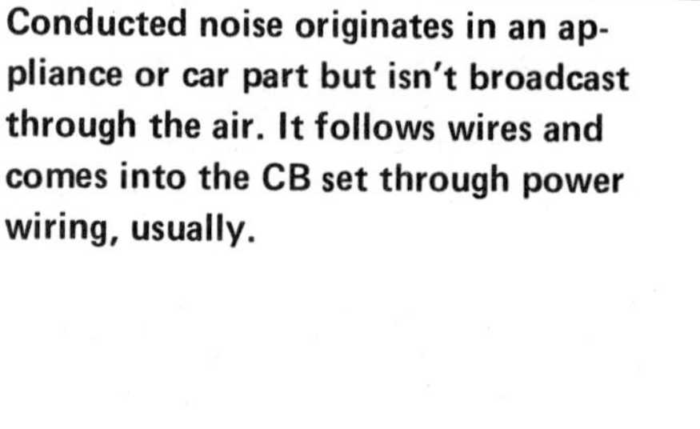

Conducted noise originates in an appliance or car part but isn't broadcast through the air. It follows wires and comes into the CB set through power wiring, usually.

HOW TO IDENTIFY AND FIX SOURCES OF CB NOISE IN VEHICLES AND BOATS

What you hear in the loudspeaker	Likely source	How to be sure	Steps to reduce noise
Popping	Ignition system	Listen with engine idling, speed up and listen for faster popping. Shut off ignition; popping should stop immediately.	1. Major tune-up of engine including new plugs, points, ignition capacitor. 2. Be sure original factory-recommended suppressor cables and resistor plugs are in ignition system. 3. Improve grounding of antenna base and coax shielding. 4. Connect transceiver directly across battery to avoid intercoupling with vehicle wiring. 5. Improve bonding around engine compartment. 6. Fully shield ignition system.
Whining	Alternator or generator	Pitch of whining varies with engine speed. Shut off ignition; decreasing whine will be heard as alternator or generator slows to stop.	1. Install 0.5 mfd coaxial capacitor in series with alternator output or generator armature lead. 2. Check alternator for dirty slip rings or worn brushes. Recondition. 3. Check generator for worn commutator or worn brushes. Recondition.
Raspy sound	Voltage regulator	Irregular sound. Watch ammeter on instrument panel with engine idling; accelerate and listen for rasp as needle moves to "charge."	1. Install 0.5 mfd coaxial capacitors. 2. Shield field wire. 3. Install tuned-circuit filter in series with field wire.
Crackles, hisses	Instrument gauges	Tap each gauge while listening to speaker.	1. Install 0.5 mfd coaxial capacitor in series with each gauge terminal. 2. Install 0.5 mfd coaxial capacitor in series with voltage-limiter battery terminal on fuel and temperature gauges.
Clicks	Gauge senders	Listen while bouncing vehicle to slosh fuel and activate fuel sender. Speed up engine and listen; oil sender click rate varies with oil pressure and stabilizes at maximum oil pressure.	1. Install capacitors on voltage limiters as above. For stubborn cases install "hash choke."
Irregular popping	Wheel static	On dry day listen while coasting 55-30 MPH with engine off on deserted road. Apply brakes lightly; popping should stop briefly.	1. Install static-collector springs in front wheel hubs.
Sharp, regular popping	Tire static	Use coasting test above. Popping continues even when brakes are applied lightly.	1. Inject anti-static powder through valves of each tire.
Various-pitched whines	Accessories like wipers, fans, window operators	Listen while operating each accessory one at a time.	1. Install 0.25 mfd coaxial capacitor in series with hot lead to accessory motor. Ground motor casing.

of a vehicle and enters the CB receiver along these metal paths rather than through the air.

Radiated Noise – Most radiated noise originates in the engine compartment so the best place for an antenna is as far away as possible without sacrificing the use of the vehicle's metal body as a groundplane.

Shielding–A conductive enclosure, usually made of metal, is called a shield. To be effective, a shield must be grounded to the frame of the vehicle. Shielding works two ways: If you *enclose the source* of rf noise in a shielded enclosure, the noise can't get out. If you *shield the receiver* parts which are sensitive to rf noise, the noise can't get in and you don't hear it. The braided outer conductor of coax cable is a shield and must be grounded. The metal case of your transceiver is also a shield.

Suppressing Noise–Still another powerful aid to noise reduction is suppressing it at the source so it doesn't happen or at least is greatly reduced.

HOW TO REDUCE RADIATED NOISE

The coax cable shield won't work well unless it is properly grounded at both antenna and transceiver. At the antenna base, at least one screw of the mount should bite into bare metal. If you use a spring-clip mount, it should make good contact with bare metal.

Test the Antenna Ground–Fasten a length of #12 or heavier wire to the vehicle firewall or frame. Start the engine. Tune the CB to a chan-el with a weak station heard over noise. Speed up the engine to check if the noise changes in loudness or character. If so, it is coming from your engine. Turn up the volume on the set so you can hear it when you are outside. Now touch the other end of the ground wire to the metal base of the antenna. If the noise is improved, the antenna needs a better ground. If it is a magnetic-base, consider switching to a bolt-on or clip-on type so you can get good metal-to-metal contact. If you already have a bolt-on or clip-on antenna, something is wrong in the grounding system.

Chapter 17 shows making antenna-ground and continuity tests with an ohmmeter. This test-wire method works well for people without an ohmmeter.

Find the Problem–If the antenna is mounted on the trunk lid or bumper and isn't making a good ground, you can suspect a poor electrical connection or *bond* between trunk lid or bumper and the rest of the vehicle. Improving the bonding will solve the problem. Later in this chapter I'll point out the advantage of having all parts and panels of the vehicle bonded anyway, so anything you do to improve antenna ground will be useful toward bonding the entire structure.

A trunk lid may not be well grounded to the rest of the vehicle body because of paint. Examine the bolts on the trunk-lid brackets at both hinges. Toothed washers should be under the bolt heads so they bite into bare metal when the bolts are tightened. If not, install them. Braided metal bonding straps from lid to body make an even better electrical bond, but you'll have to install them yourself. Auto parts stores and electronic parts stores stock bonding straps –sometimes called *grounding straps.*

Bumpers may not be well grounded to the chassis. Add a heavy-duty bonding strap between bumper and chassis. Be sure to clean the metal at the two contact points before installing.

Check Out the Transceiver Ground –Be sure the ground side of the transceiver power lead makes good contact with battery ground or an equivalent ground point on the frame. A poor contact here can cause arcing which makes noise in your set. If the transceiver has two wires for DC power leads and a third wire or terminal for ground, be sure the third wire or terminal is also properly connected to the vehicle frame or equivalent. A metal firewall is usually a good equivalent to the frame.

Check the Grounding of the Battery to the Vehicle Frame–Battery ground runs directly to the engine block. Be sure the bolt holding this terminal is tight. Engine shock mounts are electrical insulators, so another connection is made from the engine to the vehicle frame. It's there somewhere–usually a flexible bonding strap to a point on the firewall or frame. If you don't see a ground connection from engine to vehicle frame, install a braided bonding strap from engine block to firewall or frame using existing bolts and adding toothed washers.

Thanks to Champion Spark Plug company and Cornell Dubilier Electric Corporation for drawings and photos used in this chapter.

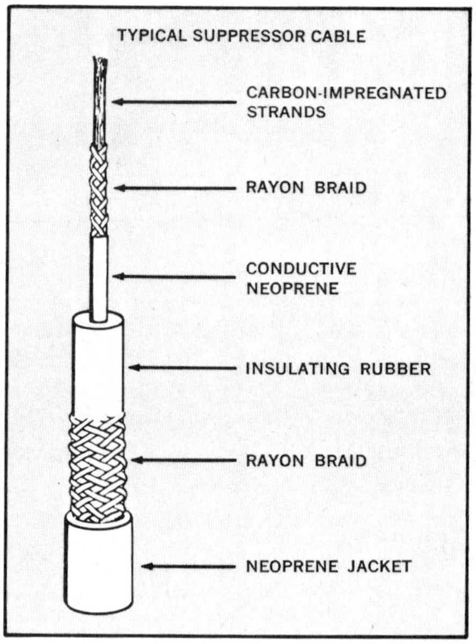

One of the most common types of suppressor cable uses a carbon center-conductor.

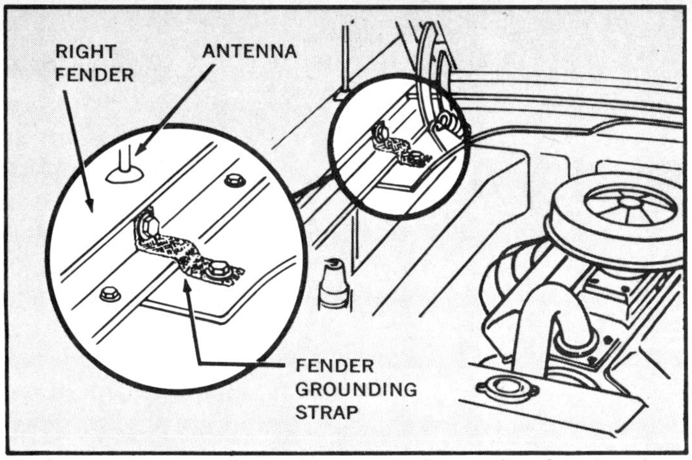

If the antenna is mounted on any body panel such as hood, trunk lid, bumper or fender, that part should be bonded to the frame of the car.

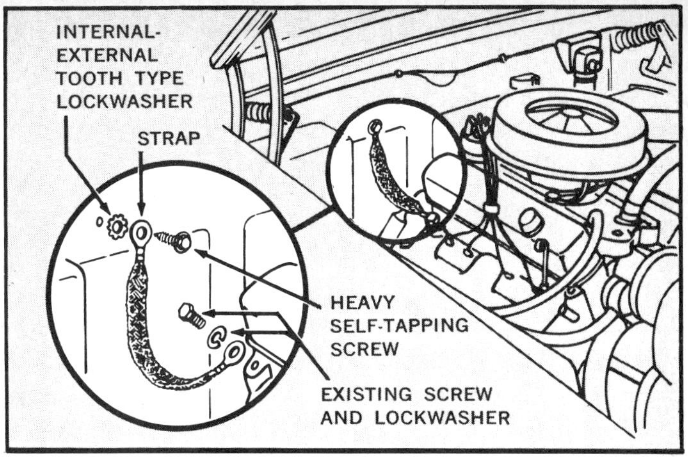

If you can't find a grounding strap between engine and frame or firewall, install one as shown.

Check Grounding of the Hood—Especially if you have a cowl-mounted antenna, you'll want to check hood grounding. American-made 1977 cars feature CB as an option—either factory-installed or dealer-installed, depending upon which brand you buy. You're likely to get a cowl-mounted antenna for factory-installed CB's. A properly-grounded hood is a shield against radiated noise from the engine compartment. Check the hood hinges for toothed washers or braided bonding straps. Some manufacturers ground the hood with an extension which acts as a wiping contact when the hood is closed.

NOISE SUPPRESSION

There's no need to tackle additional bonding or shielding until you've examined the noise-suppression equipment on the vehicle and perhaps improved it.

Vehicles are made with suppression equipment to reduce radiated and conducted noise. Most noise suppression is aimed at the generating system and the ignition. As a first step, get a major tune-up for two reasons. Cars in need of a tune-up usually make more electrical noise, so the tune-up itself will make an improvement. Also, ignitions in poor condition can't perform well anyway and are weakened by noise-suppression measures. If you add on more gadgets to make the ignition quieter, and the engine won't run at all, you've reached the ultimate in noise suppression.

When getting that tune-up, mention you are trying to reduce electrical noise. State emphatically that you want all of the original factory noise-suppression measures left intact or restored if they had been removed before. There are several combinations of ignition-noise suppression equipment. You don't want too much or too little, so the best guide is to start with whatever is supposed to be there. Here's how to identify the parts:

Resistor spark plugs—check the type number and refer to manufacturer's literature or ask the parts man at an auto store.

Resistor ignition cable—Wipe clean and read the label. It should say *radio resistance* or use codes such as *HTHR, HTLR,* or *TVR.* Again, ask the parts man if in doubt.

Ignition-suppressor resistors are individual round parts that clip onto the top of a spark plug or fit into the towers on top of the distributor. If you see something that isn't wire, it's a resistor.

Types of Suppressor Cable—*Resistor cable,* or *resistance cable,* or *suppressor cable* usually means ignition cables with a conductor which is made of carbon instead of metal. Some resistor cable uses a copper

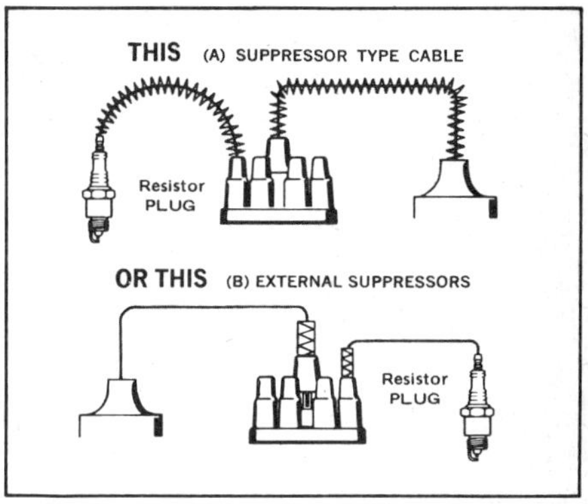

Two examples of what to look for: Suppressor cable with resistor plugs or ordinary cable and external suppressors in the distributor towers. Other combinations are described in text.

center conductor but has a built-in resistor which makes a bulge or lump in the insulation. Some use a copper center conductor with a spiral-wound shield to suppress noise called MSW, Magnetic Suppression Wire.

How It Works—The theory is complicated but the practice is simple. Adding electrical resistance in the spark-plug circuit reduces ignition noise. It can be in the spark plug itself, inside the cable, or separate little resistors.

Typical Combinations—By checking numbers and identifying parts, you should figure out what is installed in your car. Typical combinations are:

Resistor spark plugs and resistance cable or magnetic-suppression cable (MSW).

Resistor spark plugs with non-resistor cable and individual suppressors in the distributor towers.

Non-resistor spark plugs with resistors clipped onto the top of each plug, used with resistance cable.

Do not add more suppression resistance than stock for the engine or performance and gas mileage may suffer.

What To Check—Breaks or cracks in the ignition-wire insulation are a danger signal indicating the conductor inside may have deteriorated too. A qualified mechanic can make the proper resistance and continuity checks. If you do it yourself, use an ohmmeter. HTLR wire should measure between 3000 and 7000 ohms *per foot.* Disconnect each lead, *one at a time,* from a spark plug and follow it back to the distributor cap where you disconnect the other end. Caution: Never pull directly on the wire or you can break the carbonized linen thread conductor inside the insulation. Always grasp the end fitting.

Measure and reconnect each wire before going on to the next so you don't mix them up.

HTHT cable is OK if it measures between 6000 and 12,000 ohms per foot. Replace the cable when resistance is about 3 times the correct value.

If an end terminal is bad, replace the entire lead instead of trying to repair or replace the terminal. If you buy one lead or an entire ignition-cable set, get an *exact* replacement for the year and model of your engine.

Most vehicles come equipped with resistor plugs when new. Replacements should be resistor plugs,

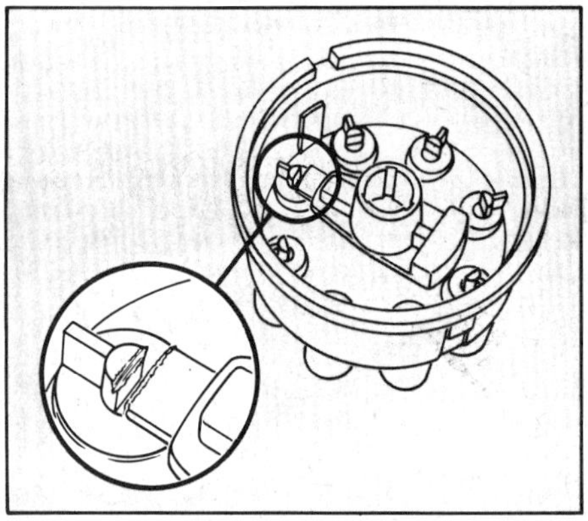

If the rotor tip or stationary contacts in the distributor cap show electrical erosion, replace the parts.

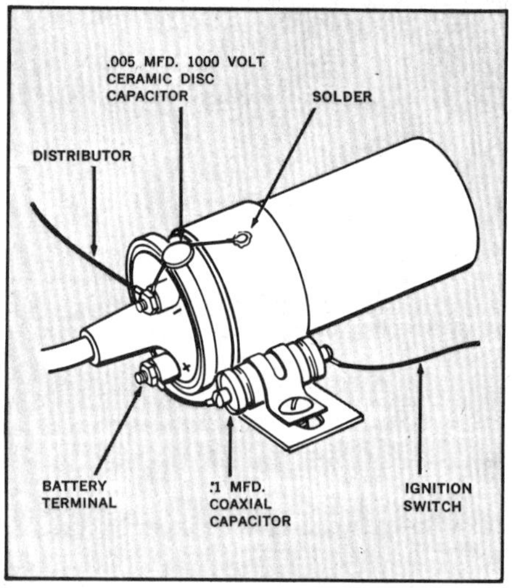

These two capacitors reduce noise from a conventional battery-coil ignition. If your car has electronic ignition, check your dealer about noise suppression.

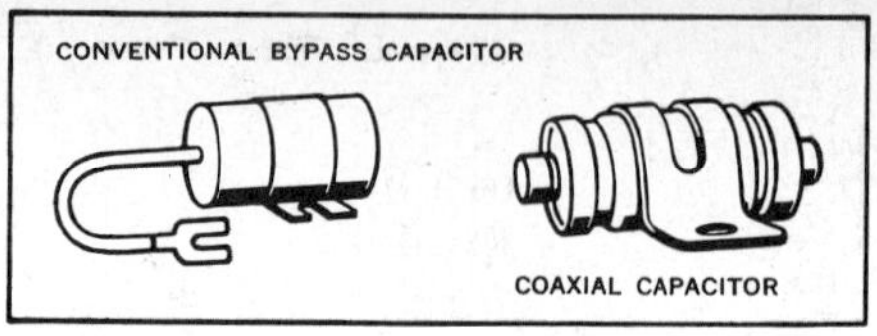

Conventional bypass capacitors have one lead and are grounded by the mounting bracket. Coaxial units surround a center conductor so DC current can flow through from end to end. They are also grounded through the mounting bracket.

too, or you'll wind up with more noise. If you use external suppressors, the resistance when new is around 10,000 ohms. Replace when they approach 3 times that value.

Distributor and Coil—Other ignition system noise sources include the distributor. Point bounce or pitted points can cause noise. Replace distributor cap and rotor when the rotor tip and contacts inside the cap appear eroded, or if the inside of the cap is rough or carbon-tracked.

The ignition coil may transfer noise to the vehicle wiring system. Check to see if capacitors are connected to two small terminals adjacent to the center tower or connector on the coil as shown in the accompanying drawing.

Caution: If the vehicle has electronic ignition, consult your dealer before adding anything!

When installing capacitors (sometimes called condensers) use a *coaxial capacitor* for most effect at CB frequencies, rather than the conventional type. The difference is shown in the accompanying drawing.

While you are at it, tag the connecting wires so you know where they go, then remove coil and mounting bracket. File and clean the mating surfaces where parts touch each other and the engine block. Replace and tighten down securely.

HOW TO MAKE A CB NOISE SNIFFER

So far I have identified the most likely offenders producing noise in your transceiver. If you still have noise in your CB ears, the sources may be hard to find without a "sniffer."

Put a male coax plug on about 15 ft. of coaxial cable. Strip back 1-1/2 to 2 inches of braided shielding from the other end of the cable which you will use as a probe or "sniffer." *Do not* remove the insulation from the center conductor. Cover the very tip of the center conductor with electrical tape.

How To Use a Noise Probe—Disconnect the antenna lead from your set and attach the probe. Start the engine. Turn on the transceiver. Turn off the following controls, if you have them: Squelch, automatic noise limiter, noise blanker. Bring up the audio volume on a channel where you get weak signals or none at all. I suggest channel 9 as the least busy.

Hold the probe near, but not touching the suspected noise source and listen for increased noise in the loudspeaker when the probe is close and the suspected noise source is operating. The probe will usually make noises seem a lot louder than they are in normal operation. If you hear a noise with your probe that you *never* hear when driving around making chin music with your good buddies, there's not much point in quieting the noise. When you find the source of some noise that has been bothering you, rejoice and fix it!

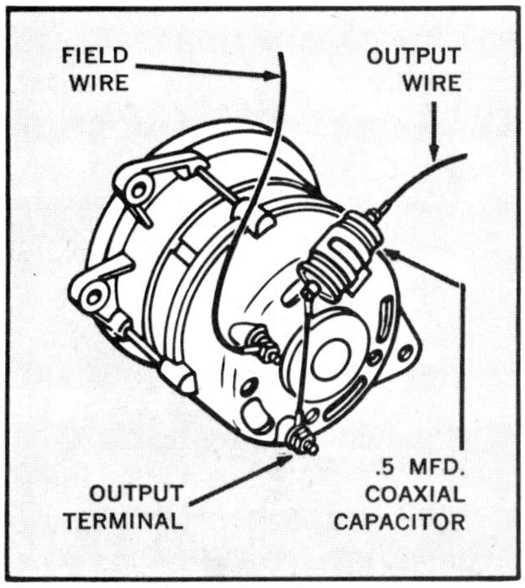

Here's how to suppress alternator noise getting into your CB set.

ALTERNATOR OR GENERATOR NOISE

Sound is a high-pitched whine which changes with engine speed. The probe may exaggerate the noise. If the whining ordinarily doesn't bother you, forget about it. Otherwise fix as follows:

An alternator may be quieted by installation of a 0.5-mfd (microfarad) coaxial capacitor at the output terminal. Be sure to buy a capacitor rated to handle the maximum alternator current. The service department of a dealer for your car can give you that information. Or, just buy the largest coaxial capacitor available—200-ampere handling capacity. Clean the output terminal—it may be marked BATTERY (B) – before connecting the capacitor in series with the output wire.

Caution: Never connect a capacitor to the alternator FIELD (F) terminal.

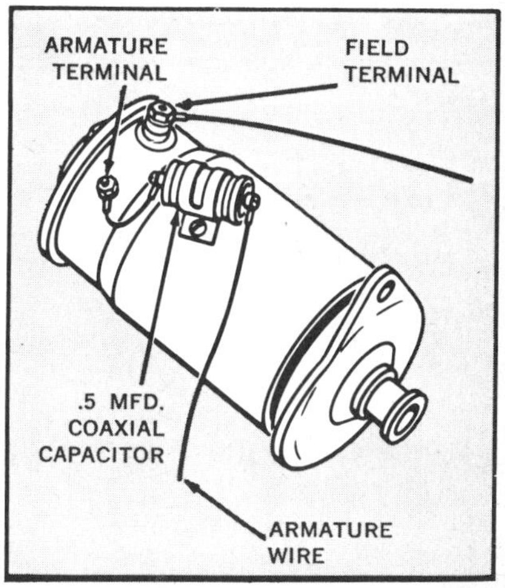

If your car has a generator, it may have a conventional capacitor installed at the factory. For CB noise suppression, use a coaxial capacitor as shown.

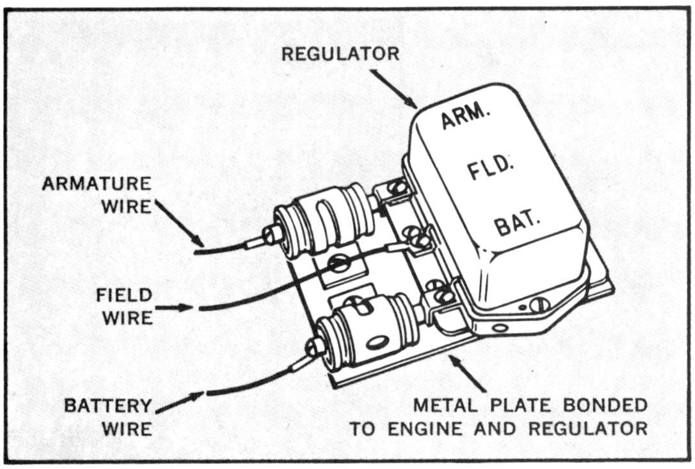

Typical noise suppression suitable for CB sets applied to the voltage regulator used with a DC *generator* in a car. The fix for an alternator regulator is different—see text.

This Cornell Dubilier L-C Tuned Filter removes frequencies near 27 MHz to which CB sets are most sensitive. Mount unit in the DC power lead as close as possible to offending noise sources.

DON'T GET YOURSELF HURT

People who are not used to working on cars often make mistakes resulting in serious physical injury or property damage.

When you are running the engine with the car stationary, BLOCK THE WHEELS SO THE CAR CAN'T MOVE!

Don't depend on the car being in neutral or remaining in neutral. Don't depend on the parking brake holding. People have depended on these and ended up mashed flat against a wall! The only way you can be sure is to place large stones, building blocks or equivalent directly in front of the wheels and close against them. Even then, don't stand in front of the car!

When the engine is running and the hood is up, MOVING ENGINE PARTS CAN INJURE YOU SERIOUSLY! The fan can cut flesh and break bones. If you get your finger or necktie caught under a fan belt, you can be mangled! The only safeguard is extreme care and awareness. Of course you won't wear a necktie. *Be aware* of moving parts under the hood.

When you are working under the hood, take the ignition key out of the lock and put it in your pocket. Then you know where it is, for sure!

If you are not experienced and handy with tools and don't have a well-equipped tool box, consider having the work done by a professional. Improvised tools used by an inexperienced mechanic rarely do good work.

If you still get excessive alternator noise, have a qualified auto-electric shop handle the problem.

A *generator* will have a factory-installed capacitor attached to the armature terminal. If it's not a coaxial capacitor, replace it with one of 0.5 mfd, rated to handle the maximum generator current. Be careful not to connect it to the generator field terminal by mistake. If the commutator and brushes of the generator are dirty and worn, you may still get generator noise. Have the generator reconditioned.

VOLTAGE-REGULATOR NOISE

A rasping sound imposed irregularly on generator or alternator noise is probably the voltage regulator.

If it is causing interference when driving, you'll hear it intermittently whenever the generated voltage changes due to changed engine speed or changed load, for example when you switch the air conditioner on. The voltage regulator works to bring the voltage back to normal—13.8 volts.

To test for voltage regulator noise, bring the noise probe near the regulator with the engine running and the car stationary. Change engine speed up and down and you should hear regulator noise if it is making any.

As the accompanying drawing of a typical *generator* regulator shows, there are three sets of contacts. Install 0.5-mfd coaxial capacitors *close* to the armature and battery terminals; they should be rated to handle the maximum generator current.

On an *alternator* regulator, the terminal marked IGNITION or IGN requires a 0.5-mfd. coaxial capacitor in series with the lead and another of the same size in series with the lead to the terminal marked BATTERY or BATT.

If you still experience excessive voltage-regulator noise, shield the field wire with a sleeve of copper braid; be sure to ground *both* ends of the braid to vehicle ground. Be sure the braid cannot slide toward the terminals to cause a short.

A last resort for stubborn noise in generators, alternators or voltage regulators is a tuned-circuit filter. Some kinds are made with pigtail leads, others are made with terminals; installation instructions are usually in the package. This is the *only* suppression accessory that is connected to a terminal marked FIELD.

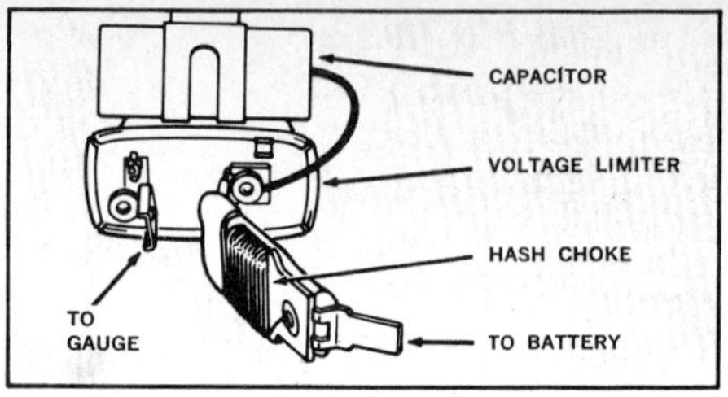

Voltage limiters for gauges are quieted by capacitors and hash chokes.

NOISE FROM INSTRUMENTS, ACCESSORIES, WHEELS AND TIRES

Instruments, gauges and sender units, may produce a hissing or crackling noise in the speaker. Tune to a quiet channel. With the vehicle parked and engine running, hit the dashboard to simulate road shock while listening for a sudden change in hissing, crackles and clicks. Bounce the car to test the fuel gauge and sender. Tap each gauge.

Isolating the offending gauges or senders requires the tedious work of disconnecting them one by one. I don't think it's worth the effort of identifying which. As long as you are getting to the instruments by crawling under the dashboard or taking apart the instrument panel, noise-proof all of them. Connect a 0.5-mfd coaxial capacitor in series with each gauge terminal. Keep the capacitor lead to the gauge as short as possible. Be sure to ground the metal mounting bracket of the capacitor to the dashboard metal—not metallic-looking plastic—or to vehicle ground.

Fuel and temperature gauges have voltage limiters mounted on or near them. Install a 0.5-mfd coaxial capacitor in series with the battery terminal at each voltage limiter and ground the capacitor mounting bracket. An alternative is to connect a pig-tail 0.1-mfd radio-type capacitor *across* the limiter terminals. If that doesn't work, ask your electronic parts dealer for a *hash-choke*—a small wire coil—and install it in series with the battery terminal of each voltage limiter.

WHEELS AND TIRES

These are heard as popping and rushing in the loudspeaker. You're likely to hear it in dry weather at higher vehicle speeds. To identify each source, find a deserted road for coasting tests with the engine off.

On a dry day, accelerate to highway speed, then coast in neutral with engine off, CB set on.

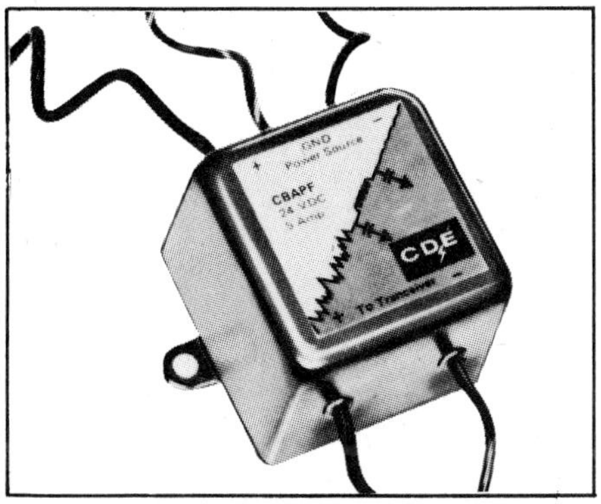

Mount this all-purpose filter as close as possible to your CB set. Feed DC power through this Cornell Dubilier filter which is tuned to 27 MHz, blocking interference traveling along the DC power wires no matter what the source.

Caution: Don't turn the ignition switch to LOCK because you can't steer.

Apply the brakes lightly. If the popping disappears you have *wheel static* build-up between the front axles and front wheels. To cure it, buy a pair of static-collector springs from an auto parts store. Remove each front hub cap and axle dust-cover. Press the small end of the spring against the tip of the axle and put back the dust-cover to hold it in place. Replace the hub cap. Check springs for wear when you have the wheel bearings repacked.

Tire static can be identified by the same coasting test. The popping is a bit more regular. It varies with road speeds but does not disappear when the brakes are applied. Both auto and electronic parts stores have kits for injecting anti-static powder into the tires. Do the work at a service station where you have access to an air hose. The kit has a small container threaded at one end to screw onto the tire valve. Fill the container with a packet of the powder, attach it to a tire valve, then blow the contents into the tire with a few blasts of the air hose. Don't over-inflate your tires in the process. Do the same for all tires on the vehicle, including the spare if you periodically include it in tire rotation.

ACCESSORIES

Operate each accessory separately while listening to the speaker on a quiet channel. You can usually identify the sound of electric windshield wipers, turn-signals, blower fans and window drives. Only wipers and blowers are worth noise-proofing because they get continuous use when they're turned on. Install a 0.25-mfd coaxial capacitor in series with the hot lead to the blower motor. Ground the capacitor mounting bracket and ground the motor casing or mounting bracket with a braided bond strap.

BONDING THE WHOLE THING

By now, if you still need to eliminate or suppress noise, you should be ready for heroic measures. Bonding to assure a common ground among all parts of the vehicle is a lot of work but is sometimes necessary. Bond at each recommended location with braided ground straps, then listen to the loudspeaker, for apparent differences. The method is cut-and-try. Location can be critical at times—One bonding location may not reduce noise, but a few inches away you might get excellent results. Here's a check list of bonding locations, including those priority points previously mentioned:

- Battery ground to frame
- Both sides of trunk lid
- Bumpers to frame
- Both sides of hood
- Corners of engine to frame
- Coil and distributor to engine and firewall
- Exhaust pipe to frame and engine
- Air cleaner to engine block
- Tail pipe to frame
- Steering column and gauge-line tubing where they pass through firewall
- Radiator to supports

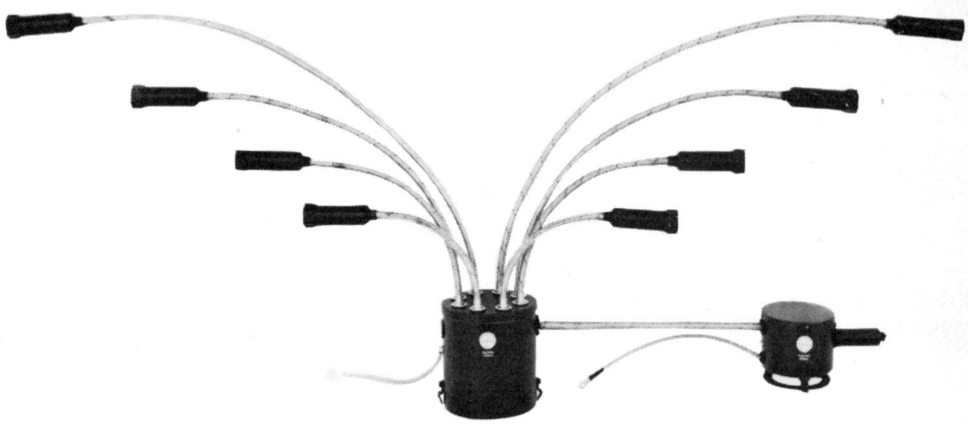

Estes Engineering Company, Port Angeles, WA, supplies complete ignition-shielding kits like this for do-it-yourself installation. This kit, for a V-8 engine has individual shields which fit over each spark plug.

MARINE APPLICATIONS

Shielding is extremely important for marine CB use because boats rarely have shielded engine compartments as cars do.

Outboard motors producing noise in a CB set usually can be improved by suppression techniques similar to those used on auto engines. Manufacturers provide kits to do the job. Don't install them on outboards with a tendency to foul plugs until you have corrected the causes of fouling. Ignition-noise suppression on a poorly performing engine will make the engine run worse.

Inboard engines also respond to suppression techniques similar to car engines. They'll be even quieter if you shield them. If your wooden or fiberglass boat has a marine ground plate, surround the engine with copper or bronze screening connected to the ground plate. You can buy metal engine shields made for the purpose.

An alternative, is a fully-shielded ignition system. This works well on vehicles, too. Corvettes, which have fiberglass bodies, have a fully-shielded ignition system you might look at to see how it can be done.

SHIELDED IGNITION SYSTEMS

All sources of radiated interference previously discussed–spark plugs, ignition wiring, distributor cap, coils–are enclosed with

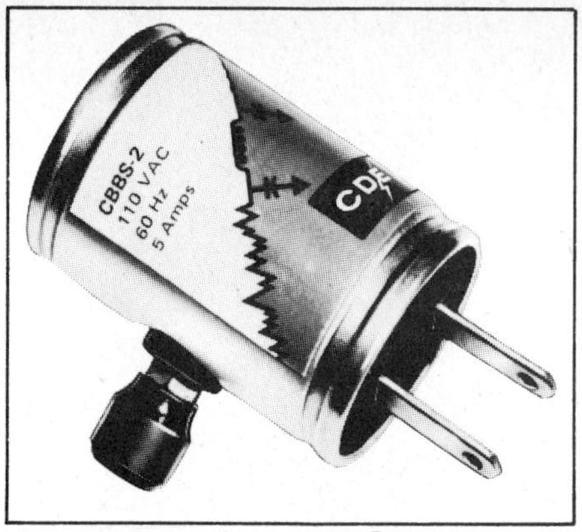

Stop noise at the source! Plug this appliance filter into the wall outlet, ground the case using the terminal on the side, plug noisy appliance into filter. This helps CB, TV and FM.

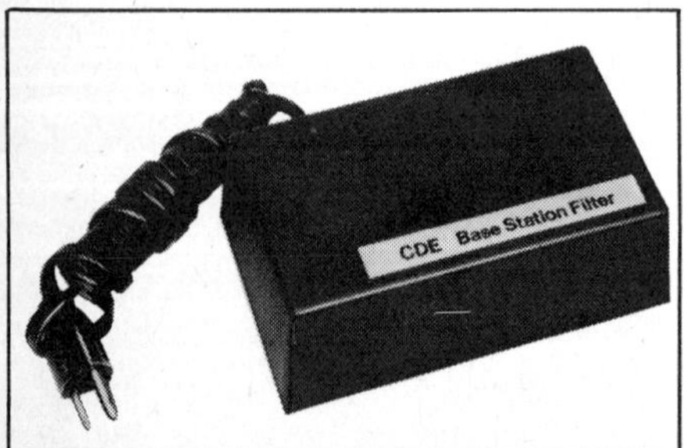

Plug this base-station filter into the wall outlet, then plug your base station into the filter. Filter removes interfering signals from power system before they can get into your CB base station.

an electrically-conductive shield grounded to the engine.

Shielding significantly reduces the voltage delivered to the spark plugs by the coil. Keep cables as short as possible for maximum coil voltage. You'll have to regap spark plugs more frequently. If engine performance is still below par with fresh plugs, install a coil with higher output.

BASE-STATION NOISE AND INTERFERENCE

When you're really reaching out with your base station for that distant contact, it's dreadful to suddenly get zapped by noise from an electric shaver, blender, vacuum cleaner, hair dryer, sewing machine, furnace, air conditioning, or other exasperating bedevilment.

You can also be bothered by noise from fluorescent lamps. One cure is to move the transceiver a little farther from the lamp. A better cure is to turn off power to the fixture and install one or more 0.01-mfd 600 V capacitors across the AC line cord inside the fixture. Then forget about it and do something about the appliance noises you now hear better.

Appliance noise usually gets into your CB through the building wiring. Connect your base station through a noise filter plugged into the wall receptacle you use for power. Noise filters of various types can be plugged into wall outlets or wired into the appliance to filter out electrical noise before it gets into the building power circuits.

I advise starting with the base-station filter. If that doesn't reduce noise enough, start putting filters on offending appliances. Buy these at CB or electronics stores. Check the current rating of a filter to be sure it can handle the current drawn by the appliance you'll connect to it.

WHEN SUNSPOTS INTERFERE

Sunspot activity will increase to a peak around 1980, then gradually taper off again. Sunspots follow roughly an 11-year cycle, producing intense magnetic storms to which the Citizens Band is especially sensitive. The magnetic storms affect a layer in the upper atmosphere from which the 27 MHz signals bounce. Welcome to skipland!

Working skip is illegal, but when the sunspots are active skip will put your signals all over the map. You can tell how strong skip propagation is by watching the S-meter on your set. Your regular contact with a good buddy who normally comes in with a solid S-8 may be overshadowed by a skip from hundreds of miles away. One day when I couldn't work more than a mile in New Jersey, I suddenly found myself talking to a trapper with a walkie-talkie tromping through the Canadian woods.

Nobody has a cure for sunspots. I just wanted to prepare you so you wouldn't think your dazzling talents gave your CB set worldwide range on only 4 powerful watts.

TELEVISION AND FM INTERFERENCE

Howboutcha, neighbors? You've got the one Tennessee Valley Indian on your television channels 2, 3, 4, 5, and 6. How do I sound? Why aren't we good buddies any more?

TV interference (TVI) can ruin your neighborhood social life. So can FM interference because the FM band is right above TV channel 6 and its frequencies coincide with CB harmonic emissions. Your station may not be at fault but the very sight of your base or mobile antenna makes you a suspect.

Your CB transmitter not only puts out a signal on the carrier frequency to which it is tuned, but also radiates signals—harmonics—at multiples of that frequency. These harmonics are not as strong as the carrier, but may be received on nearby television sets. Lots of TV receivers simply weren't designed with circuitry to reject these signals; they respond to CB signal harmonics and sometimes to the CB carrier itself.

Symptoms are herringbone patterns on the TV screen. You may even make thick horizontal lines, called sound bars, across the picture; these vary with the modulation of your voice, sometimes heard on the TV loudspeaker.

In most cases, this CB invasion of TV is not your fault but it's hard to tell that to your neighbors and still remain friends.

The FCC investigates complaints about it and checks suspected CB stations for illegal levels of harmonic radiation. Until September, 1976, if a CBer was within legal limits, the problem was tossed back at the TV owner with the advice to get a TVI filter. New regulations—paragraph 95.49 of the FCC rules—tighten the limit on spurious radiation. Transmitters FCC type-accepted after September 10, 1976, meet new harmonic suppression requirements of 60 dB below the average AM power output.

If your CB is broadcasting TVI, install a low-pass filter between set and antenna. This filter reduces your radiation of 54 MHz (TV channel 2) by more than 60 dB. This filter makes you a good neighbor.

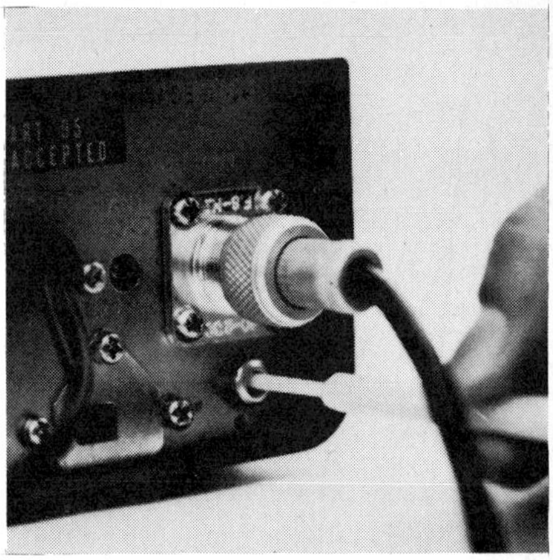

If your CB set has a built-in tunable TVI filter like this Lafayette, here's how to adjust it. Use a non-metallic screwdriver, available at radio-electronics stores. Key your transmitter and adjust for minimum interference on your own TV set.

Here are steps you can take to keep the local peace:

If your base-station antenna is on a guyed mast or tower, insert strain insulators in the guy wires at 3-ft. intervals.

Install a *low-pass* TVI filter on your rig. Ground the CB case to a cold-water pipe as described earlier.

If your CB set has a built-in TVI trap, tune it for minimum interference on your own set—channels 2 or 3—while transmitting.

Suggest your neighbors install *high-pass* TVI filters or *wave traps* on their TV or FM sets. These can be purchased at electronic stores and connect to the antenna terminals at the TV or FM set. The wave trap is usually more effective because it has a tuning adjustment to be set for minimum interference. These traps work in about 50% of TVI cases. When they don't, it's best to call a TV serviceman.

I advise you not to lay a hand on your neighbor's TV set or antenna in any way. If the set subsequently develops any other kind of trouble, you may be accused of causing them.

CHAPTER 19

ACCESSORIES THAT CAN HELP

CB-land is flooded with accessories pouring out of factories faster than transceivers. Many are enormously useful for simplifying your CB life. Some have already been discussed.

Here are guidelines to the selection and use of various types of accessories:

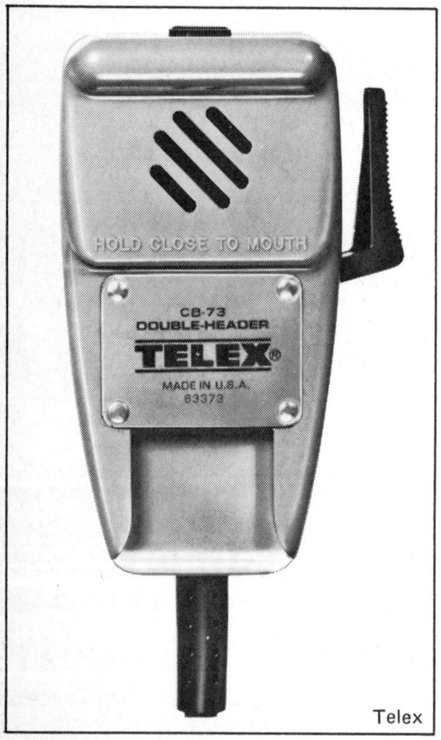

Telex

Noise-cancelling microphones **reduce wind and background noise. Useful in trucks, tractors and at construction sites. It's a good idea to get one that doubles as a power mike for more conventional operating environments.**

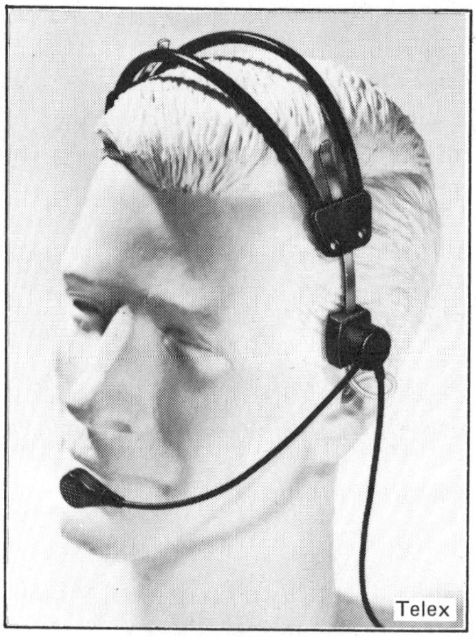

Telex

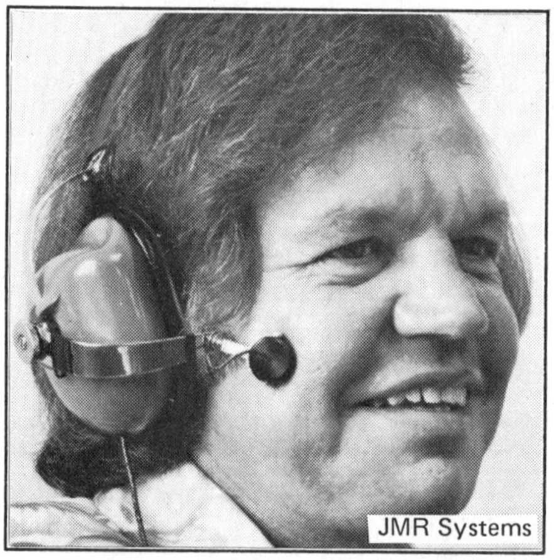

JMR Systems

Headsets **leave your hands free to work, drive or fly. Look for lightest weight to avoid discomfort. You need only one earphone. Lip mikes cancel most noise. A cheek contact mike is even better for severely noisy conditions. You'll need a push-to-talk switch. Attach it to your clothing with a clip or to a foot-operated pedal located away from vehicle pedals.**

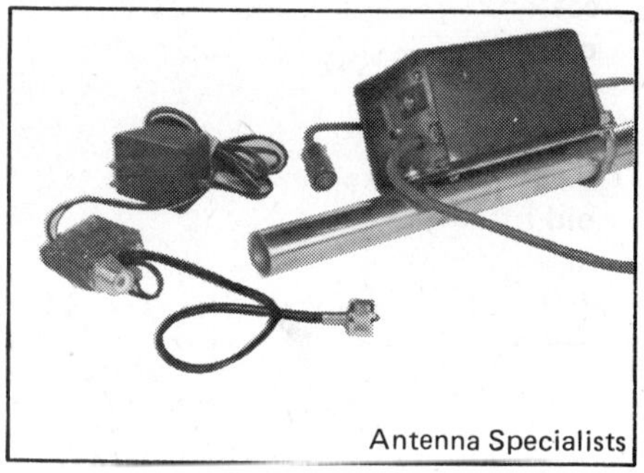

Antenna Specialists

Telco

Boosters **are used on mobile or base stations to amplify received signals. Be sure it is made for CB frequencies and has a circuit that senses when you are transmitting so it can disconnect the amplifier. Most boosters are installed near the transceiver and have about 15 to 30 dB gain. A gain control is useful to prevent receiver overloading by strong signals. Some units are mounted up on a base-station mast where they amplify more signal and less noise. Many CB sets have such a low noise in the input stages that a booster preamplifier merely adds its own circuit noise and makes the situation worse. Buy one on a money-back guarantee that it will make reception better. Recheck SWR after installing.**

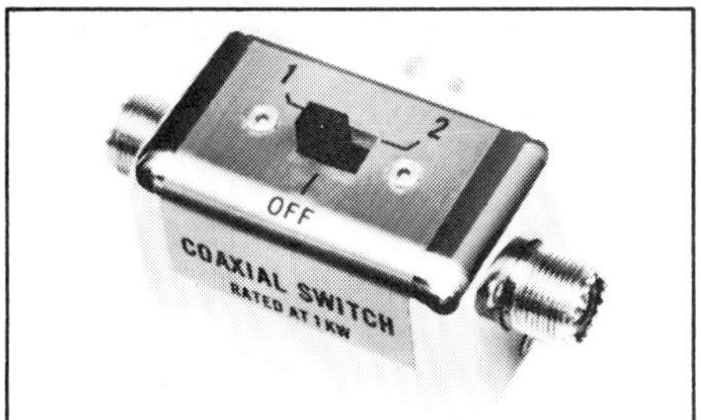

Gold Line

Coaxial switches **make it simple to switch antennas. When you graduate to a beam, keep your omni for communication with randomly located stations. Connect your transmitter to either antenna by a coax switch at the transmitter output. More expensive switches have lower losses. A switch that grounds the unused antenna will reduce noise. Unless the coax switch is a make-before-break type, avoid making changeover with the transmitter keyed or arcing will burn out switch and possibly damage transmitter.**

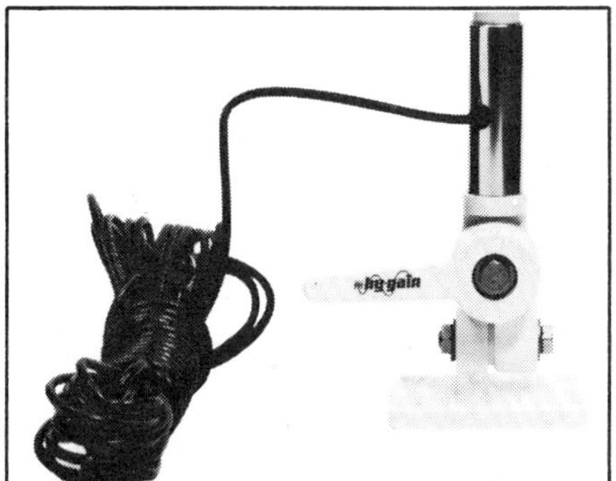

Hy-Gain

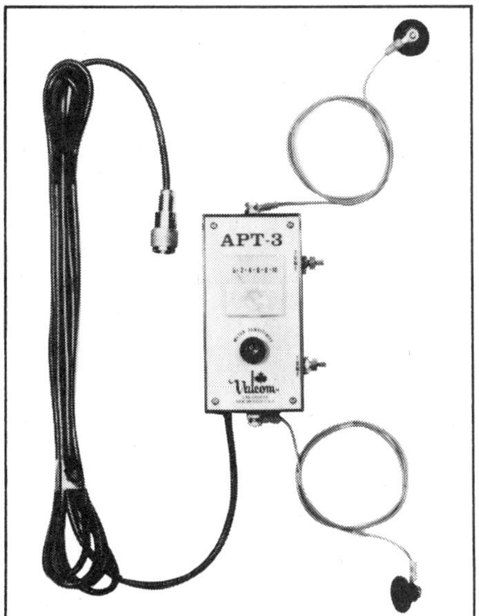

Valcom

Apartment house antennas **are for CBers who can't put anything on the roof. Try a half-wave dipole you can mount on your window sill or an antenna you can attach to window glass with suction cups.**

GC Electronics

Co-phasers **connect a pair of antennas to the same transceiver for improved radiation pattern. If you already have one antenna, you can add a second by means of a coaxial cable co-phase harness kit. Or you can install a co-phasing circuit in a little black box. Connect the box to the CB set and feed both antennas out of the box.**

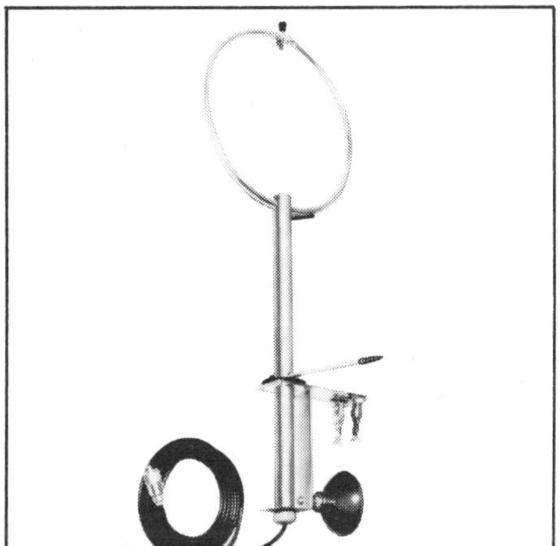
Gold Line

Direction-finding antennas **help track CB stations involved in emergencies. CB clubs use the pointing feature to track channel hogs or noisy power lines. CB jamborees hold treasure hunts with them. To each his own. I'm joining Omnis Anonymous.**

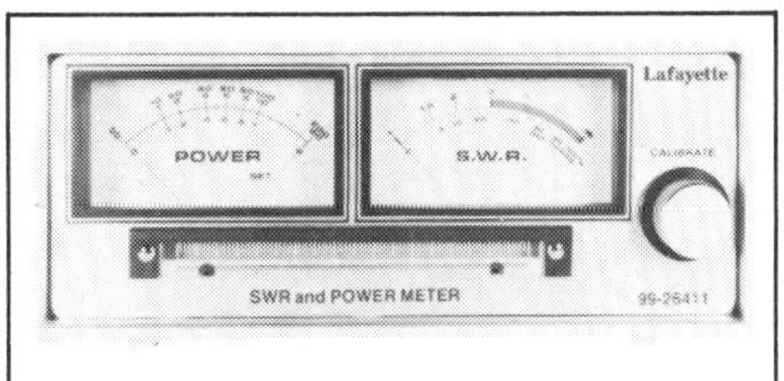

Lafayette

SWR meters **measure standing wave ratio at the point where they are connected between transceiver output and antenna. Use shortest possible coax lead between accessory SWR meter and set. Combination SWR/Power meters measure SWR and also *relative* power output of your set.**

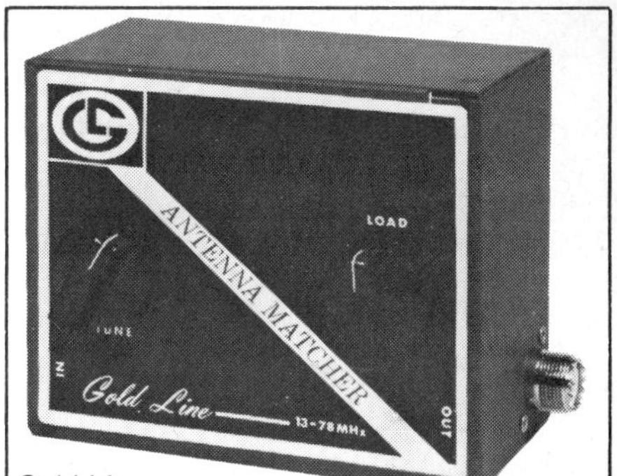

Gold Line

Antenna matchers **may help if you can't get SWR down to 1.5:1 or lower. Put one at the transceiver end of the transmission line, adjust for lowest SWR. Mount a "matchbox" out of sight of knob twiddlers. Check SWR periodically and adjust matcher as necessary.**

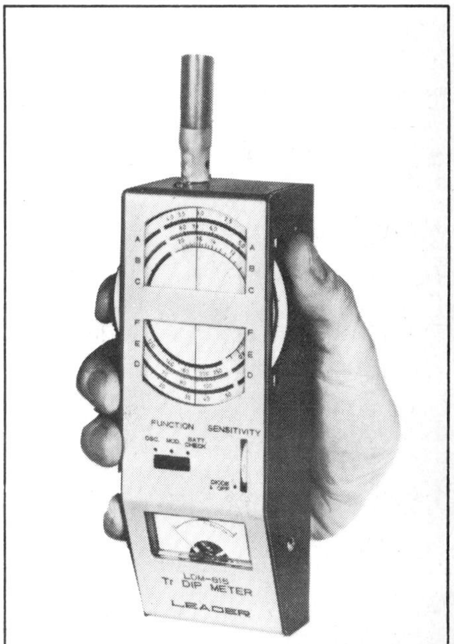

Leader

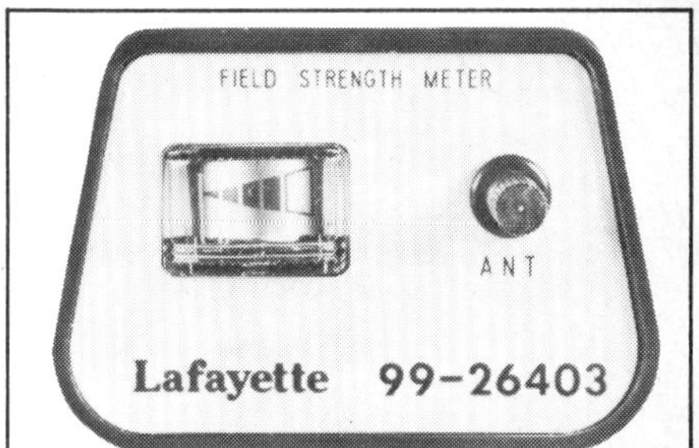

Lafayette

Field strength meters **measure rf power in the vicinity of an antenna so you can tell how your antenna is doing. Measure rf field strength in a circle around the antenna to plot the actual radiation pattern.**

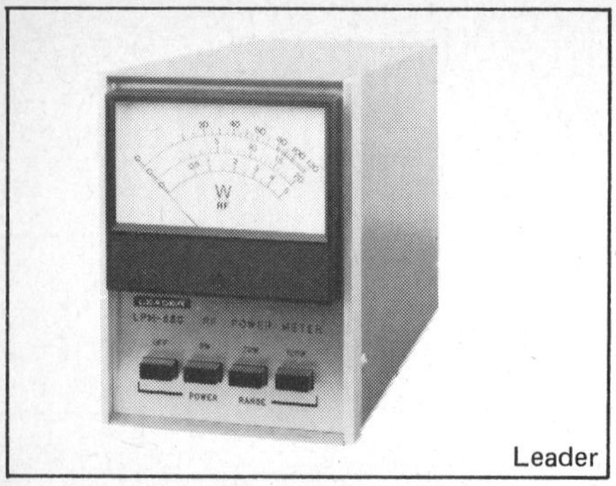

Leader

Gold Line

Power meters **measure *actual* transmitter rf output in watts. You don't need to read more than 5 watts so meters with higher ranges aren't necessary for CB. Check the meter's accuracy specification—usually a percentage of the full-scale reading. ±5% is better than ±10%, and costs more too. For most CBers, ±10% is fine.**

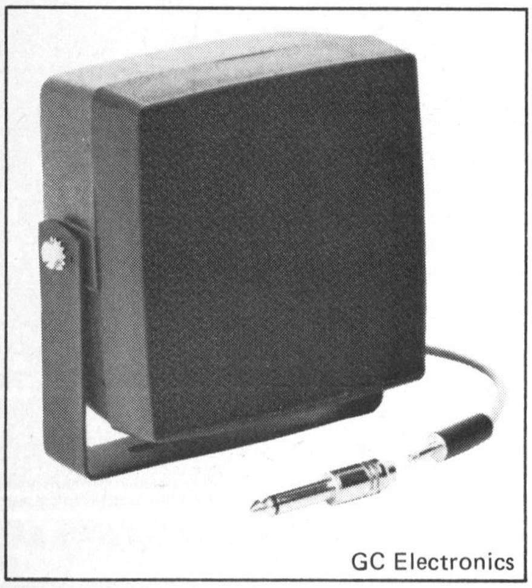

GC Electronics

Extension speakers **improve intelligibility when located and aimed better than the CB set's internal speaker. For CB use, the speaker should be restricted to voice frequencies—about 300 to 3000 Hertz. Don't use a hi-fi speaker. Ask for a *communications* speaker unit.**

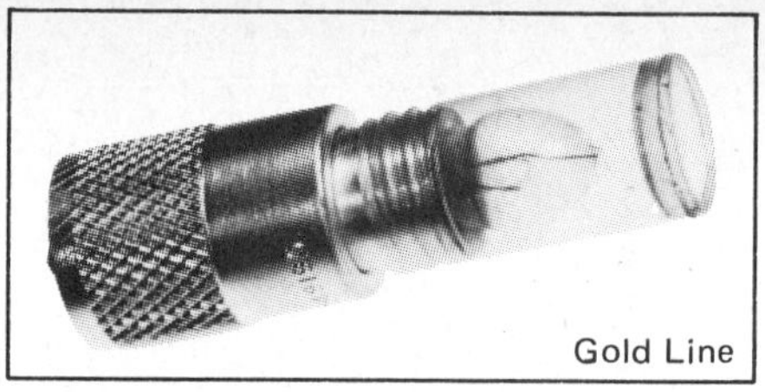

Gold Line

Dummy loads **let you test your transmitter without radiating interference. Use one that is a 50-ohm match so your power meter will read true rf output from the transceiver. If you don't have a power meter, get the type of dummy load that lights up to indicate power output and flickers to show modulation.**

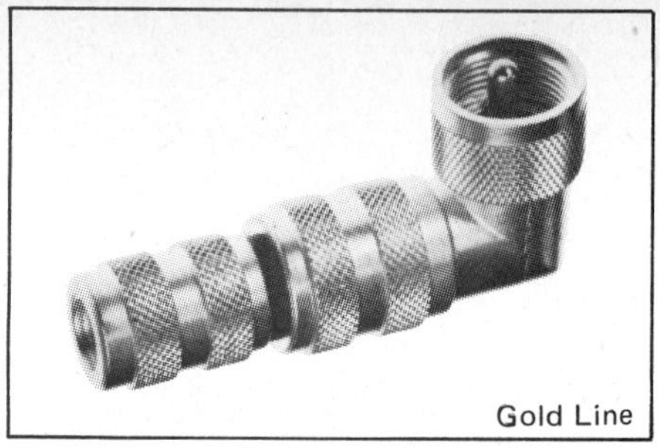

Gold Line

Connectors and adapters **are made for every cable or wire situation you'll encounter. There are double male and double female adapters for coax; splicing connectors for coax; adapters for auto radio antenna leads; clips and connectors for tapping power leads; and anything else you ever need. A good electronics catalog will show instant solutions to your connection problem.**

Communications Power, Inc.

Speech compressors **increase the average speech level to assure full modulation. Plug mike into compressor and compressor into CB mike jack. Supply power separately to compressor—12 V DC or 117 V AC. Get one that works on both power sources and use it mobile or base. A gain control and compression-meter on the unit will help you set it correctly to avoid overmodulation. Some compressors don't have controls or indicators.**

Cepco, Inc.

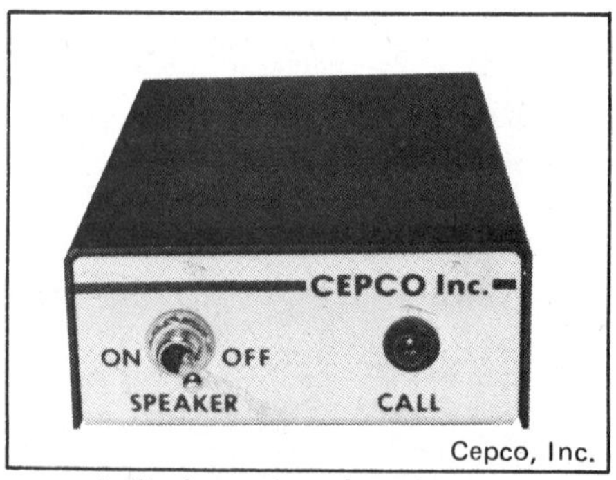

Cepco, Inc.

Tone alerts **receive or send a coded signal between similarly-equipped stations so you don't have to listen to noise and channel chatter while waiting for a call. I like the type that turns on your speaker when a call is received; then turns it off if you don't answer within a specified time and leaves a call light on instead. Look for a unit with a tone code that can't be triggered falsely by noise.**

Able Mfg.

Carry-cases with power supply give versatility to your mobile rig while you're on the road. Carry your CB in from your vehicle, plug into wall outlet in your motel room, and keep on truckin! Don't confuse these carry-cases with a regular base-station adapter that you leave in home or office.

Breaker

Burglar alarms are merely a deterrent. Amateur CB bandits will run but pros disconnect the vehicle battery and work in peace. Choose an alarm with a drop-out relay. Alarm goes off if the ground connection to antenna or CB is interrupted, a switch operated, or other trip devices. The drop-out relay keeps the alarm sounding even if thief restores the broken connection. Hide the alarm switch in the trunk, not under the dash where a wise thief can trace the cables to it.

GC Electronics

PA speakers work on the public-address output of a transceiver. Get a *weather-proof* speaker, not just weather-resistant, rugged enough to mount under a vehicle hood. Mobile PA systems may violate local laws. Check first.

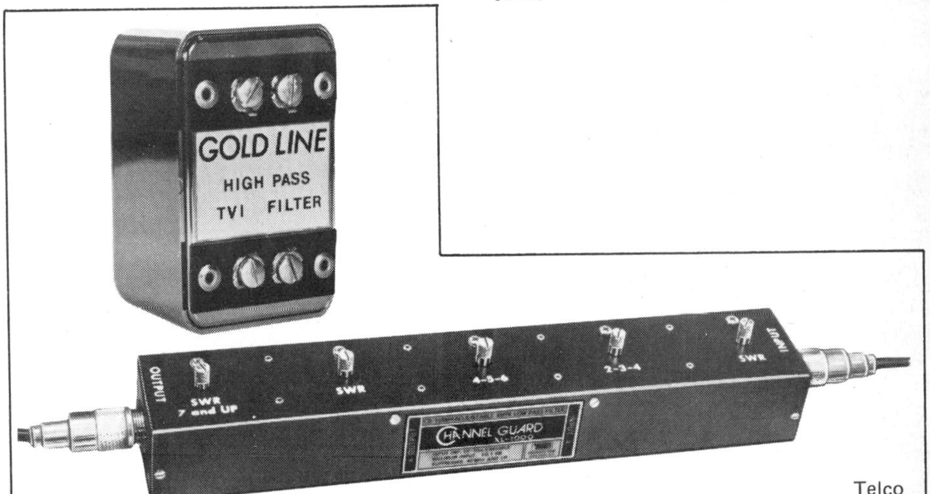

Telco

Dynamic duo solves TV interference problems. If your TV suffers from TVI, block interference by installing a *high-pass* filter such as this Gold Line in your TV lead-in. If your CB is annoying the neighbors, install a low-pass TVI filter between CB and antenna. Use a fixed filter or try this tunable *Channel Guard* from Telco.

CHAPTER 20

THE CBers SCENE

INDEX FOR THIS CHAPTER

CB SLANG AND DEFINITIONS

AC—Alternating current
AC/DC—Alternating current or direct current
Ace—An important CB'er
Advertising—A marked police car with flashers going
Align—Adjust or tune a circuit
Amplifier—Device or circuit to increase signal strength
Amplitude Modulation (AM)—Modulating a radio signal by varying its strength
Anchored Modulator—Base station operator
Antenna—Metallic rod that radiates or receives radio frequency signals
Apple—CB addict
Automatic Gain Control—A circuit which automatically holds a signal to a desired level
Automatic Noise Limiter (ANL)—Circuit which automatically reduces noise pulses
BTO—Big Time Operator
Back—Over. Go ahead
Back Door—Last vehicle in a string of three or more—all in contact with each other
Back Down—Driver slower
Back Out—Stop transmitting
Background Noise—Noise heard with the desired signal
Bad Scene—A crowded channel
Ballet Dancer—Swaying antenna
Band—A range of frequencies
Bao Bab—Wide load
Barefoot—Illegal CB transmission; operating without an amplifier to boost transmitter power higher than allowed by law
Base Station—CB set operated from a fixed location
Beam Antenna—A directional antenna that radiates or receives more energy in one direction than in others
Bear—Policeman—named for the "Smokey the Bear" hats often worn
Bear Cave—Police station on highway
Bear's Den—Any police station
Bear Report—Where are the bears?
Beat the Bushes—"Front door" (lead vehicle) looking for Smokey
Be-Bop—Radio control signals
Beer Tone—Intermittent tone signal
Big A—Atlanta
Big Daddy—The FCC
Big Orange—Snyder truck
Big Switch—Turn off CB set by pulling the big switch
Big 10-4—Acknowledged with enthusiasm
Bikini State—Florida
Bleeding—Interference from adjacent CB channels
Blessed Event—A new CB rig
Blinkin Winkin—School bus
Blood Box—Ambulance
Blow the Doors Off—Pass
Boast Toastie—A CB expert
Bodacious—Good signal
Bone Box—Ambulance
Bootlegger—Unlicensed CB'er
Boulevard—Interstate highway
Bounce Around—Next trip through
Bouncing Cardboard—Driver's license
Boy Scouts—The State Police
Break—I'd like to interrupt
Break One-Nine—I want to talk "on Channel 19" or whatever number is given after "break"
Breaker—CB'er who breaks in to ask to use a channel
Bubble Trouble—Tire problem
Bubblegum Machine – Revolving flashers used on police and emergency vehicles
Bug Out—To leave a channel
Bushel—1,000 pounds; a 20-ton load would be 40 bushels
Call Sign—Assigned station identification
Camera—Police radar unit
Can—Shell of a CB set
Carrier—Radio wave whose constant amplitude or frequency may be varied (modulated) to carry intelligence like a voice signal
Cartel—A group hogging a channel
Catch—Talk to
Cell Block—Location of the base station
Chain Gang—Members of a CB club
Channel—A specific frequency assigned a designated channel number in a band
Charlie—The FCC
Check the Seatcover—Watch for female occupants of vehicle
Chicken Choker—Poultry truck
Chicken Coop—Weigh station for trucks
Choo Choo Town—Chattanooga
Chopped Top—A short antenna
Chrome Dome—Mobile unit with a roof antenna
Circus Wagon—Monofort truck
Citizens Band—Band of radio frequencies allocated to the Citizens Radio Service
Clarifier—Fine tuning control
Class D—Citizens Radio Service which uses frequencies between 26.965 and 27.255 MHz
Clean—No Smokeys around
Clean Cut—An unmodified rig
Clear—Out, final transmission
Coax—Coaxial cable in which one conductor completely surrounds the other and acts as a shield
Coffee Break—CB social get-together, often held at a diner
Come again—Repeat your last transmission
Come on—Over. Go ahead and transmit
Comeback—Return call
Comic Book—Truckdrivers' log
Container—Chassis and shell of a CB rig
Convoy—Group of vehicles traveling together
Cornflake Machine – Consolidated Freight
County Mounty—County sheriff or highway patrol
Covered Up—Interferred with
Cradle Baby—CB'er who is afraid to ask someone to stand by
Crystal—Device whose characteristic natural vibrations maintain a transmitter (or receiver) on its assigned frequency
Cub Scouts—Sheriff's men
Curly Locks—Coils in a CB rig
Cut the Coax—Turn off CB set
Cut Out—To leave a channel
DC—Direct current
DOC—Department of Communications which regulates radio communications in Canada
Daddy-O—The FCC
Darktime—Night
Decibel (dB)—A mathematical term for stating a ratio
Delta Tune—A fine tuning control to enable reception of an off-frequency station
Despair Box—Box where spare CB components are kept
Diesel Digit—Channel 19
The Dirty Side—New York and New Jersey
Do You Copy—Do you hear me?
Don't Tense—Take it easy
Double-Conversion—A superheterodyne receiver circuit which converts the incoming signal frequency twice for greater selectivity
Double Nickel—55 MPH, legal speed for most highways
Draggin' Wagon—Wrecker
Drop the Hammer—Accelerate to top speed
Ears—CB set
Earth Ground—A metallic connection with the earth
Eatem Up—Roadside restaurant
Effective Radiated Power—Antenna input power in watts multiplied by antenna gain
Eighteen Wheeler—Any semi-tractor truck with any number of wheels
Eights (eighty-eights) – Goodbye (from amateur radio 88's meaning "love and kisses")
8's and Other Good Numbers—Signing off with all kinds of best wishes
Eyeball—Face-to-face meeting
FCC—Abbreviation for Federal Communications Commission, regulates all electrical communications systems originating in the U.S.
FET—Field effect transistor
Fat Load—Overload

Feds—Government inspectors
Feed the Bears—Pay a fine to police for speeding
Field Strength Meter—A measuring instrument for determining energy radiated from an antenna
50 Dollar Lane—The inside or passing lane on multi-lane highways
Filter—A selective network allowing only certain frequencies or signals to pass
Final—Last transmission
Fingers—A channel-hopping CB'er
Five-Five—55, the legal limit on most highways
Flag Waver—Highway worker
Flag Waver Taxi—Highway truck
Flappers—Ears
Flip Flop—Return trip
Fog Lifter—Interesting CB'er
Foot in the Carburetor—Police in pursuit
Foot Warmer—A linear amplifier (illegal)
Four Wheeler—Automobile
Fox Charlie Charlie—The FCC
Frequency Synthesizer—Circuit using one or a few crystals to generate many frequencies
Friendly Candy Company – The FCC
Front Door—First vehicle of convoy in radio contact
Fuse—A protective device against excessive current flow
Fuse Block—An insulating base carrying fuses
GRS—General Radio Service, the Canadian counterpart of Citizens Radio Service
Gain—Any increase in power
Gain Control—A device for varying the power of a circuit
Getting Out—Being heard
Glory Card—Class D License
Goldie Locks—Mobile business women
Gone—Final transmission
Good Buddy—Any other CB'er
Goodies—CB accessories
Goon Squad—Channel hoggers
Grass—Side of the road or median strip
Green Stamps—Dollars; toll roads
Groundplane Antenna—A nondirectional vertical antenna with metal radials which create an artifical ground
Ground Wave—A radio wave which travels along the earth's surface, as opposed to waves reflected from the upper atmosphere (sky waves)
Grounded—Connected to earth or to some conductor which takes the place of earth
Halloween Machine—Cooper-Jarrett truck
Ham—Slang for amateur radio operator
Hammer—Accelerator
Handle—A nickname used by a CB'er
Hertz (Hz)—Cycles per second
Holler—Call
Home-Twenty—Home location
Hot Pants—Smoke or fire
Hound Men—Policemen looking for CB'ers
How About?—Phrase used to call someone—"How boutcha Happy Booker?"
Hung Up—CB'er who can't leave set
Idiot Box—TV set
Ignition Noise—Interference produced by sparks or other ignition discharges
Impedance—Opposition to the flow of alternating current
Impedance Match—Condition in which the impedance of components or circuits are matched for maximum transfer of energy
In the Grass—Parked or pulled over on the median strip
Indian—Neighbor who has TVI from you
Integrated Circuit – Subminiature combination of interconnected circuit elements
Interference—Any disturbance in the reception of desired signals
Jack—A form of socket
Jamboree—Gathering of CB'ers sponsored by a CB club for equipment displays and entertainment
Jaw Jacking—Conversation
KHz—Kilohertz. 1000 cycles per second
Keep Your Nose Between the Ditches and Smokey Out of Your Britches—Drive safely and look out for speed traps and speeding fines
Kenosha Cadillac—Any car made by AMC
Keyboard—Transceiver controls
Kodiak with a Kodak—Police (bear) with camera
LED—Light emitting diode
Land Line—The telephone
Lay an Eye On—See
Lead—A wire to or from a circuit
Let the Channel Roll—Let others break in and use the channel
Limp Line—Rigging loose—shifted load
Line Filter—Device in power line to block noise
Linear Amplifier—Extra power amplifier used to increase CB transmitter output (strictly illegal)
Loading Coil—Coil of wire used to electrically lengthen the antenna while maintaining its same physical dimensions
Local Yokel—Small town police officer
MHz—Abbreviation for megahertz One million cycles per second
Mail—Overheard conversation
Man in Blue—Policeman
Man in Slicker—Fireman
Man in White—Doctor
Mercy—Used in place of swearing
Microphone—Device for converting sound into electrical signals
Microvolt (uV)—One-millionth of one volt
Mike—Slang for microphone
Mile Markers—Small mile post signs along highways
Mobile—CB set mounted in a vehicle
Mobile Eyeball—Check operator's rig while moving
Modulate—Slang for "to talk with." Technically it means putting voice information on a radio carrier signal
Modulation Indicator- Light or meter giving visible indication of modulation
Monster Lane—Inside lane
Moth Ball—Annual CB Convention
Move—In motion
Music Town—Nashville
Negative—No
Negative Contact—Station being called failed to respond
Negative Copy—Did not hear response by the station called
Negative Ground—The negative battery terminal is grounded to the vehicle frame.

Negatory—No, negative
Nickel-Cadmium Cell (Nicad)—A rechargeable battery cell
Noise Blanker—Circuit which silences a receiver during brief noise bursts like ignition popping
Omnidirectional Antenna—Antenna which radiates or receives signals equally well in all directions
On the Peg—Legal limit
On the Side—Standing by (listening). Also a vehicle may be parked on the side of the road
One Eyed Monster—TV set
One Time—A short contact
Other Half—Wife (usually) or husband
Over Shoulder—Behind
PTT—Press-to-talk or push-to-talk; a switch on a microphone that turns the transmitter on
Pack Set—A hand-carried portable transceiver
Panic in the Streets—Area being monitored by FCC
Part 95 Rules—FCC Rules and Regulations governing the Citizens Radio Service
Pavement Princess—Roadway happy hooker
Peak Envelope Power (p.e.p.)—Maximum power generated by an SSB transmitter when fully modulated
Peanut Butter in Ears—Not listening to CB set
Phase Lock Loop—Circuit for maximum frequency stability
Pickem-Up—Light truck; pickup truck
Picture Box—Radar set up
Picture Taker—Radar speed trap
Plain Wrapper—An unmarked police car; color may be indicated as in "plain green wrapper"
Play Dead—Stand by
Politz-eye—The police
Portable Chicken Coop—A portable truck weighing station
Positive Ground—The positive battery terminal is grounded to the vehicle frame
Pounds—Number on S-meter (S-3 is three pounds)
Power Output—Power delivered by a transmitter to the antenna, expressed in watts
Power Supply—Electrical power for a unit or circuit
Pregnant Roller Skate—Volkswagen
Prescription—FCC rules
Pull the Big One—Signing off for good
Puttin' the Hammer Down—Pushing on the accelerator; speeding
Putting On—Signal strength put out
QSL Card—Post card confirming a radio communication contact
Q-Bird—Intermittent tone generator
Q-Signals—International code of three letter abbreviations starting with the letter Q.
rf—Radio frequency
Radio Check—Reception report
Ratchet Jaw—Non-stop talker
Read—Hear
Reefer—Refrigerated trailer
Regulated Power Supply—Supply in which output voltage remains constant as the load or source is varied
Restem Up—Roadside rest area
Rig—CB set
Riot Squad—Neighbors who have TVI
Rockin' Chair—Middle vehicles in convoy
Roger—I acknowledge
Roger Rollerskate—Car going more than 20 MPH over the limit
Roller Skate—Small car
S/RF Meter—A meter that indicates the relative strength of a received signal or the relative transmitter of power output
SWR—Standing wave ratio, the lower the better for an efficient transfer of rf power between transceiver and antenna
Savages—Cb'ers who hog the channel
Scanner—Sequential samples of several channels
Seat Covers—Occupants of passenger car, usually exposed female legs in passing car
Selectivity—Ability of a receiver to reject frequencies other than the specific one to which it is tuned
Selectivity Filter—An integral component or circuit used within a CB receiver to improve selectivity
Sensitivity—The minimum input signal required by a receiver to produce a specified output at a

specified signal-to-noise ratio
Set of Dials—CB rig
Seventy-Three—Best regards. And hope to catch you later
Shake The Trees and Rake The Leaves—The first vehicle in a convoy watching for speed traps, the last vehicle looking for anything moving in from the rear
Shanty Shaker—Mobile home driver
Shout—Call
Signal-to-Noise Ratio—Magnitude of a signal compared to noise received or present in the circuit
Single-Conversion Receiver—Superheterodyne circuit in which the intercepted radio signal is converted once to another frequency for amplification
Single Sideband (SSB)—An AM radio transmission technique in which only one sideband is transmitted and the other sideband and the carrier are suppressed
Six Wheeler—Passenger car pulling a trailer
Skip—Stations from great distances heard by means of ionospheric reflection of their signals
Sky Wave—A radio wave that has been reflected from one of the layers of the ionosphere
Slave Drivers—CB'ers who take control of a channel
Slider—An illegal VFO (variable frequency oscillator)
Smokey—police
Smokey Dozing—Police, in stopped car
Smokey with Ears—Police with CB
Smokey on Four Legs—Mounted police (used in New York City and Chicago)
Smokey on the Ground—Trooper out of patrol car
Smokey on Rubber—Trooper in a patrol car
Snooperscope—An illegally high antenna
Souped Up—A rig running illegally high power
Squelch—A circuit in a receiver which shuts off the loudspeaker unless a strong signal is received, thus eliminating annoying background noise
Standby Current—Current consumed by a transceiver turned on but not receiving or transmitting
Static—Noise from man-made or atmospheric electrical disturbances
Static Discharge Unit—A grounded protective device that discharges excessive static electricity on the antenna
Stroller—CB'er with a walkie-talkie
Struggle—Trying to "break" a channel
Sucker—CB rig on the service bench
Surface Wave—A ground wave which travels along the surface of the earth
Swindle Sheets—Trucker's log book
Talk Power—A high modulation level
Tear Jerker—CB'er who always cries the blues
Ten-Code—Abbreviations used by CB'ers and other two-way radio users to minimize use of air time.
Ten Roger—I acknowledge
Tennessee Valley Indians—TV Interference
Thermos Bottle—Tanker truck
Thin—Very weak signal
Thin Man—CB'er with a weak carrier
13-Code—A zany code of put-downs for fellow CB'ers
Thirty Three—10-33. This is an emergency.
Thread—Wires in a CB rig
Threes—Seventy-three, best regards
Throwing—Transmitting
Ticker Tape—FCC rules
Tijuana Taxi—A marked police car
Train Station—Traffic court that fines everybody
Transceiver—A combination radio transmitter and receiver
Transmit Current—The amount of current drawn by a transceiver when transmitting
Trip—Strong signal
Turkey Call—Intermittent tone generator
Twenty—10-20, location
Two-Way Radar—Radar used from moving police car
Two Wheeler—Motorbike; motorcycle
Ungowa Bwana—O.K.
Walked All Over—Overpowered by a stronger signal
Walkie-Talkie—A generic term for a hand-held transceiver
Wall-to-Wall—Loud and clear
Wall-to-Wall Bears—High concentration of police with strict enforcement
Wallace Lane—Middle lane of 3-lane highway
Wallpaper—Postcard acknowledging a two-way contact
Warden—The wife, The FCC
We Gone—Stopping our sending; will listen
Wind Jammer—A long winded CB'er
Wrapped Loaf—CB rig in its original carton
XYL—Ex-young lady, wife
X-Ray Machine—Radar speed meter
YL—Young lady
Youngville—Young children using the channel

13-Codes

Nutty, zany put-downs in code are raging across CB-land. This collection is from S-9 Magazine, Port Washington, N.Y., which is almost as old as CB.

GOOD COPY

13-1 All units can copy you and think you're an idiot.
13-2 Yes, I copy you, but I'm ignoring you.
13-3 You're beautiful when you're angry.
13-4 Sorry 'bout that, Big Fella.
13-5 Same to you, Mack!
13-6 OK, so I goofed—none of us are human.
13-7 If you can't copy me it must be your fault because I'm running 3,000 watts.

OPERATORS

13-20 Is your mike clinking or are your uppers loose again?
13-21 Good grief—are you being paid by the word?
13-22 Lady, is that your voice or did you install a steam whistle?
13-23 If you had spoken for another 30 seconds you would have been eligible for a Broadcast Station license.
13-24 Y'know, you made more sense last time when you were smashed.
13-25 Some of the local operators and I have chipped in to purchase your rig from you. Have you considered stamp collecting?
13-26 Next time you eat garlic would you talk farther from the mike?

TECHNICAL

13-40 Your signal sounds great, now shut off the set and give me a land line so I can find out what you want!
13-41 Either my receiver is out of alignment or you're on channel 28.
13-42 Either my speaker cone is ripped or you better try it again when you sober up a little.
13-43 That was a beautiful 10 try it with your mike connected.
13-44 I love the way your new rig sounds, now I know why the manufacturer discontinued that model so fast.
13-45 Your transmitter must have a short circuit because there's smoke coming from my loudspeaker.
13-46 That's a new antenna? I could get a better signal out of a 6-inch piece of damp string!
13-47 What a fantastic signal—give me a few minutes to bring the mobile unit to your driveway so I can copy your message.

SIDEBANDERS

13-50 Say, fella, can you slide that thing down 250 KHz?
13-51 You've tried the upper sideband, you've tried the lower side, you've even tried both sides—hope you're satisfied. Now will you go QRT so we can use the center slot?
13-52 Only good thing about hearing you on Single Sideband is that with only 1 sideband you're only half as offensive as you were on AM!
13-53 Attention—AM station on center slot: Just because they won't talk to you on your own channels what makes you think we'll talk to you here?

United States of America
Federal Communications Commission

Form Approved
GAO No. B-180227(R01 02)

FCC FORM 505

April 1976

APPLICATION FOR CLASS C OR D STATION LICENSE IN THE CITIZENS RADIO SERVICE

INSTRUCTIONS

A. Print clearly in capital letters or use a typewriter. Put one letter or number per box. Skip a box where a space would normally appear.

B. Enclose appropriate fee with application. Make check or money order payable to Federal Communications Commission. DO NOT SEND CASH. No fee is required of governmental entities. For additional fee details see FCC Form 76-K, or Subpart G of Part 1 of the FCC Rules and Regulations, or you may call any FCC Field Office.

C. Mail application to Federal Communications Commission, P.O. Box 1010, Gettysburg, Pa. 17325

NOTICE TO INDIVIDUALS REQUIRED BY PRIVACY ACT OF 1974

Sections 301, 303 and 308 of the Communications Act of 1934 and any amendments thereto (licensing powers) authorize the FCC to request the information on this application. The purpose of the information is to determine your eligibility for a license. The information will be used by FCC staff to evaluate the application, to determine station location, to provide information for enforcement and rulemaking proceedings and to maintain a current inventory of licensees. No license can be granted unless all information requested is provided.

1. Complete **ONLY** if license is for an Individual or Individual Doing Business AS

FIRST NAME　INIT　LAST NAME

2. DATE OF BIRTH

MONTH　DAY　YEAR

3. Complete **ONLY** if license is for a business, an organization, or Individual Doing Business AS

NAME OF BUSINESS OR ORGANIZATION

4. Mailing Address

4A. NUMBER AND STREET

4B. CITY　**4C.** STATE　**4D.** ZIP CODE

(See reverse side of this form, for filling in Item 4C.)

5. If you gave a P.O. Box No., RFD No., or General Delivery in Item 4A, you must also answer items 5A, 5B, and 5C.

5A. *NUMBER AND STREET WHERE YOU OR YOUR PRINCIPLE STATION CAN BE FOUND (If your location can not be described by number and street, give other description, such as, on RT. 2, 3 mi., north of York.)*

5B. CITY　**5C.** STATE

(See reverse side of this form for filling in Item 5C.)

6. *Type of Applicant* **(Check Only One Box)**

- ☐ *Individual*
- ☐ *Association*
- ☐ *Corporation*
- ☐ *Business Partnership*
- ☐ *Governmental Entity*
- ☐ *Sole Proprietor or Individual/Doing Business As*
- ☐ *Other (Specify)* ______

7. *This application is for*

- ☐ *New License*
- ☐ *Renewal*
- ☐ *Increase in Number of Transmitters*

IMPORTANT
Give Official FCC Call Sign

8. *This application is for* **(Check Only One Box)**

- ☐ *Class C Station License (NON-VOICE—REMOTE CONTROL OF MODELS)*
- ☐ *Class D Station License (VOICE)*

9. *Indicate number of transmitters applicant will operate during the five year license period* **(Check Only One Box)**

- ☐ *1 to 5*
- ☐ *6 to 15*
- ☐ *16 or more (Specify No. and attach statement justifying need.)*

10. CERTIFICATION I certify that:

- The applicant is not a foreign government or a representative thereof.
- The applicant has or has ordered a current copy of Part 95 of the Commission's rules governing the Citizens Radio Service. See reverse side for ordering information.
- The applicant will operate his transmitter in full compliance with the applicable law and current rules of the FCC and that his station will not be used for any purpose contrary to Federal, State, or local law or with greater power than authorized.
- The applicant waives any claim against the regulatory power of the United States relative to the use of a particular frequency or the use of the medium of transmission of radio waves because of any such previous use, whether licensed or unlicensed.

THIS APPLICATION WILL NOT BE PROCESSED UNLESS SIGNED AND DATED.

WILLFUL FALSE STATEMENTS MADE ON THIS FORM OR ATTACHMENTS ARE PUNISHABLE BY FINE AND IMPRISONMENT. U.S. CODE, TITLE 18, SECTION 1001.

11. SIGNATURE	12. DATE

Signature of: Individual applicant, partner, or authorized person on behalf of a governmental entity, or an officer of a corporation or association

Sometimes it becomes necessary to return an application. By putting your name and address in the area below, you will enable us to return quickly any application which needs correction or clarification: 1) Put your name on the first line in regular order (for example, Joe Doe); 2) Put your number and street on the second line; 3) Put your city, state, and zip code on the third line.

If necessary, use abbreviations to stay within the guidemarks provided.

ORDERING PART 95 OF THE FCC RULES AND REGULATIONS

The 1976 edition of Part 95, Citizens Radio Service, is for sale at $1.50 per copy. Order from: Superintendent of Documents, U.S. Government Printing Office, Washington, D. C. 20402. Order by Stock Number: 004-000-00324-1 PLEASE NOTE: Part 95, Citizens Radio Service, is revised annually. The prices of government publications are subject to change without notice. Therefore, the price charged for 1977 and later editions may differ from that shown for the 1976 editions.

For Items 4C and 5C: use the two-letter state abbreviations below for filling in these items.

Alabama	AL	Kentucky	KY	Ohio	OH
Alaska	AK	Louisiana	LA	Oklahoma	OK
Arizona	AZ	Maine	ME	Oregon	OR
Arkansas	AR	Maryland	MD	Pennsylvania	PA
California	CA	Massachusetts	MA	Puerto Rico	PR
Colorado	CO	Michigan	MI	Rhode Island	RI
Connecticut	CT	Minnesota	MN	South Carolina	SC
Delaware	DE	Mississippi	MS	South Dakota	SD
District of Columbia	DC	Missouri	MO	Tennessee	TN
Florida	FL	Montana	MT	Texas	TX
Georgia	GA	Nebraska	NB	Utah	UT
Guam	GU	Nevada	NV	Vermont	VT
Hawaii	HI	New Hampshire	NH	Virginia	VA
Idaho	ID	New Jersey	NJ	Virgin Islands	VI
Illinois	IL	New Mexico	NM	Washington	WA
Indiana	IN	New York	NY	West Virginia	WV
Iowa	IA	North Carolina	NC	Wisconsin	WI
Kansas	KS	North Dakota	ND	Wyoming	WY

TEMPORARY PERMIT

(Do not use for license renewals)

If you will use a Temporary Permit, FCC Form 555-B, complete the following:

I hereby request a temporary permit pursuant to my certification on FCC Form 555-B.
My temporary call sign will be [K] [ ] [ ]   [ ] [ ] [ ] [ ] [ ]

☆U.S.Government Printing Office: 1976—205-411/16

United States of America
Federal Communications Commission

FCC FORM 555–B
April 1976

Temporary Permit

Class D Citizens Radio Station

1 Instructions

- Use this form only if you want a temporary permit while your regular application , FCC Form 505, is being processed by the FCC.
- Do not use this form if you already have a Class D license.
- Do not use this form when renewing your Class D license.

2 Certification

Read, Fill In Blanks, and Sign

I Hereby Certify:

☐ I am at least 18 years of age.

☐ I am not a representative of a foreign government.

☐ I have applied for a Class D Citizens Radio Station License by mailing a completed Form 505 and $4.00 filing fee to the Federal Communications Commission, Box 1010, Gettysburg, PA. 17325.

☐ I have not been denied a license or had my license revoked by the FCC.

☐ I am not the subject of any other legal action concerning the operation of a radio station.

____
Name

____
Signature

If you cannot certify to the above, you are not eligible for a temporary permit.

Willful false statements void this permit and are punishable by fine and/or imprisonment.

____
Address

____
Date Form 505 mailed to FCC

3 Temporary Call Sign

- **Complete the blocks as indicated.**

Use this temporary call sign until given a call sign by the Federal Communications Commission.

K	☐	☐	☐☐☐☐☐
	Initial of Applicant's First Name	Initial of Applicant's Last Name	Applicant's Zip Code

4 Limitations

Your authority under this permit is subject to all applicable laws, treaties and regulations and is subject to the right of use or control by the Government of the United States.

This permit is valid for 60 days from the date the Form 505 is mailed to the FCC.

You must have a temporary permit or a license from the FCC to operate your Citizens Band radio transmitter.

Do Not Mail this form, it is your Temporary Permit.

See the reverse side of this form for a summary of operating instructions.

Using Your Citizens Radio Station

(See Part 95 of FCC Rules & Regulations for complete instructions on authorized station use.)

Welcome to the Citizens Radio Service

Citizens Band Radio is a shared communications service with many people using the same frequencies and channels.

The guidelines provided in this form are not intended as a substitute for FCC Rules, but as a general reference to those operating practices and procedures which will benefit you and other users of Citizens Radio.

Your compliance with these guidelines and your consideration for the rights of others in your radio service is necessary if the full potential and enjoyment of Citizens Radio is to be realized.

1 Who May Operate Your Citizens Radio Station?

You, members of your immediate family living with you, and your employees while on the job.

2 How Many Transmitters Does this Permit Authorize?

A maximum of five (5).

3 Can the FCC Inspect My Station?

Your station and station records must be available for inspection by an authorized agent of the FCC.

4 Where Should I Keep This Permit?

Keep it in a safe place. Post photocopies at all fixed station locations. Indicate on photocopies the location of this permit. Attach a card with your name, address and temporary call sign to each transmitter .

5 How Shall I Identify My Station?

Identify transmissions in English with your temporary call sign.

6 How Can I Use My Station?

Use it for private short-distance radio-communications for your personal or business activities. Channel 9 is reserved solely for emergency communications and to assist motorists.

Prohibited Communications Include:

- Activities contrary to law
- Transmitting obscene, indecent or profane messages
- Communicating with non-Class D stations
- Intentional interference to other radio stations
- Transmitting for amusement, entertainment, or over a public address system
- Transmitting false distress messages
- Advertising, selling, or for hire

7 How High Can My Fixed Station Antenna Be?

See Section 95.37 if your antenna will be over 20 feet above ground. Additional information is available in SS Bulletin 100l-h.

8 May Amplifiers Be Used With My Transmitter?

'Linear' amplifiers are absolutely prohibited. 'Power' microphones may require adjustments to your transmitter.

9 Who Can Make Adjustments to My Transmitters?

Adjustments affecting proper operation may be made only by, or under the supervision of a licensed first or second-class radio operator.

FCC Form 410
January 1976

United States of America
FEDERAL COMMUNICATIONS COMMISSION

Form Approved
Budget Bureau No. 52-R0138

PART I - REGISTRATION OF CANADIAN RADIO STATION LICENSEE AND APPLICATION FOR PERMIT TO OPERATE

(See Instructions on the Reverse Side)

.... SERIAL NO.

Name of Canadian licensee	Mailing address (including U. S. address, if any)

The above named person requests registration and permit for operation in the United States of the radio station(s) licensed to him by the Government of Canada as follows:

License No.	Expiration date	Call sign	Authorized communication service, and frequency if not Amateur

Specify below the area of proposed operation in the United States, and the approximate period or periods of time.

Date signed ....

....
(Signature of Applicant Licensee or Common Carrier Subscriber)

FCC Form 410
January 1976

United States of America
FEDERAL COMMUNICATIONS COMMISSION

SERIAL NO. ....

PART II - CERTIFICATE OF REGISTRATION OF CANADIAN RADIO STATION LICENSEE AND PERMIT FOR OPERATION IN THE UNITED STATES

Name of Canadian licensee	Mailing address

The above named person holding a radio station license issued by the Government of Canada as follows:

License No.	Expiration date	Call sign	Authorized communication service, and frequency if not Amateur

has registered with the Federal Communications Commission, and this certificate, together with evidence of applicability of the above described license, shall constitute authority for use of the radio station in the United States under the terms of the treaty between the United States and Canada relating to mutual recognition of certain radio station and operator licenses issued by either country, effective May 15, 1952.

Registration Expires:

FEDERAL
COMMUNICATIONS
COMMISSION

INSTRUCTIONS

1. This form is to be used by the licensee of a Canadian radio station requesting registration to permit operation of land mobile units or an amateur station in the United States. If applying for the Amateur Service, you must enter a U. S. address on the reverse.
2. The applicant is to complete all items on the reverse side and be sure to insert the applicant's name and address on the bottom of this page. Fold on dotted lines, enclose in envelope, and mail to the Federal Communications Commission, Washington, D.C. 20554.
3. Upon approval of the application, the "Certificate of Registration" will be authenticated and returned to the licensee. It may be photo-copied to the extent necessary to provide a copy in each vehicle covered by the license.
4. The following applicable Federal Communications Commission Rules and Regulations should be obtained from the Superintendent of Documents, Government Printing Office, Washington, D. C. 20402. They are available for the amount shown, in United States currency or International money order:

TITLE	ORDER BY VOLUME NO.	PRICE OF VOLUME*
Part 87 - Aviation Services Part 89 - Public Safety Radio Services Part 91 - Industrial Radio Services Part 93 - Land Transportation Radio Services	V	$22.75
Part 97 - Amateur Radio Service	VI	6.70
Part 21 - Domestic Public Radio Services (Other than Maritime Mobile)	VII	9.65

* SOLD IN VOLUMES ONLY - INDIVIDUAL PARTS ARE NOT AVAILABLE. This price includes basic volume plus all amendments. Be sure to give address to which you wish to have future amendments mailed.

NOTICE TO INDIVIDUALS REQUIRED BY PRIVACY ACT OF 1974

Sections 301, 303, and 308 of the Communications Act of 1934, as amended, *(licensing powers)* authorize the FCC to request the information on this application. The purpose of the information is to determine your eligibility for a license. The information will be used by FCC staff to evaluate the application, to determine station location, to provide information for enforcement and rulemaking proceedings and to maintain a current inventory of licensees. No license can be granted unless all information requested is provided.

NO

7530-21-029-5268

16-52
871

DEPARTMENT OF COMMUNICATIONS

TELECOMMUNICATIONS REGULATION BRANCH

CANADA

APPLICATION FOR REGISTRATION OF RADIO STATION LICENSEE OF UNITED STATES OF AMERICA

▶ Before completing Note instructions on reverse side

FULL NAME OF U.S.A. LICENSEE

MAILING ADDRESS (Including Canadian, if any)

Requests authority to operate in Canada a radio station(s) licensed in his name by the Government of the United States of America as follows:

Class of F.C.C. Licence	Date Licences Expires	Call Sign on F.C.C. Licence

Authorized communication services in U.S.A. Indicate frequency if not Amateur or Citizen Band

Specify area of proposed operation in Canada and approximate period or period(s) of time

Date signed	Signature of Applicant Licensee or common carrier subscriber

▶ NOTE REVERSE SIDE ◀

DO NOT DETACH THIS PORTION

DEPARTMENT OF COMMUNICATIONS NO.

CANADA

CERTIFICATE OF REGISTRATION OF RADIO STATION LICENSEE OF UNITED STATES OF AMERICA AND AUTHORITY FOR OPERATION IN CANADA

Class of FCC Licence	Date licence expires	Call sign

Authorized Communication service in U.S.A. Indicate frequency if not amateur or Citizens Band

Full name and address of U.S. licensee

This certificate expires	For the Minister of Communications

▶ NOTE REVERSE SIDE ◀

UNITED STATES OF AMERICA
FEDERAL COMMUNICATIONS COMMISSION
TRANSMITTER IDENTIFICATION CARD
FCC Form 452-C (July 1972)

1. Station call sign:
2. Name and Address of Permittee or Licensee:

UNITED STATES OF AMERICA
FEDERAL COMMUNICATIONS COMMISSION
TRANSMITTER IDENTIFICATION CARD
FCC Form 452-C (July 1972)

1. Station call sign:
2. Name and Address of Permittee or Licensee:

UNITED STATES OF AMERICA
FEDERAL COMMUNICATIONS COMMISSION
TRANSMITTER IDENTIFICATION CARD
FCC Form 452-C (July 1972)

1. Station call sign:
2. Name and Address of Permittee or Licensee:

UNITED STATES OF AMERICA
FEDERAL COMMUNICATIONS COMMISSION
TRANSMITTER IDENTIFICATION CARD
FCC Form 452-C (July 1972)

1. Station call sign:
2. Name and Address of Permittee or Licensee:

INSTRUCTIONS

1. This form is to be used by the licensee of a radio station in the United States of America requesting authority to operate such station(s) in Canada under the relevant USA/Canada Treaty or Convention.
2. The applicant is to complete all items and fill in all blanks of both sections of this form with the exception of the spaces provided for authentication and expiry date and return to Regional Superintendent, Telecommunications Regulation, Department of Communications, nearest to the area of proposed operation. Regional Offices are located at Vancouver, B.C. Winnipeg, Man., Toronto, Ont., Montréal, P.Q. and Moncton, N.B.
3. Upon approval of the application, the Certificate of Registration will be authenticated and returned to the licensee. It may be photo-copied to the extent necessary to provide a copy with each Unit covered by the licence.
4. Operation of Radio Stations in Canada are required to conform with the provisions of the Radio Act, and Regulations made thereunder, copies of which may be obtained from the Queen's Printer, Ottawa.
5. Application should be made at least 60 days prior to the date the applicant will enter Canada.

The person whose name appears on the reverse side of this certificate, holder of a radio station licence issued by the United States of America as stipulated on this certificate, has registered with the Department of Communications. This Certificate, together with evidence of validity of the aforementioned licence, shall constitute authority for the use of the radio Station(s) in Canada under the terms of the relevant USA/Canada Treaty or Convention.

NOTE: You can get Part 95 relating to the Citizens Band by ordering "FCC Part 95, Stock Number 004-000-00324-1" at a cost of $1.50. Those rules, revised as of the date indicated, appear elsewhere in this chapter.

SIDEBANDERS Q-SIGNALS

Q-signals are standard radio operating abbreviations adopted by international agreement. They mean the same thing in any language. Sidebanders often set up networks and pop Q-signals the way people pop bubble-gum.

CODE	MEANING
QRA	What is the name of your station?
QRB	How far approximately are you from my station?
QRD	Where are you bound and where are you from?
QRG	Will you tell me my exact frequency?
QRH	Does my frequency vary?
QRK	What is the readability of my signals (1-5)?
QRL	Are you busy?
QRM	Are you being interferred with?
QRN	Are you troubled by static?
QRT	Shall I stop sending?
QRU	Have you anything for me?
QRV	Are you ready?
QRW	Shall I tell––that you are calling him on channel –?
QRX	When will you call again?
QRY	What is my turn?
QRZ	Who is calling me?
QSA	What is the strength of my signals (1-5)?
QSB	Are my signals fading?
QSL	Can you acknowledge receipt?
QSO	Can you communicate with ––direct or by relay?
QSW	Do you wish to transmit on this frequency?
QSX	Will you listen to––on channel–?
QSY	Shall I change to another frequency?
QTE	What is my TRUE bearing from you? or What is my TRUE bearing from––?
QTH	What is your position in latitude and longitude?
QTI	What is your TRUE course?
QTJ	What is your speed?
QTL	What is your TRUE heading?
QTR	What is the exact time?
QTU	What are the hours during which your station is open?
QTX	Will you keep your station open for further communication with me until further notice (or until––hours)?
QUA	Have you news of––?
QUD	Have you received the urgent signal sent by––?
QUF	Have you received the distress signal sent by––?

WHERE TO GET HARD-TO-FIND CB PRODUCTS

Most of the items mentioned in this book are nationally distributed. But some will be hard to get because they are new, or are not the kind of products ordinarily stocked by a CB store, or are mail-order items.

CB Climate Barrier for corrosion protection of antenna and electrical connections–$2.95 (plus sales tax for New Jersey residents only) from:

Bob-Cat Corporation
567 Deal Parkway
West Allenhurst, New Jersey
07711

Owen Low-Profile CB Antenna for recreational vehicles with metal roofs. Model RV 269 may be stocked by some dealers or order from manufacturer at $59.95 plus shipping and tax where applicable:

Bill Owen Inc.
7934 Napoleon Road
Jackson, Michigan 49201

Boatlife Liquid Tape and *Boatlife Marine Silicone Sealer*–non-corrosive waterproofing for cable entries, antenna and electrical connections is nationally distributed through marine supply stores, but such outlets may be sparse in certain parts of the country. For address of nearest supplier, write the manufacturer:

Boatlife, Inc.
65 Bloomingdale Road
Hicksville, New York 11801

Cylindrical-head locking screws to protect transceiver mounting brackets from theft are hard to find. The toughest I know of are available by mail-order only. Specify set when requesting price from:

Art Brown
R 1 Box 22M
Boone, N.C. 28607

Electrical contact cleaner is found in electrical and electronic parts stores and some automotive stores. I found major differences among them. If you can't easily locate the following non-residue types, write these manufacturers for the supplier nearest you.:

CO Contact Cleaner
Corrosion Reaction Consultants
Dresher, Pa. 18025

LPS Contact Cleaner
LPS Research Laboratories
2050 Cotner Avenue
Los Angeles, Cal. 90025

Shielded ignition kits may not be in all auto parts stores. If you can't locate a source write:

Estes Engineering Company
930 Marine Drive
Port Angeles, Washington 98362

Portable hump mounts are handy but perhaps hard to find. The SMA Port-A-Mount may not be found in all CB stores. For the store nearest you write:

SMA Products
Box 152
Springfield, Illinois 62705

FEDERAL COMMUNICATIONS COMMISSION RULES AND REGULATIONS
PART 95/CITIZENS RADIO SERVICE

Revised July 27, 1976

SUBPART A – GENERAL

95.1 Basis and pupose.

The rules and regulations set forth in this part are issued pursuant to the provisions of Title III of the Communications Act of 1934, as amended, which vests authority in the Federal Communications Commission to regulate radio transmissions and to issue licenses for radio stations. These rules are designed to provide for private short-distance radiocommunications service for the business or personal activities of licensees, for radio signaling, for the control of remote objects or devices by means of radio; all to the extent that these uses are not specifically prohibited in this part. They also provide for procedures whereby manufacturers of radio equipment to be used or operated in the Citizens Radio Service may obtain type acceptance and/or type approval of such equipment as may be appropriate.

95.3 Definitions.

For the purpose of this part, the following definitions shall be applicable. For other definitions, refer to Part 2 of this chapter.

(a) Definitions of services.

Citizens Radio Service. A radiocommunications service of fixed, land, and mobile stations intended for short-distance personal or business radiocommunications, radio signaling, and control of remote objects or devices by radio; all to the extent that these uses are not specifically prohibited in this part.

Fixed Service. A service of radiocommunication between specified fixed points.

Mobile service. A service of radiocommunication between mobile and land stations or between mobile stations.

(b) Definitions of stations.

Base Station. A land station in the land mobile service carrying on a service with land mobile stations.

Class A station. A station in the Citizens Radio Service licensed to be operated on an assigned frequency in the 460—470 MHz band with a transmitter output power of not more than 50 watts.

Class B station. (All operations terminated as of November 1, 1971.)

Class C station. A station in the Citizens Radio Service licensed to be operated on an authorized frequency in the 26.96— 27.23 MHz band, or on the frequency 27.255 MHz, for the control of remote objects or devices by radio, or for the remote actuation of devices which are used solely as a means of attracting attention, or on an authorized frequency in the 72—76 MHz band for the radio control of models used for hobby purposes only.

Class D station. A station in the Citizens Radio Service licensed to be operated for radiotelephony, only, on authorized frequencies in the 26.96MHz to 27.41 MHz band.

Fixed station. A station in the fixed service.

Land station. A station in the mobile service not intended for operation while in motion. (Of the various types of land stations, only the base station is pertinent to this part.)

Mobile station. A station in the mobile service intended to be used while in motion or during halts at unspecified points. (For the purposes of this part, the term includes hand-carried and pack-carried units.)

(c) Miscellaneous definitions.

Antenna structures. The term "antenna structures" includes the radiating system its supporting structures and any appurtenances mounted thereon.

Assigned frequency. The frequency appearing on a station authorization from which the carrier frequency may deviate by an amount not to exceed that permitted by the frequency tolerance.

Authorized bandwidth. The maximum permissible bandwidth for the particular emission used. This shall be the occupied bandwidth or necessary bandwidth, whichever is greater.

Carrier power. The average power at the output terminals of a transmitter (other than a transmitter having a suppressed, reduced or controlled carrier) during one radio frequency cycle under conditions of no modulation.

Control point. A control point is an operating position which is under the control and supervision of the licensee, at which a person immediately responsible for the proper operation of the transmitter is stationed, and at which adequate means are available to aurally monitor all transmissions and to render the transmitter inoperative.

Dispatch point. A dispatch point is any position from which messages may be transmitted under the supervision of the person at a control point.

Double sideband emission. An emission in which both upper and lower sidebands resulting from the modulation of a particular carrier are transmitted. The carrier, or a portion thereof, also may be present in the emission.

External radio frequency power amplifiers. As defined in 2.815 (a) and as used in this part, an external radio frequency power amplifier is any device which, (1) when used in conjunction with a radio transmitter as a signal source is capable of amplification of that signal, and (2) is not an integral part of a radio transmitter as manufactured.

Harmful interference. Any emission, radiation or induction which endangers the functioning of a radionavigation service or other safety service or seriously degrades, obstructs or repeatedly interrupts a radiocommunication service operating in accordance with applicable laws, treaties, and regulations.

Man-made structure. Any construction other than a tower, mast or pole.

Mean power. The power at the output terminals of a transmitter during normal operation, averaged over a time sufficiently long compared with the period of the lowest frequency encountered in the modulation. A time of 1/10 second during which the mean power is greatest will be selected normally.

Necessary bandwidth. For a given class of emission, the minimum value of the occupied bandwidth sufficient to ensure the transmission of information at the rate and with the quality required for the system employed, under specified conditions. Emissions useful for the good functioning of the receiving equipment, as for example, the emission corresponding to the carrier of reduced carrier systems, shall be included in the necessary bandwidth.

Occupied bandwidth. The frequency bandwidth such that, below its lower and above its upper frequency limits, the mean powers radiated are each equal to 0.5% of the total mean power radiated by a given emission.

Omnidirectional antenna. An antenna designed so the maximum radiation in any horizontal direction is within 3 dB of the minimum radiation in any horizontal direction.

Peak envelope power. The average power at the output terminals of a transmitter during one radio frequency cycle at the highest crest of the modulation envelope, taken under conditions of normal operation.

Person. The term "person" includes an individual, partnership, association, joint-stock company, trust or corporation.

Remote control. The term "remote control" when applied to the use or operation of a citizens radio station means control of the transmitting equipment of that station from any place other than the location of the transmitting equipment, except that direct mechanical control or direct electrical control by wired connections of transmitting equipment from some other point on the same premises, craft or vehicle shall not be considered to be remote control.

Single sideband emission. An emission in which only one sideband is transmitted. The carrier, or a portion thereof, also may be present in the emission.

Station authorization. Any construction permit, license, or special temporary authorization issued by the Commission.

95.5 Policy governing the assignment of frequencies.

(a) The frequencies which may be assigned to Class A stations in the Citizens Radio Service, and the frequencies which are available for use by Class C or Class D stations are listed in Subpart C of this part. Each frequency available for assignment to, or use by, stations in this service is available on a shared basis only, and will not be assigned for the exclusive use of any one applicant; however, the use of a particular frequency may be restricted to (or in) one or more specified geographical areas.

(b) In no case will more than one frequency be assigned to Class A stations for the use of a single applicant in any given area until it has been demonstrated conclusively to the Commission that the assignment of an additional frequency is essential to the operation proposed.

(c) All applicants and licensees in this service shall cooperate in the selection and use of the frequencies assigned or authorized, in order to minimize interference and thereby obtain the most effective use of the authorized facilities.

(d) Simultaneous operation on more than one frequency in the 72—76 MHz band by a transmitter or transmitters of a single licensee is prohibited whenever such operation will cause harmful interference to the operation of other licensees in this service.

95.6 Types of operation authorized.

(a) Class A stations may be authorized as mobile stations, as base stations, as fixed stations, or as base or fixed stations to be operated at unspecified or temporary locations.

(b) Class C and Class D stations are authorized as mobile stations only; however, they may be operated at fixed locations in accordance with other provisions of this part.

95.7 General citizenship requirements.

A station license shall not be granted to or held by a foreign government or a representative thereof.

[95.7 revised eff. 2-5-75; VI(75)-1]

SUBPART A – GENERAL
SUBPART B – APPLICATIONS AND LICENSES

95.11 Station authorization required.

No radio station shall be operated in the Citizens Radio Service except under and in accordance with an authorization granted by the Federal Communications Commission.

95.13 Eligibility for station license.

(a) Subject to the general restrictions of 95.7, any person is eligible to hold an authorization to operate a station in the Citizens Radio Service: *Provided,* That if an applicant for a Class A or Class D station authorization is an individual or partnership, such individual or each partner is eighteen or more years of age; or if an applicant for a Class C station authorization is an individual or partnership, such individual or each partner is twelve or more years of age. An unincorporated association, when licensed under the provisions of this paragraph, may upon specific prior approval of the Commission provide radiocommunications for its members.

NOTE: While the basis of eligibility in this service includes any state, territorial, or local governmental entity, or any agency operating by the authority of such governmental entity, including any duly authorized state, territorial, or local civil defense agency, it should be noted that the frequencies available to stations in this service are shared without distinction between all licensees and that no protection is afforded to the communications of any station in this service from interference which may be caused by the authorized operation of other licensed stations.

(b) [Reserved]

(c) No person shall hold more than one Class C and one Class D station license.

95.14 Mailing address furnished by licensee.

Except for applications submitted by Canadian citizens pursuant to agreement between the United States and Canada (TIAS No. 2508 and No. 6931), each application shall set forth and each licensee shall furnish the Commission with an address in the United States to be used by the Commission in serving documents or directing correspondence to that licensee. Unless any licensee advises the

Commission to the contrary, the address contained in the licensee's most recent application will be used by the Commission for this purpose.

[*95.14 amended eff. 12-17-75; VI (75)-2*]

95.15 Filing of applications.

(a) To assure that necessary information is supplied in a consistent manner by all persons, standard forms are prescribed for use in connection with the majority of applications and reports submitted for Commission consideration. Standard numbered forms applicable to the Citizens Radio Service are discussed in 95.19 and may be obtained from the Washington, D.C., 20554, office of the Commission, or from any of its engineering field offices.

(b) All formal applications for Class C or Class D new, modified, or renewal station authorizations shall be submitted to the Commission's office, Box 1010, Gettysburg, Pa. 17325. An application for a temporary permit shall be made by completing and making the certifications required by FCC Form 555-B. Applications for Class A station authorizations, applications for consent to transfer of control of a corporation holding any citizens radio station authorization, requests for special temporary authority or other special requests, and correspondence relating to an application for any class citizens radio station authorization shall be submitted to the Commission's Office at Washington, D.C. 20554, and should be directed to the attention of the Secretary. Applicants for Class A stations in the Chicago Regional Area, defined in 95.19, shall submit their application to the Commission's Chicago Regional Office. Applications involving Class A or Class D station equipment which is neither type approved nor crystal controlled, whether of commercial or home construction, shall be accompanied by supplemental data describing in detail the design and construction of the transmitter and methods employed in testing it to determine compliance with the technical requirements set fourth in Subpart Co of this part.

(c) Unless otherwise specified, an application shall be filed at least 60 days prior to the date on which it is desired that Commission action thereon be completed. In any case where the applicant has made timely and sufficient application for renewal of license, in accordance with the Commission's rules, no license with reference to any activity of a continuing nature shall expire until such application shall have been finally determined.

(d) A temporary permit may not be held by an applicant already holding a Class D station license.

(e) Failure on the part of the applicant to provide all the information required by the application form, or to supply the ncessary exhibits or supplementary statements may constitute a defect in the application.

(f) Applicants proposing to construct a radio station on a site located on land under the jurisdiction of the U.S. Forest Service, U.S. Department of Agriculture, or the Bureau of Land Management, U.S. Department of the Interior, must supply the information and must follow the procedure prescribed by 170 of this chapter.

95.17 Who may sign applications.

(a) Except as provided in paragraph (b) of this section, applications, amendments thereto, and related statements of fact required by the Commission shall be personally signed by the applicant, if the applicant is an individual; by one of the partners, if the applicant is a partnership; by an officer, if the applicant is a corporation; or by a member who is an officer, if the applicant is an unincorporated association. Applications, amendments, and related statements of fact filed on behalf of eligible government entities, such as states and territories of the United States and political subdivisions thereof, the District of Columbia, and units of local government, including incorporated municipalities, shall be signed by such duly elected or appointed officials as may be competent to do so under the laws of the applicable jurisdiction.

(b) Applications, amendments thereto, and related statements of fact required by the Commission may be signed by the applicant's attorney in case of the applicant's physical disability or of his absence from the United States. The attorney shall in that event separately set forth the reason why the application is not signed by the applicant. In addition, if any matter is stated on the basis of the attorney's belief only (rather than his knowledge), he shall separately set forth his reasons for believing that such statements are true.

(c) Only the original of applications, amendments, or related statements of fact need be signed; copies may be conformed.

(d) Applications, amendments, and related statements of the fact need not be signed under oath. Willful false statements made therein, however, are punishable by fine and imprisonment. U.S. Code, Title 18, section 1001, and by appropriate administrative sanctions, including revocation of station license pursuant to section 312 (a) (1) of the Communications Act of 1934, as amended.

95.19 Standard forms to be used.

(a) *FCC Form 505, Application for Class C or D Station License in the Citizens Radio Service.* This form shall be used when:

(1) Application is made for a new Class C or Class D authorization. A separate application shall be submitted for each proposed class of station.

(2) Application is made for modification of any existing Class C or Class D station authorization in those cases where prior Commission approval of certain changes is required (see 95.35).

(3) Application is made for renewal of an existing Class C or Class D station authorization, or for reinstatement of such an expired authorization.

(b) FCC Form 555-B, *Temporary Permit, Class D Citizens Radio Station.* This form shall be used when application is made for a temporary permit.

(c) *FCC Form 400, Application for Radio Station Authorization in the Safety and Special Radio Services.* Except as provided in paragraph (d) of this section, this form shall be used when:

(1) Application is made for a new Class A base station or fixed station authorization. Separate applications shall be submitted for each proposed base or fixed station at different fixed locations; however, all equipment intended to be operated at a single fixed location is considered to be one station which may, if necessary, be classed as both a base station and a fixed station.

(2) Application is made for a new Class A station authorization for any required number of mobile units (including hand-carried and pack-carried units) to be operated as a group in a single radiocommunication system in a particular area. An application for Class A mobile station authorization may be combined with the application for a single Class A base station authorization when such mobile units are to be operated with that base station only.

(3) Application is made for station license of any Class A base station or fixed station upon completion of construction or installation in accordance with the terms and conditions set forth in any construction permit required to be issued for that station, or application for extension of time within which to construct such a station.

(4) Application is made for modification of any existing Class A station authorization in those cases where prior Commission approval of certain changes is required (see 95.35).

(5) Application is made for renewal of an existing Class A station authorization, or for reinstatement of such an expired authorization.

(6) Each applicant in the Safety and Special Radio Services (1) for modification of a station license involving a site change or a substantial increase in tower height or (2) for a license for a new station must, before commencing construction, supply the environmental information, where required, and must follow the procedure prescribed by Subpart I of Part 1 of this chapter (1.1301 through 1.1319) unless Commission action authorizing such construction would be a minor action within the meaning of Subpart I of Part 1.

(7) Application is made for an authorization for a new Class A base or fixed station to be operated at unspecified or temporary locations. When one or more individual transmitters are each intended to be operated as a base station or as a fixed station at unspecified or temporary locations for indeterminate periods, such transmitters may be considered to comprise a single station intended to be operated at temporary location. The application shall specify the general geographic area within which the operation will be confined. Sufficient data must be submitted to show the need for the proposed area of operation.

(d) *FCC Form 703, Application for Consent to Transfer of Control of Corporation Holding Construction Permit or Station License.* This form shall be used when application is made for consent to transfer control of a corporation holding any citizens radio station authorization.

(e) Beginning April 1, 1972, FCC Form 425 shall be used in lieu of FCC Form 400, applicants for Class A stations located in the Chicago Regional Area defined to consist of the counties listed below:

ILLINOIS

1. Boone.
2. Bureau.
3. Carroll.
4. Champaign.
5. Christian
6. Clark.
7. Coles.
8. Cook.
9. Cumberland.
10. De Kalb.
11. De Witt.
12. Dougles.
13. Du Page.
14. Edgar.
15. Ford.
16. Fulton.
17. Grundy.
18. Henry.
19. Iroquois.
20. Jo Daviess.
21. Kane.
22. Kankaee
23. Kendall.
24. Knox.
25. Lake.
26. La Salle.
27. Lee.
28. Livingston.
29. Logan.
30. Macon.
31. Marshall.
32. Mason.
33. McHenry.
34. McLean.
35. Menard.
36. Mercer.
37. Moultrie
38. Ogle.
39. Peoria
40. Piatt.
41. Putnam.
42. Rock Island.
43. Sangamon.
44. Shelby.
45. Stark.
46. Stephenson.
47. Tazewell.
48. Vermillion.
49. Warren.
50. Whiteside.
51. Will.
52. Winnebago.
53. Woodford.

INDIANA

1. Adams.
2. Allen.
3. Benton.
4. Blackford.
5. Boone.
6. Carroll.
7. Cass.
8. Clay.
9. Clinton.
10. De Kalb.
11. Delaware.
12. Elkhart.
13. Fountain.
14. Fulton.
15. Grant.
16. Hamilton.
17. Hancock.
18. Hendricks.
19. Henry.
20. Howard.
21. Huntington.
22. Jasper.
23. Jay.
24. Kosciusko.
25. Lake.
26. Lagrange.
27. La Porte.
28. Madison.
29. Marion.
30. Marshall.
31. Miami
32. Montgomery.
33. Morgan.
34. Newton.
35. Noble.
36. Owen.
37. Parke.
38. Porter.
39. Pulaski
40. Putnam.
41. Randolph.
42. St. Joseph.
43. Starke.
44. Steuben.
45. Tippecanoe.
46. Tipton.
47. Vermillion.
48. Vigo.
49. Wabash.
50. Warren.
51. Wells.
52. White.
53. Whitley.

IOWA

1. Cedar.
2. Clinton.
3. Dubuque.
4. Jackson.
5. Jones.
6. Muscatine.
7. Scott.

MICHIGAN

1. Allegan.
2. Barry.
3. Berrien.
4. Branch.
5. Calhoun.
6. Cass.
7. Clinton.
8. Eaton.
9. Hillsdale.
10. Ingham.
11. Ionia.
12. Jackson.
13. Kalamazoo.
14. Kent.
15. Lake.
16. Mason.
17. Mecosta.
18. Montcalm.
19. Muskegon.
20. Newaygo.
21. Oceana.
22. Ottawa.
23. St. Joseph.
24. Van Buren.

OHIO

1. Defiance.
2. Mercer.
3. Paulding.
4. Van Wert.
5. Williams.

WISCONSIN

1. Adams.
2. Brown.
3. Calumet.
4. Columbia.
5. Dane.
6. Dodge.
7. Door.
8. Fond du Lac.
9. Grant.
10. Green.
11. Green Lake.
12. Iowa.
13. Jefferson.
14. Juneau.
15. Kenosha.
16. Kewaunee.
17. Lafayette.
18. Manitowoc.
19. Marquette
20. Milwaukee
21. Outagamie.
22. Ozaukee.
23. Racine.
24. Richland.
25. Rock.
26. Sauk.
27. Sheboygan.
28. Walworth.
29. Washington.
30. Waukesha.
31. Waupaca.
32. Waushara.
33. Winnebago.

95.25 Amendment or dismissal of application.

(a) Any application may be amended upon request of the applicant as a matter of right prior to the time the application is granted or designated for hearing. Each amendment to an application shall be signed and submitted in the same manner and with the same number of copies as required for the original application.

(b) Any application may, upon written request signed by the applicant or his attorney, be dismissed without prejudice as a matter of right prior to the time the application is granted or designated for hearing.

95.27 Transfer of license prohibited.

A station authorization in the Citizens Radio Service may not be transferred or assigned. In lieu of such transfer or assignment, an application for new station authorization shall be filed in each case, and the previous authorization shall be forwarded to the Commission for cancellation.

95.29 Defective applications.

(a) If an applicant is requested by the Commission to file any documents or information not included in the prescribed application form, a failure to comply with such request will constitute a defect in the application.

(b) When an application is considered to be incomplete or defective, such application will be returned to the applicant, unless the Commission may otherwise direct. The reason for return of the applications will be indicated, and if appropriate, necessary additions or corrections will be suggested:

95.31 Partial grant.

Where the Commission, without a hearing, grants an application in part, or with any privileges, terms, or conditions other than those requested, the action of the Commission shall be considered as a grant of such application unless the applicant shall, within 30 days from the date on which such grant is made, or from its effective date if a later date is specified, file with the Commission a written rejection of the grant as made. Upon receipt of such rejection, the Commission will vacate its original action upon the application and, if appropriate, set the application for hearing.

95.33 License term.

Licenses for stations in the Citizens Radio Service will normally be issued for a term of 5 years from the date of original issuance, major modification, or renewal.

95.35 Changes in transmitters and authorized stations.

Authority for certain changes in transmitters and authorized stations must be obtained from the Commission before the changes are made, while other changes do not require prior Commission approval. The following paragraphs of this section describe the conditions under which prior Commission approval is or is not necessary.

(a) Proposed changes which will result in operation inconsistent with any of the terms of the current authorization require that an application for modification of license be submitted to the commission. Application for modification shall be submitted in the same manner as an application for a new station license, and the licensee shall forward his existing authorization to the Commission for cancellation immediately upon receipt of the superseding authorization. Any of the following changes to authorized stations may be made only upon approval by the Commission:

(1) Increase the overall number of transmitters authorized.

(2) Change the presently authorized location of a Class A fixed or base station or control point.

(3) Move, change the height of, or erect a Class A station antenna structure.

(4) Make any change in the type of emission or any increase in bandwidth of emission or power of a Class A station.

(5) Addition or deletion of control point(s) for an authorized transmitter of a Class A station.

(6) Change or increase the area of operation of a Class A mobile station or a Class A base or fixed station authorized to be operated at temporary locations.

(7) Change the operating frequency of a Class A station.

(b) When the name of a licensee is changed (without changes in the ownership, control, or corporate structure), or when the mailing address of the licensee is changed (without changing the authorized location of the base or fixed Class A station) a formal application for modification of the license is not required. However, the licensee shall notify the Commission promptly of these changes. The notice, which may be in letter form, shall contain the name and address of the licensee as they appear in the Commission's records, the new name and/or address, as the case may be, and the call signs and classes of all radio stations authorized to the licensee under this part. The notice concerning Class C or D radio stations shall be sent to Federal Communications Commission, Gettysburg, Pa. 17325, and a copy shall be maintained with the records of the station. The notice concerning Class A stations shall be sent to (1) Secretary, Federal Communications Commission, Washington, D.C. 20554, and (2) to Engineer in Charge of the Radio District in which the station is located, and a copy shall be maintained with the license of the station until a new license is issued.

(c) Proposed changes which will not depart from any of the terms of the outstanding authorization for the station may be made without prior Commission approval. Included in such changes is the substitution of transmitting equipment at any station, provided that the equipment employed is included in the Commission's "Radio Equipment List," and is listed as acceptable for use in the appropriate class of station in this service. Provided it is crystal-controlled and otherwise complies with the power, frequency tolerance, emission and modulation percentage limitations prescribed, non-type accepted equipment may be substituted at:

(1) Class C stations operated on frequencies in the 26.99—27.26 MHz band;

(2) Class D stations until November 22, 1974.

(d) Transmitting equipment type accepted for use in Class D stations shall not be modified by the user. Changes which are specifically prohibited include:

(1) Internal or external connection or addition of any part, device or accessory not included by the manufacturer with the transmitter for its type acceptance. This shall not prohibit the external connection of antennas or antenna transmission lines, antenna switches, passive networks for coupling transmission lines or antennas to transmitters, or replacement of microphones.

(2) Modification in any way not specified by the transmitter manufacturer and not approved by the Commission.

(3) Replacement of any transmitter part by a part having different electrical characteristics and ratings from that replaced unless such part is specified as a replacement by the transmitter manufacturer.

(4) Substitution or addition of any transmitter oscillator crystal unless the crystal manufacturer or transmitter manufacturer has made an express determination that the crystal type, as installed in the specific transmitter type, will provide that transmitter type with the capability of operating within the frequency tolerance specified in Section 95.45 (a).

(5) Addition or substitution of any component, crystal or combination of crystals, or any other alteration to enable transmission on any frequency not authorized for use by the licensee.

(e) Only the manufacturer of the particular unit of equipment type accepted for use in Class D stations may make the permissive changes allowed under the provisions of Part 2 of this chapter for type acceptance. However, the manufacturer shall not make any of the following changes to the transmitter without prior written authorization from the Commission:

(1) Addition of any accessory or device not specified in the application for type acceptance and approved by the Commission in granting said type acceptance.

(2) Addition of any switch, control, or external connection.

(3) Modification to provide capability for an additional number of transmitting frequencies.

95.37 Limitations on antenna structures.

(a) Except as provided in paragraph (b) of this section, an antenna for a Class A station which exceeds the following height limitations may not be erected or used unless notice has been filed with both the FAA on FAA Form 7460—1 and with the Commission on Form 714 or on the license application form, and prior approval by the Commission has been obtained for:

(1) Any construction or alteration of more than 200 feet in height above ground level at its site (17.7(a) of this chapter).

(2) Any construction or alteration of greater height than an imaginary surface extending outward and upward at one of the following slopes (17.7(b) of this chapter):

95.37 Limitations on antenna structures

(i) 100 to 1 for a horizontal distance of 20,000 feet from the nearest point of the nearest runway of each airport with at least one runway more than 3,200 feet in length, excluding heliports, and seaplane bases without specified boundaries, if that airport is either listed in the Airport Directory of the current Airman's Information Manual or is operated by a Federal military agency.

(ii) 50 to 1 for a horizontal distance of 10,000 feet from the nearest point of the nearest runway of each airport with its longest runway no more than 3,200 feet in length, excluding heliports, and seaplane bases without specified boundaries, if that airport is either listed in the Airport Directory or is operated by a Federal military agency.

(iii) 25 to 1 for a horizontal distance of 5,000 feet from the nearest point of the nearest landing and takeoff area of each heliport listed in the Airport Directory or operated by a Federal military agency.

(3) Any construction or alteration on any airport listed in the Airport Directory of the current Airman's Information Manual (17.7(c) of this chapter).

(b) A notification to the Federal Aviation Administration is not required for any of the following construction or alteration of Class A station antenna structures.

(1) Any object that would be shielded by existing structures of a permanent and substantial character or by natural terrain or topographic features of equal or greater height, and would be located in the congested area of a city, town, or settlement where it is evident beyond all reasonable doubt that the structure so shielded will not adversely affect safety in air navigation. Applicants claiming such exemption shall submit a statement with their application to the Commission explaining the basis in detail for their finding (17.14(a) of this chapter).

(2) Any antenna structure of 20 feet or less in height except one that would increase the height of another antenna structure (17.14(b) of this chapter).

(c) All antennas (both receiving and transmitting) and supporting structures associated or used in conjunction with a Class C or D Citizens Radio Station operated from a fixed location must comply with at least one of the following:

(1) The antenna and its supporting structure does not exceed 20 feet in height above ground level; or

(2) The antenna and its supporting structure does not exceed by more than 20 feet the height of any natural formation, tree or man-made structure on which it is mounted; or

NOTE: A man-made structure is any

construction other than a tower, mast, or pole.

(3) The antenna is mounted on the transmitting antenna structure of another authorized radio station and exceeds neither 60 feet above ground level nor the height of the antenna supporting structure of the other station; or

(4) The antenna is mounted on and does not exceed the height of the antenna structure otherwise used solely for receiving purposes, which structure itself complies with subparagraph (1) or (2) of this paragraph.

(5) The antenna is omnidirectional and the highest point of the antenna and its supporting structure does not exceed 60 feet above ground level and the highest point also does not exceed one foot in height above the established airport elevation for each 100 feet of horizontal distance from the nearest point of the nearest airport runway.

NOTE: A work sheet will be made available upon request to assist in determining the maximum permissible height of an antenna structure.

(d) Class C stations operated on frequencies in the 72—76 MHz band shall employ a transmitting antenna which complies with all of the following:

(1) The gain of the antenna shall not exceed that of a half-wave dipole;

(2) The antenna shall be immediately attached to, and an integral part of, the transmitter; and

(3) Only vertical polarization shall be used.

(e) Further details as to whether an aeronautical study and/or obstruction marking and lighting may be required, and specifications for obstruction marking and lighting when required, may be obtained from Part 17 of this chapter, "Construction, Marking, and Lighting of Antenna Structures."

(f) Subpart I of Part 1 of this chapter contains procedures implementing the National Environmental Policy Act of 1969. Applications for authorization of the construction of certain classes of communications facilities defined as "major actions" in 1.305 thereof, are required to be accompanied by specified statements. Generally these classes are:

(1) Antenna towers or supporting structures which exceed 300 feet in height and are not located in areas devoted to heavy industry or to agriculture.

(2) Communications facilities to be located in the following areas:

(i) Facilities which are to be located in an officially designated wilderness area or in an area whose designation as a wilderness is pending consideration;

(ii) Facilities which are to be located in an officially designated wildlife preserve or in an area whose designation as a wildlife preserve is pending consideration;

(iii) Facilities which will affect districts, sites, buildings, structures or objects, significant in American history, architecture, archaeology or culture, which are listed in the National Register of Historic Places or are eligible for listing (see 36 CFR 800.2 (d) and (f) and 800.10); and

(iv) Facilities to be located in areas which are recognized either nationally or locally for their special scenic or recreational value.

(3) Facilities whose construction will involve extensive change in surface features (e.g. wetland fill, deforestation or water diversion).

NOTE: The provisions of this paragraph do not include the mounting of FM, television or other antennas comparable thereto in size on an existing building or antenna tower. The use of existing routes, buildings and towers is an environmentally desirable alternative to the construction of new routes or towers and is encouraged.

If the required statements do not accompany the application, the pertinent facts may be brought to the attention of the Commission by any interested person during the course of the license term and considered de novo by the Commission.

[95.37 (c) & (c) (3) amended eff. 9-15-75; VI (75)-2]

SUBPART C—TECHNICAL REGULATIONS

95.41 Frequencies Available.

(a) Frequencies available for assignment to Class A stations:

(1) The following frequencies or frequency pairs are available primarily for assignment to base and mobile stations. They may also be assigned to fixed stations as follows:

(i) Fixed stations which are used to control base stations of a system may be assigned the frequency assigned to the mobile units associated with the base station. Such fixed stations shall comply with the following requirements if they are located within 75 miles of the center of urbanized areas of 200,000 or more population.

(a) If the station is used to control one or more base stations located within 45 degrees of azimuth, a directional antenna having a front-to-back ratio of at least 15 dB shall be used at the fixed station. For other situations where such a directional antenna cannot be used, a cardioid, bidirectional or omnidirectional antenna may be employed. Consistent with reasonable design, the antenna used must, in each case, produce a radiation pattern that provides only the coverage necessary to permit satisfactory control of each base station and limit radiation in other directions to the extent feasible.

(b) The strength of the signal of a fixed station controlling a single base station may not exceed the signal strength produced at the antenna terminal of the base receiver by a unit of the associated mobile station, by more than 6 dB. When the station controls more than one base station, the 6 dB control-to-mobile signal difference need be verified at only one of the base station sites. The measurement of the signal strength of the mobile unit must be made when such unit is transmitting from the control station location or, if that is not practical, from a location within one-fourth mile of the control station site.

(c) Each application for a control station to be authorized under the provisions of this paragraph shall be accompanied by a statement certifying that the output power of the proposed station transmitter will be adjusted to comply with the foregoing signal level limitation. Records of the measurements used to determine the signal ratio shall be kept with the station records and shall be made available for inspection by Commission personnel upon request.

(d) Urbanized areas of 200,000 or more population are defined in the U.S. Census of Population, 1960, Vol. 1, table 23, page 50. The centers of urbanized areas are determined from the Appendix, page 226 of the U.S. Commerce publication "Air Line Distance Between Cities in the United States."

(ii) Fixed stations, other than those used to control base stations, which are located 75 or more miles from the center of an urbanized area of 200,000 or more population. The centers of urbanized areas of 200,000 or more population are listed on page 226 of the Appendix to the U.S. Department of Commerce publication "Air Line Distance Between Cities in the United States." When the fixed station is located 100 miles or less from the center of such an urbanized area, the power output may not exceed 15 watts. All fixed systems are limited to a maximum of two frequencies and must employ directional antennas with a front-to-back ratio of at least 15 dB. For two-frequency systems, separation between transmit-receive frequencies is 5 MHz.

Base and Mobile (MHz)	Mobile Only (MHz)
462.550	467.550
462.575	467.575
462.600	467.600
462.625	467.625
462.650	467.650
462.675	467.675
462.700	467.700
462.725	467.725

(2) Conditions governing the operation of stations authorized prior to March 18, 1968:

(i) All base and mobile stations authorized to operate on frequencies other than those listed in subparagraph (1) of this paragraph may continue to operate on those frequencies only until January 1, 1970.

(ii) Fixed stations located 100 or more miles from the center of any urbanized area of 200,000 or more population authorized to operate on frequencies other than those listed in subparagraph (1) of this paragraph will not have to change frequencies provided no interference is caused to the operation of stations in the land mobile service.

(iii) Fixed stations, other than those used to control base stations, located less than 100 miles (75 miles if the transmitter power output does not exceed 15 watts) from the center of any urbanized area of 200,000 or more population must discontinue operation by November 1, 1971. However, any operation after January 1, 1970, must be on frequencies listed in subparagraph (1) of this paragraph.

(iv) Fixed stations, located less than 100 miles from the center of any urbanized area of 200,000 or more population, which are used to control base stations and are authorized to operate on frequencies other than those listed in subparagraph (1) of this paragraph may continue to operate on those frequencies only until January 1, 1970.

(v) All fixed stations must comply with the applicable technical requirements of subparagraph (1) relating to antennas and radiated signal strength of this paragraph by November 1, 1971.

(vi) Notwithstanding the provisions of subdivisions (i) through (v) of this subparagraph, all stations authorized to operate on frequencies between 465.000 and 465.500 MHz and located within 75 miles of the center of the 20 largest urbanized areas of the United States, may continue to operate on these frequencies only until January 1, 1969. An extension to continue operation on such frequencies until January 1, 1970, may be granted to such station licensees on a case by case basis if the Commission finds that continued operation would not be inconsistent with planned usage of the particular frequency for police purposes. The 20 largest urbanized areas can be found in the U.S. Census of Population, 1960, vol. 1, table 23, page 50. The centers of urbanized areas are determined from the appendix, page 226, of the U.S. Commerce publication, "Air Line Distance Between Cities in the United States."

(b) [Reserved]

(c) Class C mobile stations may employ only amplitude tone modulation or on-off keying of the unmodulated carrier, on a shared basis with other stations in the Citizens Radio Service on the frequencies and under the conditions specified in the following tables:

(1) For the control of remote objects or devices by radio, or for the remote actuation of devices which are used solely as a means of attracting attention and subject to no protection from interference due to the operation of industrial, scientific, or medical devices within the 26.96—27.28 MHz band, the following frequencies are available:

(MHz)	(MHz)	(MHz)
26.995	27.095	27.195
27.045	27.145	[1]27.255

[1]The frequency 27.255 MHz also is shared with stations in other services.

(2) Subject to the conditions that interference will not be caused to the remote control of industrial equipment operating on the same or adjacent frequencies and to the reception of television transmissions on Channels 4 or 5; and that no protection will be afforded from interference due to the operation of fixed and mobile stations in other services assigned to the same or adjacent frequencies in the band, the following frequencies are available solely for the radio remote control of models used for hobby pby purposes:

(i) For the radio remote control of any model used for hobby purposes:

MHz	MHz	MHz
72.16	72.32	72.96

(ii) For the radio remote control of aircraft models only:

MHz	MHz	MHz
72.08	72.24	72.40
75.64		

(d) The frequencies listed in the following paragraphs are available for use by Class D stations and are subject to no protection from terference resulting from the operation of industrial, scientific, or medical devices in the 26.96 MHz to 27.28 MHz band.

(1) The following frequencies may be used for communications between Class D stations:

MHz	MHz
26.965	27.115
26.975	27.125
26.985	27.135
27.005	27.155
27.015	27.165
27.025	27.175
27.035	27.185
27.055	27.205
27.075	27.214
27.085	27.215
27.105	27.225
	27.255

(2) Effective January 1, 1977, the following frequencies may be used for communications between Class D stations:

MHz	MHz
26.965	27.225
26.975	27.235
26.985	27.245
27.005	27.255
27.015	27.265
27.025	27.275
27.035	27.285
27.055	27.295
27.075	27.305
27.085	27.315
27.105	27.325
27.115	27.335
27.125	27.345
27.135	27.355
27.155	27.365
27.165	27.375
27.175	27.385
27.185	27.395
27.205	27.405
27.215	

(3) The frequency 27.065 MHz shall be used solely for:

(i) Emergency communications involving the immediate safety of life of individuals or the immediate protection of property, or

(ii) Communications necessary to render assistance to a motorist.

NOTE—A licensee, before using 27.065 MHz must make a determination that his communication is either or both (a) an emergency communication or (b) is necessary to render assistance to a motorist. To be an emergency communication, the message must have some direct relation to the immediate safety of life or immediate protection of property. If no immediate action is required, it is not an emergency. What may not be an emergency under one set of circumstances may be an emergency under different circumstances. There are many worthwhile public service communications that do not qualify as emergency communications. In the case of motorist assistance, the message must be necessary to assist a particular motorist and not, except in a valid emergency, motorists in general. If the communications are to be lengthy, the exchange should be shifted to another frequency, if feasible, after contact is established. No nonemergency or nonmotorist assistance communications are permitted on 27.065 MHz even for the limited purpose of calling a licensee monitoring a frequency to ask him to switch to another frequency. Although 27.065 MHz may be used for marine emergencies, it should not be considered a substitute for the authorized marine distress system. The Coast Guard has stated it will not "participate directly in the Citizens Radio Service by fitting with and/or providing a watch on any Citizens Band Channel. (Coast Guard Commandant Instructions 2302.6)"

The following are examples of permitted and prohibited types of communications. They are guidelines and are not intended to be all inclusive.

Permitted	*Example message*
Yes	A tornado is sighted six miles north of town.
No	This is observation post number 10. No tornados sighted.
Yes	I am out of gas on Interstate 95.
No	I am out of gas in my driveway.
Yes	There is a four-car collision at Exit 10 on the Beltway.
No	Traffic is moving smoothly on the Beltway.
Yes	Base to Unit 1, the Weather Bureau has just issued a thunderstorm warning. Bring the sailboat into port.
No	Attention all motorists. The Weather Bureau advises that the snow tomorrow will accumulate 4 to 6 inches.
Yes	There is a fire in the building on the corner of 6th and Main Streets.
No	This is Halloween patrol unit number 3. Everything is quiet here.

The following priorities should be observed in the use of 27.065 MHz:

1. Communications relating to an existing situation dangerous to life or property, i.e., fire, automobile accident.

2. Communications relating to a potentially hazardous situation, i.e., car stalled in a dangerous place, lost child, boat out of gas.

3. Road assistance to a disabled vehicle on the highway or street.

4. Road and street directions.

(e) Upon specific request accompanying application for renewal of station authorization, a Class A station in this service, which was authorized to operate on a frequency in the 460—461 MHz band until March 31, 1967, may be assigned that frequency for continued use until not later than March 31, 1968, subject to all other provisions of this part.

[95.41 (d) & (d) (1) amended, (d) (2) deleted, (d) (3) redesig. (d) (2), and new (d) (3) added eff. 9-15-75; VI (75)-2]

95.42 Special provisions.

Effective September 10, 1976 station authorizations for the use of frequencies between 26.96 MHz and 27.41 MHz will be issued only to applicants in the Citizens Radio Service. Any license in a radio service other than the Citizens Radio Service authorizing the use of frequencies between 26.96 MHz and 27.41 MHz shall remain valid until December 31, 1979.

95.43 Transmitter power.

(a) Transmitter power is the power at the transmitter output terminals and delivered to the antenna, antenna transmission line, or any other impedance matched, radio frequency load.

(1) For single sideband transmitters and other transmitters employing a reduced carrier, a suppressed carrier or a controlled carrier, used at Class D stations, transmitter power is the peak envelope power.

(2) For all transmitters other than those covered by paragraph (a) (1) of this section, the transmitter power is the carrier power.

(b) The transmitter power of a station shall not exceed the following values under any condition of modulation or other circumstances.

Class of station:	Transmitter power in watts
A	50
C—27.255 MHz	25
C—26.995—27.195 MHz	4
C—72—76 MHz	0.75
D—Carrier (where applicable)	4
D—Peak envelope power (where applicable)	12

95.44 External radio frequency power amplifiers prohibited.

No external radio frequency power amplifier shall be used or attached, by connection, coupling attachment or in any other way at any Class D station.

NOTE: An external radio frequency power amplifier at a Class D station will be presumed to have been used where it is in the operator's possession or on his premises and there is extrinsic evidence of any operation of such Class D station in excess of power limitations provided under this rule part unless the operator of such equipment holds a station license in another radio service under which license the use of the said amplifier at its maximum rated output power is permitted.

95.45 Frequency tolerance.

(a) Except as provided in paragraphs (b) and (c) of this section, the carrier frequency of a transmitter in this service shall be maintained within the following percentage of the authorized frequency:

Class of station	Frequency tolerance	
	Fixed and base	Mobile
A	0.0025	0.0005
C		.005
D		.005

(b) Transmitters used at Class C stations operating on authorized frequencies between 26.99 and 27.26 MHz with 2.5 watts or less mean output power, which are used solely for the control of remote objects or devices by radio (other than devices used solely as a means of attracting attention), are permitted a frequency tolerance of 0.01 percent.

(c) Class A stations operated at a fixed location used to control base stations, through use of a mobile only frequency, may operate with a frequency tolerance of 0.0005 percent.

95.47 Types of emission.

(a) Except as provided in paragraph (e) of this section, Class A stations in this service will normally be authorized to transmit radiotelephony only. However, the use of tone signals or signaling devices solely to actuate receiver circuits, such as tone operated squelch or selective calling circuits, the primary function of which is to establish or establish and maintain voice communications, is permitted. The use of tone signals solely to attract attention is prohibited.

(b) [Reserved]

(c) Class C stations in this service are authorized to use amplitude tone modulation or on-off unmodulated carrier only, for the control of remote objects or devices by radio or for the remote actuation of devices which are used solely as a means of attracting attention. The transmission of any form of telegraphy, telephony or record communications by a Class C station is prohibited. Telemetering, except for the transmission of simple, short duration signals indicating the presence or absence of a condition or the occurrence of an event, is also prohibited.

(d) Transmitters used at Class D stations in this service are authorized to use amplitude voice modulation, either single or double sideband. Tone signals or signalling devices may be used only to actuate receiver circuits, such as tone operated squelch or selective calling circuits, the primary function of which is to establish or maintain voice communications. The use of any signals solely to attract attention or for the control of remote objects or devices is prohibited.

(e) Other types of emission not described in paragaph (a) of this section may be authorized for Class A citizens radio stations upon a showing of need therefor. An application requesting such authorization shall fully describe the emission desired, shall indicate the bandwidth required for satisfactory communication, and shall state the purpose for which such emission is required. For information regarding the classification of emissions and the calculation of bandwidth, reference should be made to Part 2 of this chapter.

95.49 Emission limitations.

(a) Each authorization issued to a Class A citizens radio station will show, as a prefix to the classification of the authorized emission, a figure specifying the maximum bandwidth to be occupied by the emission.

(b) [Reserved]

(c) The authorized bandwidth of the emission of any transmitter employing amplitude modulation shall be 8 kHz for double sideband and 4 kHz for single sideband. The authorized bandwidth of the emission of any transmitter employing frequency or phase modulation (Class F2 or F3) shall be 20 kHz. The use of F2 and F3 emissions in the frequency band 26.96 MHz—27.41 is not authorized.

(d) The mean power of emissions shall be attenuated below the mean power of the transmitter in accordance with the following schedule:

(1) When using emissions other than single sideband:

(i) On any frequency removed from the center of the authorized bandwidth by more than 50 percent up to and including 100 percent of the authorized bandwidth: at least 25 decibels;

(ii) On any frequency removed from the center of the authorized bandwidth by more than 100 percent up to and including 250 percent of the authorized bandwidth; At least 35 decibels;

(2) When using single sideband emissions:

(i) On any frequency removed from the center of the authorized bandwidth by more than 50 percent up to and including 150 percent of the authorized bandwidth: At least 25 decibels:

(ii) On any frequency removed from the center of the authorized bandwidth by more than 150 percent up to and including 250 percent of the authorized bandwidth: At least 35 decibels;

(3) On any frequency removed from the center of the authorized bandwidth by more than 250 percent of the authorized bandwidth: at least 43+10 $\log_{10}$10 (mean power in watts) decibels, for Class D transmitters type accepted before September 10, 1976 and all Class A transmitters.

(4) On any frequency removed from the center of the authorized bandwidth by more than 250 percent of the authorized bandwidth up to a frequency of twice the fundamental frequency; at least 53+10 $\log_{10}$ (mean power in watts) decibels, for Class D transmitters type accepted after September 10, 1976.

(5) On any frequency twice or greater than twice the fundamental frequency: at least 60 decibels (mean power in watts) for Class D transmitters type accepted after September 10, 1976.

NOTE—The requirements of paragraph (d) must be met both with and without connection of all attachments acceptable for use with such transmitters. External speakers, microphones, power cords, and antennas are among the devices included in this requirement. Additionally, if it is shown that a licensee causes interference to television reception because of insufficient harmonic attenuation, he may be required to insert a low pass filter between the transmitter RF output terminal and the antenna feedline.

(e) When an unauthorized emission results in harmful interference, the Commission may, in its discretion, require appropriate technical changes in equipment to alleviate the interference.

95.51 Modulation requirements.

(a) When double sideband, amplitude modulation is used for telephony, the modulation percentage shall be sufficient to provide efficient communication and shall not exceed 100 percent.

(b) Each transmitter for use in Class D stations, other than single sideband, suppressed carrier, or controlled carrier, for which type acceptance is requested after May 24, 1974, having more than 2.5 watts maximum output power shall be equipped with a device which automatically prevents modulation in excess of 100 percent on positive and negative peaks.

(c) The maximum audio frequency required for satisfactory radiotelephone intelligibility for use in this service is considered to be 3000 Hz.

(d) Transmitters for use at Class A stations shall be provided with a device which automatically will prevent greater than normal audio level from causing modulation in excess of that specified in this subpart; *Provided, however,* That the requirements of this paragraph shall not apply to transmitters authorized at mobile stations and having an output power of 2.5 watts or less.

(e) Each transmitter of a Class A station which is equipped with a modulation limiter in accordance with the provisions of paragraph (d) of this section shall also be equipped with an audio low-pass filter. This audio low-pass filter shall be installed between the modulation limiter and the modulated stage and, at audio frequencies between 3 kHz and 20 kHz, shall have an attenuation greater than the attenuation at 1 kHz by at least:

$$60 \log_{10} (f/3) \text{ decibels}$$

where "f" is the audio frequency in kHz. At audio frequencies above 20 kHz, the attenuation shall be at least 50 decibels greater than the attenuation at 1 kHz.

(f) Simultaneous amplitude modulation and frequency or phase modulation of a transmitter is not authorized.

(g) The maximum frequency deviation of frequency modulated transmitters used at Class A stations shall not exceed +5 kHz.

95.53 Compliance with technical requirements.

(a) Upon receipt of notification from the Commission of a deviation from the technical requirements of the rules in this part, the radiations of the transmitter involved shall be suspended immediately, except for necessary tests and adjustments, and shall not be resumed until such deviation has been corrected.

(b) When any citizens radio station licensee receives a notice of violation indicating that the station has been operated contrary to any of the provisions contained in Subpart C of this part, or where it otherwise appears that operation of a station in this service may not be in accordance with applicable technical standards, the Commission may require the licensee to conduct such tests as may be necessary to determine whether the equipment is capable of meeting these standards and to make such adjustments as may be necessary to assure compliance therewith. A licensee who is notified that he is required to conduct such tests and/or make adjustments must, within the time limit specified in the notice, report to the Commission the results thereof.

(c) All tests and adjustments which may be required in accordance with paragraph (b) of this section shall be made by, or under the immediate supervision of, a person holding a first- or second-class commercial operator license, either radiotelephone or radio telegraph as may be appropriate for the type of emission employed. In each case, the report which is submitted to the Commission shall be signed by the licensed commercial operator. Such report shall describe the results of the tests and adjustments, the test equipment and procedures used, and shall state the type, class, and serial number of the operator's license. A copy of this report shall also be kept with the station records.

95.55 Acceptability of transmitters for licensing.

Transmitters type approved or type accepted for use under this part are included in the Commission's Radio Equipment List. Copies of this list are available for public reference at the Commission's Washington, D.C., offices and field offices. The requirements for transmitters which may be operated under a license in this service are set forth in the following paragraphs.

(a) Class A stations: All transmitters shall be type accepted.

(b) Class C stations:

(1) Transmitters operated in the band 72—76 MHz shall be type accepted.

(2) All transmitters operated in the band 26.99—27.26 MHz shall be type approved, type accepted or crystal controlled.

(c) Class D stations:

(1) All transmitters first licensed, or marketed as specified in 2.805 of this chapter, prior to November 22, 1974 shall be type accepted or crystal controlled.

(2) All transmitters first licensed, or marketed as specified in 2.803 of this chapter, on or after November 22, 1974, shall be type accepted.

(3) Effective November 23, 1978, all transmitters shall be type accepted.

(4) Prior to January 1, 1977 transmitters which are equipped to operate on any frequency not included in 95.41(d) (1) may not be installed at, or used by, any Class D station unless there is a station license posted at the transmitter location, or a transmitter identification card (FCC Form 452—C) attached to the transmitter, which indicates that operation of the transmitter on such frequency has been authorized by the Commission.

(5) Effective January 1, 1977 transmitters which are equipped to operate on any frequency not included in 95.41 may not be installed at or used by any Class D station unless there is a station license posted at the transmitter location, or a transmitter identification card (FCC Form 452—C) attached to the transmitter, which indicates that operation of the transmitter on such frequency has been authorized by the Commission.

NOTE— A "transmitter" is defined to include any radio frequency (RF) power amplifier.

(d) With the exception of equipment type approved for use at a Class C station, all transmitting equipment authorized in this service shall be crystal controlled.

(e) No controls, switches or other functions which can cause operation in violation of the technical regulations of this part shall be accessible from the operating panel or exterior to the cabinet enclosing a transmitter authorized in this service.

95.57 Procedure for type acceptance of equipment.

(a) Any manufacturer of a transmitter built for use in this service, except noncrystal controlled transmitters for use at Class C stations, may request type acceptance for such transmitter in accordance with the type acceptance requirements of this part, following the type acceptance procedure set forth in Part 2 of this chapter.

(b) Type acceptance for an individual transmitter may also be requested by an applicant for a station authorization by following the type acceptance procedures set forth in Part 2 of this chapter. Such transmitters, if accepted, will not normally be included on the Commission's "Radio Equipment List", but will be individually enumerated on the station authorization.

(c) Additional rules with respect to type acceptance are set forth in Part 2 of this chapter. These rules include information with respect to withdrawal of type acceptance, modification of type-accepted equipment, and limitations on the findings upon which type acceptance is based.

(d) Transmitters equipped with a frequency or frequencies not listed in 95.41(d) (1) will not be type accepted for use at Class D stations unless the transmitter is also type accepted for use in the service in which the frequency is authorized, if type acceptance in that service is required.

95.58 Additional requirements for type acceptance.

(a) All tansmitters shall be crystal controlled.

(b) Except for transmitters type accepted for use at Class A stations, transmitters shall not include any provisions for increasing power to levels in excess of the pertinent limits specified in Section 95.43.

(c) In addition to all other applicable technical requirements set forth in this part, transmitters for which type acceptance is requested after May 24, 1974, for use at Class D stations shall comply with the following:

(1) Single sideband transmitters and other transmitters employing reduced, suppressed or controlled carrier shall include a means for automatically preventing the transmitter power from exceeding either the maximum permissible peak envelope power or the rated peak envelope power of the transmitter, whichever is lower.

(2) Multi-frequency transmitters shall be capable of operation only on those frequencies authorized by 95.41.

(3) All transmitter frequency determining circuitry (including crystals), other than the frequency selection mechanism, employed in Class D station equipment shall be internal to the equipment and shall not be accessible from the exterior of the equipment cabinet or operating panel. Add-on devices, whether internal or external to the equipment, the function of which is to extend the frequency coverage capability of a Class D unit beyond its original frequency coverage capability, shall not be sold, manufactured, or attached to any transmitter capable of operation on Class D Citizens Radio Service frequencies.

(4) Single sideband transmitters shall be capable of transmitting on the upper sideband. Capability for transmission also on the lower sideband is permissible.

(5) The total dissipation ratings, established by the manufacturer of the electron tubes or semiconductors which supply radio frequency power to the antenna terminals of the transmitter, shall not exceed 10 watts. For electron tubes, the rating shall be the Intermittent Commercial and Amateur Service (ICAS) plate dissipation value if established. For semiconductors, the rating shall be the collector or device dissipation value, whichever is greater, which may be temperature de-rated to not more than 50°C.

(d) Only the following external transmitter controls, connections or devices will normally be permitted in transmitters for which type acceptance is requested after May 24, 1974, for use at Class D stations. Approval of additional controls, connections or devices may be given after consideration of the function to be performed by such additions.

(1) Primary power connection. (Circuitry or devices such as rectifiers, transformers, or inverters which provide the nominal rated transmitter primary supply voltage may be used without voiding the transmitter type acceptance.)

(2) Microphone connection.

(3) Radio frequency output power connection.

(4) Audio frequency power amplifier output connector and selector switch.

(5) On-off switch for primary power to transmitter. May be combined with receiver controls such as the receiver on-off switch and volume control.

(6) Upper-lower sideband selector; for single sideband transmitters only.

(7) Selector for choice of carrier level; for single sideband transmitters only. May be combined with sideband selector.

(8) Transmitting frequency selector switch.

(9) Transmit-receive switch.

(10) Meter(s) and selector switch for monitoring transmitter performance.

(11) Pilot lamp or meter to indicate the presence of radio frequency output power or that transmitter control circuits are activated to transmit.

(e) An instruction book for the user shall be furnished with each transmitter sold and one copy (a draft or preliminary copy is acceptable providing a final copy is furnished when completed) shall be forwarded to the Commission with each request for type acceptance or type approval. The book shall contain all information necessary for the proper installation and operation of the transmitter including:

(1) Instructions concerning all controls, adjustments and switches which may be operated or adjusted without causing violation of technical regulations of this part;

(2) Warnings concerning any adjustment which, according to the rules of this part, may be made only by, or under the immediate supervision of, a person holding a commercial first or second class radio operator license;

(3) Warnings concerning the replacement or substitution of crystals, tubes or other components which could cause violation of the technical regulations of this part and of the type acceptance or type approval requirements of Part 2 of this chapter.

(4) Warnings concerning licensing requirements and details concerning the application procedures for licensing.

(f) A Class D Citizens Radio Service application form (FCC Form 505), a Temporary Permit, Class D Citizens Radio Station (FCC Form 555—B), and a copy of Part 95 of the Commission's Rules and Regulations, each to be current at the time of packing of the transmitter, shall be furnished with each transmitter sold after January 1, 1977.

(g) The serial number of each new Class D unit sold after January 1, 1977 shall be engraved on the unit's chassis.

95.59 Submission of noncrystal controlled Class C station transmitters for type approval.

Type approval of noncrystal controlled transmitters for use at Class C stations in this service may be requested in accordance with the procedure specified in Part 2 of this chapter.

95.61 Type approval of receiver-transmitter combinations.

Type approval will not be issued for transmitting equipment for operation under this part when such equipment is enclosed in the same cabinet, is constructed on the same chassis in whole or in part, or is identified with a common type or model number with a radio receiver; unless such receiver has been certificated to the Commission as complying with the requirements of Part 15 of this chapter.

sion as complying with the requirements of Part 15 of this chapter.

95.63 Minimum equipment specifications.

Transmitters submitted for type approval in this service shall be capable of meeting the technical specifications contained in this part, and in addition, shall comply with the following:

(a) Any basic instructions concerning the proper adjustment, use, or operation of the equipment that may be necessary shall be attached to the equipment in a suitable manner and in such positions as to be easily read by the operator.

(b) A durable nameplate shall be mounted on each transmitter showing the name of the manufacturer, the type or model designation, and providing suitable space for permanently displaying the transmitter serial number, FCC type approval number, and the class of station for which approved.

(c) The transmitter shall be designed, constructed, and adjusted by the manufacturer to operate on a frequency or frequencies available to the class of station for which type approval is sought. In designing the equipment, every reasonable precaution shall be taken to protect the user from high voltage shock and radio frequency burns. Connections to batteries (if used) shall be made in such a manner as to permit replacement by the user without causing improper operation of the transmitter. Generally accepted modern engineering principles shall be utilized in the generation of radio frequency currents so as to guard against unnecessary interference to other services. In cases of harmful interference arising from the design, construction, or operation of the equipment, the Commission may require appropriate technical changes in equipment to alleviate interference.

(d) Controls which may effect changes in the carrier frequency of the transmitter shall not be accessible from the exterior of any unit unless such accessibility is specifically approved by the Commission.

95.65 Test procedure.

Type approval tests to determine whether radio equipment meets the technical specifications contained in this part will be conducted under the following conditions:

(a) Gradual ambient temperature variations from 0° to 125° F.

(b) Relative ambient humidity from 20 to 95 percent. This test will normally consist of subjecting the equipment for at least three consecutive periods of 24 hours each, to a relative ambient humidity of 20, 60, and 95 percent, respectively, at a temperature of approximately 80° F.

(c) Movement of transmitter or objects in the immediate vicinity thereof.

(d) Power supply voltage variations normally to be encountered under actual operating conditions.

(e) Additional tests as may be prescribed, if considered necessary or desirable.

95.67 Certificate of type approval.

A certificate or notice of type approval, when issued to the manufacturer of equipment intended to be used or operated in the Citizens Radio Service, constitutes a recognition that on the basis of the test made, the particular type of equipment appears to have the capability of functioning in accordance with the technical specifications and regulations contained in this part: *Provided,* That all such additional equipment of the same type is properly constructed, maintained, and operated: *And provided further,* That no change whatsoever is made in the design or construction of such equipment except upon specific approval by the Commission.

SUBPART D—STATION OPERATING REQUIREMENTS

95.81 Permissible Communications.

Stations licensed in the Citizens Radio Service are authorized to transmit the following types of communications:

(a) Communications to facilitate the personal or business activities of the licensee.

(b) Communication relating to:

(1) The immediate safety of life or the immediate protection of property in accordance with 95.85.

(2) The rendering of assistance to a motorist, mariner or other traveler.

(3) Civil defense activities in accordance with 95.121.

(4) Other activities only as specifically authorized pursuant to 95.87.

(c) Communications with stations authorized in other radio services except as prohibited in 95.83(a) (3).

[95.81 added eff. 9-15-75; VI (75)-2]

95.83 Prohibited communications.

(a) A citizens radio station shall not be used:

(1) For any purpose, or in connection with any activity, which is contrary to Federal, State, or local law.

(2) For the transmission of communications containing obscene, indecent, profane words, language, or meaning.

(3) To communicate with an Amateur Radio Service station, an unlicensed station, or foreign stations (other than as provided in Subpart E of this part) except for communications pursuant to 95.85(b) and 95.121.

(4) To convey program material for retransmission, live or delayed, on a broadcast facility. *NOTE:* A Class A or Class D station may be used in connection with administrative, engineering, or maintenance activities of a broadcasting station; a Class A or Class C station may be used for control functions by radio which do not involve the transmission of program material; and a Class A or Class D station may be used in the gathering of news items or preparation of programs: Provided, that the actual or recorded transmissions of the Citizens radio station are not broadcast at any time in whole or in part.

(5) To intentionally interfere with the communications of another station.

(6) For the direct transmission of any material to the public through a public address system or similar means.

(7) For the transmission of music, whistling, sound effects, or any material for amusement or entertainment purposes, or solely to attract attention.

(8) To transmit the word "MAYDAY" or other international distress signals, except when the station is located in a ship, aircraft, or other vehicle which is threatened by grave and imminent danger and requests immediate assistance.

(9) For advertising or soliciting the sale of any goods or services.

(10) For transmitting messages in other than plain language. Abbreviations including nationally or internationally recognized operating signals, may be used only if a list of all such abbreviations and their meaning is kept in the station records and made available to any Commission representative on demand.

(11) To carry on communications for hire, whether the remuneration or benefit received is direct or indirect.

[95.83(a) & headnote amended eff. 9-15-75; VI (75)-2]

(b) A Class D station may not be used to communicate with, or attempt to communicate with, any unit of the same or another station over a distance of more than 150 miles.

(c) A licensee of a Citizens radio station who is engaged in the business of selling Citizens radio transmitting equipment shall not allow a customer to operate under his station license. In addition, all communications by the licensee for the purpose of demonstrating such equipment shall consist only of brief messages addressed to other units of the same station.

95.85 Emergency and assistance to motorist use.

(a) All Citizens radio stations shall give priority to the emergency communications of other stations which involve the immediate safety of life of individuals or the immediate protection of property.

(b) Any station in this service may be utilized during an emergency involving the immediate safety of life of individuals or the immediate protection of property for the transmission of emergency communications. It may also be used to transmit communications necessary to render assistance to a motorist.

(1) When used for transmission of emergency communications certain provisions in this part concerning use of frequencies (95.41(d)); prohibited uses (95.83(a) (3)); operation by or on behalf of persons other than the licensee (95.87); and duration of transmissions (95.91 (a) and (b) shall not apply.

(2) When used for transmissions of communications necessary to render assistance to a traveler, the provisions of this Part concerning duration of transmission (95.91(b)) shall not apply.

(3) The exemptions granted from certain rule provisions in subparagraphs (1) and (2) of this paragraph may be rescinded by the Commission at its discretion.

(c) If the emergency use under paragraph (b) of this section extends over a period of 12 hours or more, notice shall be sent to the Commission in Washington, D.C., as soon as it is evident that the emergency has or will exceed 12 hours. The notice should include the identity of the stations participating, the nature of the emergency, and the use made of the stations. A single notice covering all participating stations may be submitted.

[95.85(b) (1) & (2) amended eff. 9-15-75; VI (75)-2]

95.87 Operation by, or on behalf of, persons other than the licensee.

(a) Transmitters authorized in this service must be under the control of the licensee at all times. A licensee shall not transfer, assign, or dispose of, in any manner, directly or indirectly, the operating authority under his station license, and shall be responsible for the proper operation of all units of the station.

(b) Citizens radio stations may be operated only by the following persons, except as provided in paragraph (c) of this section:

(1) The licensee;

(2) Members of the licensee's immediate family living in the same household;

(3) The partners, if the licensee is a partnership, provided the communications relate to the business of the partnership;

(4) The members, if the licensee is an unincorporated association, provided the communications relate to the business of the association;

(5) Employees of the licensee only while acting within the scope of their employment;

(6) Any person under the control or supervision of the licensee when the station is used solely for the control of remote objects or devices, other than devices used only as a means of attracting attention; and

(7) Other persons, upon specific prior approval of the Commission shown on or attached to the station license, under the following circumstances:

(i) Licensee is a corporation and proposes to provide private radiocommunication facilities for the transmission of messages or signals by or on behalf of its parent corporation, another subsidiary of the parent corporation, or its own subsidiary. Any remuneration or compensation received by the licensee for the use of the radiocommunication facilities shall be governed by a contract entered into by the parties concerned and the total of the compensation shall not exceed the cost of providing the facilities. Records which show the cost of service and its nonprofit or cost-sharing basis shall be maintained by the licensee.

(ii) Licensee proposes the shared or cooperative use of a Class A station with one or more other licensees in this service for the purpose of communicating on a regular basis with units of their respective Class A stations, or with units of other Class A stations if the communications transmitted are otherwise permissible. The use of these private radiocommunication facilities shall be conducted pursuant to a written contract which shall provide that contributions to capital and operating expense shall be made on a nonprofit, cost-sharing basis, the cost to be divided on an equitable basis among all parties to the agreement. Records which show the cost of service and its nonprofit, cost-sharing basis shall be maintained by the licensee. In any case, however, licensee must show a separate and independent need for the particular units proposed to be shared to fulfill his own communications requirements.

(iii) Other cases where there is a need for other persons to operate a unit of licensee's radio station. Requests for authority may be made either at the time of the filing of the application for station license or thereafter by letter. In either case, the licensee must show the nature of the proposed use and that it relates to an activity of the licensee, how he proposes to maintain control over the transmitters at all times, and why it is not appropriate for such other person to obtain a station license in his own name. The authority, if granted, may be specific with respect to the names of the persons who are permitted to operate, or may authorize operation by unnamed persons for specific purposes. This authority may be revoked by the Commission, in its discretion, at any time.

(c) An individual who was formerly a citizens radio station licensee shall not be permitted to operate any citizens radio station of the same class licensed to another person until such time as he again has been issued a valid radio station license of that class, when his license has been:

(1) Revoked by the Commission.

(2) Surrendered for cancellation after the institution of revocation proceedings by the Commission.

(3) Surrendered for cancellation after a notice of apparent liability to forfeiture has been served by the Commission.

95.89 Telephone answering services.

(a) Notwithstanding the provisions of 95.87, a licensee may install a transmitting unit of his station on the premises of a telephone answering service. The same unit may not be operated under the authorization of more than one licensee. In all cases, the licensee must enter into a written agreement with the answering service. This agreement must be kept with the licensee's station records and must provide, as a minimum, that:

(1) The licensee will have control over the operation of the radio unit at all times;

(2) The licensee will have full and unrestricted access to the transmitter to enable him to carry out his responsibilities under his license;

(3) Both parties understand that the licensee is fully responsible for the proper operation of the citizens radio station; and

(4) The unit so furnished shall be used only for the transmission of communications to other units belonging to the licensee's station.

(b) A citizens radio station licensed to a telephone answering service shall not be used to relay messages or transmit signals to its customers.

95.91 Duration of transmissions.

(a) All communications or signals, regardless of their nature, shall be restricted to the minimum practicable transmission time. The radiation of energy shall be limited to transmissions modulated or keyed for actual permissible communications, tests, or control signals. Continuous or uninterrupted transmissions from a single station or between a number of communicating stations is prohibited, except for communications involving the immediate safety of life or property.

(b) All communications between Class D stations (interstation) shall be restricted to not longer than five (5) continuous minutes. At the conclusion of this 5 minute period, or the exchange of less than 5 minutes, the participating stations shall remain silent for at least one minute.

(c) All communication between units of the same Class D station (intrastation) shall be restricted to the minimum practicable transmission.

(d) The transmission of audible tone signals or a sequence of tone signals for the operation of the tone operated squelch or selective calling circuits in accordance with 95.47 shall not exceed a total of 15 seconds duration. Continuous transmission of a subaudible tone for this purpose is permitted. For the purposes of this section, any tone or combination of tones having no frequency above 150 hertz shall be considered subaudible.

(e) The transmission of permissible control signals shall be limited to the minimum practicable time necessary to accomplish the desired control or actuation of remote objects or devices. The continuous radiation of energy for periods exceeding 3 minutes duration for the purpose of transmission of control signals shall be limited to control functions requiring at least one or more changes during each minute of such transmission. However, while it is actually being used to control model aircraft in flight by means of interrupted tone modulation of its carrier, a citizens radio station may transmit a continuous carrier without being simultaneously modulated if the presence or absence of the carrier also performs a control function. An exception to the limitations contained in this paragraph may be authorized upon a satisfactory showing that a continuous control signal is required to perform a control function which is necessary to insure the safety of life or property.

[95.91Ib) amended & (c) added, present par. (c) & (d) redesig. par. (d) & (e) eff. 9-15-75; VI (75)-2]

95.93 Tests and adjustments.

All tests or adjustments of citizens radio transmitting equipment involving an external connection to the radio frequency output circuit shall be made using a nonradiating dummy antenna. However, a brief test signal, either with or without modulation, as appropriate, may be transmitted when it is necessary to adjust a transmitter to an antenna for a new station installation or for an existing installation involving a change of antenna or change of transmitters, or when necessary for the detection, measurement, and suppression of harmonic or other spurious radiation. Test transmissions using a radiating antenna shall not exceed a total of 1 minute during any 5-minute period, shall not interfere with communications already in progress on the operating frequency, and shall be properly identified as required by 95.95, but may otherwise be unmodulated as appropriate.

95.95 Station identification.

(a) The call sign of a citizens radio station shall consist of three letters followed by four digits.

(b) Each transmission of the station call sign shall be made in the English language by each unit, shall be complete, and each letter and digit shall be separately and distinctly transmitted. Only standard phonetic alphabets, nationally or internationally recognized, may be used in lieu of pronunciation of letters for voice transmission of call signs. A unit designator or special identification may be used in addition to the station call sign but not as a substitute therefor.

(c) Except as provided in paragraph (d) of this section, all transmission from each unit of a citizens radio station shall be identified by the transmission of its assigned call sign at the beginning and end of each transmission or series of transmissions, but at least at intervals not to exceed ten (10) minutes.

(d) Unless specifically required by the station authorization, the transmissions of a citizens radio station need not be identified when the station (1) is a Class A station which automatically retransmits the information received by radio from another station which is properly identified or (2) is not being used for telephony emission.

(e) In lieu of complying with the requirements of paragraph (c) of this section, Class A base stations, fixed stations, and mobile units when communicating with base stations may identify as follows:

(1) Base stations and fixed stations of a Class A radio system shall transmit their call signs at the end of each transmission or exchange of transmissions, or once each 15-minute period of a continuous exchange of communications.

(2) A mobile unit of a class A station communicating with a base station of a Class A radio system on the same frequency shall transmit once during each exchange of transmissions any unit identifier which is on file in the station records of such base station.

(3) A mobile unit of Class A stations communicating with a base station of a Class A radio system on a different frequency shall transmit its call sign at the end of each transmission or exchange of transmissions, or once each 15-minute period of a continuous exchange of communications.

[95.95(c) amended eff. 9-15-75; VI (75)-2]

95.97 Operator license requirements.

(a) No operator license is required for the operation of a citizens radio station except that stations manually transmitting Morse Code shall be operated by the holders of a third or higher class radiotelegraph operator license.

(b) Except as provided in paragraph (c) of this section, all transmitter adjustments or tests while radiating energy during or coincident with the construction, installation, servicing, or maintenance of a radio station in this service, which may affect the proper operation of such stations, shall be made by or under the immediate supervision and responsibility of a person holding a first- or second-class commercial radio operator license, either radiotelephone or radio telegraph, as may be appropriate for the type of emission employed, and such person shall be responsible for the proper functioning of the station equipment at the conclusion of such adjustments or tests. Further, in any case where a transmitter adjustment which may affect the proper operation of the transmitter has been made while not radiating energy by a person not the holder of the required commercial radio operator license or not under the supervision of such licensed operator, other than the factory assembling or repair of equipment, the transmitter shall be checked for compliance with the technical requirements of the rules by a commercial radio operator of the proper grade before it is placed on the air.

(c) Except as provided in 95.53 and in paragraph (d) of this section, no commercial radio operator license is required to be held by the person performing transmitter adjustments or test during or coincident with the construction, installation, servicing, or maintenance of Class C transmitters, or Class D transmitters used at stations authorized prior to May 24, 1974: *Provided*, That there is compliance with all of the following conditions:

(1) The transmitting equipment shall be crystal controlled with a crystal capable of maintaining the station frequency within the prescribed tolerance;

(2) The transmitting equipment either shall have been factory assembled or shall have been provided in kit form by a manufacturer who provided all components together with full and detailed instructions for their assembly by nonfactory personnel;

(3) The frequency determining elements of the transmitter, including the crystal(s) and all other components of the crystal oscillator circuit, shall have been preassembled by the manufacturer, pretuned to a specific available frequency, and sealed by the manufacturer so that replacement of any component or any adjustment which might cause off-frequency operation cannot be made without breaking such seal and thereby voiding the certification of the manufacturer required by this paragraph;

(4) The transmitting equipment shall have been so designed that none of the transmitter adjustments or tests normally performed during or coincident with the installation, servicing, or maintenance of the station, or during the normal rendition of the service of the station, or during the final assembly of kits or partially preassembled units, may reasonably be expected to result in off-frequency operation, excessive input power, overmodulation, or excessive harmonics or other spurious emissions; and

(5) The manufacturer of the transmitting equipment or of the kit from which the transmitting equipment is assembled shall have certified in writing to the purchaser of the equipment (and to the Commission upon request) that the equipment has been designed, manufactured, and furnished in accordance with the specifications contained in the foregoing subparagraphs of this paragraph. The manufacturer's certification concerning design and construction features of Class C or Class D station transmitting equipment, as required if the provisions of this paragraph are invoked, may be specific as to a particular unit of transmitting equipment or general as to a group or model of such equipment, and may be in any form adequate to assure the purchase of the equipment or the Commission that the conditions described in this paragraph have been fulfilled.

(d) Any tests and adjustments necessary to correct any deviation of a transmitter of any Class of station in this service from the technical requirements of the rules in this part shall be made by, or under the immediate supervision of, a person holding a first- or second-class commercial operator license, either radiotelephone or radiotelegraph, as may be appropriate for the type of emission employed.

95.101 Posting station license and transmitter identification cards or plates.

(a) The current authorization, or a clearly legible photocopy thereof, for each station (including units of a Class C or Class D station) operated at a fixed location shall be posted at a conspicuous place at the principal fixed location from which such station is controlled, and a photocopy of such authorization shall also be posted at all other fixed locations from which the station is controlled. If a photocopy of the authorization is posted at the principal control point, the location of the original shall be stated on that photocopy. In addition, an executed Transmitter Identification Card (FCC Form 452-C) or a plate of metal or other durable substance, legibly indicating the call sign and the licensee's name and address, shall be affixed, readily visible for inspection, to each transmitter operated at a fixed location when such transmitter is not in view of, or is not readily accessible to, the operator of at least one of the locations at which the station authorization or a photocopy thereof is required to be posted.

(b) The current authorization for each station operated as a mobile station shall be retained as a permanent part of the station records, but need not be posted. In addition, an executed Transmitter Identification Card (FCC Form 452—C) or a plate of metal or other durable substance, legibly indicating the call sign and the licensee's name and address, shall be affixed, readily visible for inspection, to each of such transmitters: *Provided*, That, if the transmitter is not in view of the location from which it is controlled, or is not readily accessible for inspection, then such card or plate shall be affixed to the control equipment at the transmitter operating position or posted adjacent thereto.

95.103 Inspection of stations and station records.

All stations and records of stations in the Citizens Radio Service shall be made available for inspection upon the request of an authorized representative of the Commission made to the licensee or to his representative (see 1.6 of this chapter). Unless otherwise stated in this part, all required station records shall be maintained for a period of at least 1 year.

95.105 Current copy of rules required.

Each licensee in this service shall maintain as a part of his station records a current copy of Part 95, Citizens Radio Service, of this chapter,

95.107 Inspection and maintenance of tower marking and lighting, and associated control equipment.

The licensee of any radio station which has an antenna structure required to be painted and illuminated pursuant to the provisions of section 303(q) of the Communications Act of 1934, as amended, and Part 17 of this chapter. shall perform the inspection and maintain the tower marking and lighting, and associated control equipment, in accordance with the requirements set forth in Part 17 of this chapter.

95.111 Recording of tower light inspections.

When a station in this service has an antenna structure which is required to be illuminated, appropriate entries shall be made in the station records in conformity with the requirements set forth in Part 17 of this chapter.

95.113 Answers to notices of violations.

(a) Any licensee who appears to have violated any provision of the Communications Act or any provision of this chapter shall be served with a written notice calling the facts to his attention and requesting a statement concerning the matter. FCC Form 793 may be used for this purpose.

(b) Within 10 days from receipt of notice or such other period as may be specified, the licensee shall send a written answer, in duplicate, direct to the office of the Commission originating the notice. If an answer cannot be sent nor an acknowledgment made within such period by reason of illness or other unavoidable circumstances, acknowledgment and answer shall be made at the earliest practicable date with a satisfactory explanation of the delay.

(c) The answer to each notice shall be complete in itself and shall not be abbreviated by reference to other communications or answers to other notices. In every instance the answer shall contain a statement of the action taken to correct the condition or omission complained of and to preclude its recurrence. If the notice relates to violations that may be due to the physical or electrical characteristics of transmitting apparatus, the licensee must comply with the provisions of 95.53, and the answer to the notice shall state fully what steps, if any, have been taken to prevent future violations, and, if any new apparatus is to be installed, the date such apparatus was ordered, the name of the manufacturer, and the promised date of delivery. If the installation of such apparatus requires a construction permit, the file number of the application shall be given, or if a file number has not been assigned by the Commission, such identification shall be given as will permit ready identification of the application. If the notice of violation relates to lack of attention to or improper operation of the transmitter, the name and license number of the operator in charge, if any, shall also be given.

95.115 False signals.

No person shall transmit false or deceptive communications by radio or identify the station he is operating by means of a call sign which has not been assigned to that station.

95.117 Station location.

(a) The specific location of each Class A base station and each Class A fixed station and the specific area of operation of each Class A mobile station shall be indicated in the application for license. An authorization may be granted for the operation of a Class A base station or fixed station in this service at unspecified temporary fixed locations within a specified general area of operation. However, when any unit or units of a base station or fixed station authorized to be operated at temporary locations actually remains or is intended to remain at the same location for a period of over a year, application for separate authorization specifying the fixed location shall be made as soon as possible but not later than 30 days after the expiration of the 1-year period.

(b) A Class A mobile station authorized in this service may be used or operated anywhere in the United States subject to the provisions of paragraph (d) of this section: *Provided*, That when the area of operation is changed for a period exceeding 7 days, the following procedure shall be observed:

(1) When the change of area of operation occurs inside the same Radio District, the Engineer in Charge of the Radio District involved and the Commission's office, Washington, D.C., 20554, shall be notified.

(2) When the station is moved from one Radio District to another, the Engineers in Charge of the two Radio Districts involved and the Commission's office, Washington, D.C., 20554, shall be notified.

(c) A Class C or Class D mobile station may be used or operated anywhere in the United States subject to the provisions of paragraph (d) of this section.

(d) A mobile station authorized in this service may be used or operated on any vessel, aircraft, or vehicle of the United States: *Provided*, That when such vessel, aircraft, or vehicle is outside the territorial limits of the United States, the station, its operation, and its operator shall be subject to the governing provisions of any treaty concerning telecommunications to which the United States is a party, and when within the territorial limits of any foreign country, the station shall be subject also to such laws and regulations of that country as may be applicable.

95.119 Control points, dispatch points, and remote control.

(a) A control point is an operating position which is under the control and supervision of the licensee, at which a person immediately responsible for the proper operation of the transmitter is stationed, and at which adequate means are available to aurally monitor all transmissions and to render the transmitter inoperative. Each Class A base or fixed station shall be provided with a control point, the location of which will be specified in the license. The location of the control point must be the same as the transmitting equipment unless the application includes a request for a different location. Exception to the requirement for a control point may be made by the Commission upon specific request and justification therefor in the case of certain unattended Class A stations employing special emissions pursuant to 95.47(e). Authority for such exception must be shown on the license.

(b) A dispatch point is any position from which messages may be transmitted under the supervision of the person at a control point who is responsible for the proper operation of the transmitter. No authorization is required to install dispatch points.

(c) Remote control of a Citizens radio station means the control of the transmitting equipment of that station from any place other than the location of the transmitting equipment, except that direct mechnical control or direct electrical control by wired connections of transmitting equipment from some other point on the same premises, craft, or vehicle shall not be considered remote control. A Class A base or fixed station may be authorized to be used or operated by remote control from another fixed location or from mobile units: *Provided,* That adequate means are available to enable the person using or operating the station to render the transmitting equipment inoperative from each remote control position should improper operation occur.

(d) Operation of any Class C or Class D station by remote control is prohibited except remote control by wire upon specific authorization by the Commission when satisfactory need is shown.

[95.119(d) amended eff. 9-15-75; VI(75)-2]

95.121 Civil defense communications.

A licensee of a station authorized under this part may use the licensed radio facilities for the transmission of messages relating to civil defense activities in connection with official tests or drills conducted by, or actual emergencies proclaimed by, the civil defense agency having jurisdiction over the area in which the station is located: *Provided,* That:

(a) The operation of the radio station shall be on a voluntary basis.

(b) [Reserved]

(c) Such communications are conducted under the direction of civil defense authorities.

(d) As soon as possible after the beginning of such use, the licensee shall send notice to the Commission in Washington, D.C., and to the Engineer in Charge of the Radio District in which the station is located, stating the nature of the communications being transmitted and the duration of the special use of the station. In addition, the Engineer in Charge shall be notified as soon as possible of any change in the nature of or termination of such use.

(e) In the event such use is to be a series of pre-planned tests or drills of the same or similar nature which are scheduled in advance for specific times or at certain intervals of time, the licensee may send a single notice to the Commission in Washington, D.C., and to the Engineer in Charge of the Radio District in which the station is located, stating the nature of the communications to be transmitted, the duration of each such test, and the times scheduled for such use. Notice shall likewise be given in the event of any change in the nature of or termination of any such series of tests.

(f) The Commission may, at any time, order the discontinuance of such special use of the authorized facilities.

SUBPART E—OPERATION OF CITIZENS RADIO STATIONS IN THE UNITED STATES BY CANADIANS

95.131 Basis, purpose and scope.

(a) The rules in this subpart are based on, and are applicable solely to the agreement (TIAS #6931) between the United States and Canada, effective July 24, 1970, which permits Canadian stations in the General Radio Service to be operated in the United States.

(b) The purpose of this subpart is to implement the agreement (TIAS #6931) between the United States and Canada by prescribing rules under which a Canadian licensee in the General Radio Service may operate his station in the United States.

95.133 Permit required.

Each Candian licensee in the General Radio Service desiring to operate his radio station in the United States, under the provisions of the agreement (TIAS #6931), must obtain a permit for such operation from the Federal Communications Commission. A permit for such operation shall be issued only to a person holding a valid license in the General Radio Service issued by the appropriate Canadian governmental authority.

95.135 Application for permit.

(a) Application for a permit shall be made on FCC Form 410—B. Form 410—B may be obtained from the Commission's Washington, D.C., office or from any of the Commission's field offices. A separate application form shall be filed for each station or transmitter desired to be operated in the United States.

(b) The application form shall be completed in full in English and signed by the applicant. The application must be filed by mail or in person with the Federal Communications Commission, Gettysburg, Pa. 17325, U.S.A. To allow sufficient time for processing, the application should be filed at least 60 days before the date on which the applicant desires to commence operation.

(c) The Commission, at its discretion, may require the Canadian licensee to give evidence of his knowledge of the Commission's applicable rules and regulations. Also the Commission may require the applicant to furnish any additional information it deems necessary.

95.137 Issuance of permit.

(a) The Commission may issue a permit under such conditions, restrictions and terms as it deems appropriate.

(b) Normally, a permit will be issued to expire 1 year after issuance but in no event after the expiration of the license issued to the Canadian licensee by his government.

(c) If a change in any of the terms of a permit is desired, an application for modification of the permit is required. If operation beyond the expiration date of a permit is desired an application for renewal of the permit is required. Application for modification or for renewal of a permit shall be filed on FCC Form 410—B.

(d) The Commission, in its discretion, may deny any application for a permit under this subpart. If an application is denied, the applicant will be notified by letter. The applicant may, within 30 days of the mailing of such letter, request the Commission to reconsider its action.

95.139 Modification or cancellation of permit.

At any time the Commission may, in its discretion, modify or cancel any permit issued under this subpart. In this event, the permittee will be notified of the Commission's action by letter mailed to his mailing address in the United States and the permittee shall comply immediately. A permittee may, within 30 days of the mailing of such letter, request the Commission to reconsider its action. The filing of a request for reconsideration shall not stay the effectiveness of that action, but the Commission may stay its action on its own motion.

95.141 Possession of permit.

The current permit issued by the Commission, or a photocopy thereof, must be in the possession of the operator or attached to the transmitter. The license issued to the Canadian licensee by his government must also be in his possession while he is in the United States.

95.143 Knowledge of rules required.

Each Canadian permittee, operating under this subpart, shall have read and understood this Part 95, Citizens Radio Service.

95.145 Operating conditions.

(a) The Canadian licensee may not under any circumstances begin operation until he has received a permit issued by the Commission.

(b) Operation of station by a Canadian licensee under a permit issued by the Commission must comply with all of the following:

(1) The provisions of this subpart and of Subparts A through D of this part.

(2) Any further conditions specified on the permit issued by the Commission.

95.147 Station identification.

The Canadian licensee authorized to operate his radio station in the United States under the provisions of this subpart shall identify his station by the call sign issued by the appropriate authority of the government of Canada followed by the station's geographical location in the United States as nearly as possible by city and state.

EXCERPTS FROM CANADIAN REGULATIONS FOR GENERAL RADIO SERVICE

GENERAL RADIO SERVICE OPERATION

70. (1) Subject to these Regulations, a licence for a station performing a General Radio Service shall be in force for a period not exceeding three fiscal years.

(2) In this section "fiscal year" means a twelve-month period commencing on the first day of April and ending on the thirty-first day of March following.

(3) In section 71 to 80 "licensed station" means a station licensed to perform a General Radio Service.

71. A license for a station performing a General Radio Service may authorize the licensed station to be operated:
(a) in any craft or vehicle;
(b) at fixed locations; or
(c) while carried on the person.

72. (1) Subject to subsection (2), the frequencies authorized to be used by each licensed station on a shared basis, subject to no protection from interference caused by the operation of industrial, scientific and medical apparatus in the frequency band 26.96 — 27.28 Mc/s, are as follows:

26.965 Mc/s	27.065 Mc/s	27.155 Mc/s
26.975 Mc/s	27.075 Mc/s	27.165 Mc/s
26.985 Mc/s	27.085 Mc/s	27.175 Mc/s
27.005 Mc/s	27.105 Mc/s	27.185 Mc/s
27.015 Mc/s	27.115 Mc/s	27.205 Mc/s
27.025 Mc/s	27.125 Mc/s	27.215 Mc/s
27.035 Mc/s	27.135 Mc/s	27.225 Mc/s
27.055 Mc/s		

(2) Licensed stations may only use the frequency of 27.065 Mc/s
(a) for radiocommunications that involve the immediate protection of lives or property; or
(b) to establish communication with other stations.

73. (1) A licensed station may carry on two-way radiotelephone communication with stations that are
(a) licensed to perform a General Radio Service;
(b) licensed by the Government of the United States as Class D stations in the Citizens Radio Service if the licensees of the Class D stations are authorized by the Minister to operate them in Canada; or
(c) exempted from licensing under subsection 6(3).

(2) the type of communications referred to in subsection (1) may include transmissions of signals for the actuation of radio receivers to establish and maintain voice communication.

73.1. The maximum power authorized for use by a licensed station is
(a) where determined on the basis of direct current input power, five watts to the anode or collector circuit of the transmitter amplifier stage supplying radio frequency energy to the antenna; or
(b) where determined on the basis of radio frequency output power,
(i) twelve watts peak envelope power for transmitters producing suppressed carrier, reduced carrier, controlled carrier or other types of single sideband emissions; or
(ii) four watts carrier power for transmitters producing other types of emissions,
as measured at the transmitter output terminals when terminated by an impedence matched load.

74. (1) A licensed station shall be used only for communication concerning the business activities and personal affairs of the licensee.

(2) A licensed station shall not be used for any of the following purposes:
(a) activity contrary to federal or provincial law or municipal by-laws;
(b) the transmission of music or other material for amusement or entertainment;
(c) broadcasting or any transmission in connection with broadcasting;
(d) transmissions of any nature to the

public through a public address system;
(e) transmission of a frivolous nature;
(f) transmissions directed to any person or station beyond the ground wave coverage range of the station;
(g) communication used in itself as a diversionary or recreational activity; or
(h) calls directed to stations generally.

(2a) Notwithstanding paragraph (h) of subsection (2), in an emergency, calls for assistance may be transmitted as a call to "GRS stations" specifying the area concerned.

(3) No tolls shall be levied or collected on account of any business transacted or messages transmitted or received by means of a licensed station.

75. (1) Each exchange of communications between licensed stations shall not exceed five consecutive minutes duration and upon the termination of an exchange of communications no further transmission shall be made until the lapse of two minutes or until interference will not be caused to other stations using the same frequency.

(2) The emission of a carrier wave is prohibited except when actual communications are being transmitted or for making brief test transmissions.

(3) A licensed station shall transmit its assigned call sign.
(a) at the beginning and at the end of each exchange of communications in which it is engaged, and
(b) at the end of each test transmission.

76. (1) No transmission by a licensed station shall
(a) cause any interference to other licensed radio stations operating outside the frequency band 26.96 — 27.28 Mc/s;
(b) cause any interference to a private receiving station; or
(c) interfere maliciously with the communications of another station operating in the General Radio Service.

(2) In the event of interference as referred to in subsection (1) being caused by licensed station, the licensee may be required by a departmental radio inspector to take such steps as are necessary for the prevention of further interference or to restrict or cease operation of the station pending a satisfactory adjustment of the radio apparatus of the station.

77. (1) Each licensee shall be responsible at all times for the control and operation of his licensed station.

(2) Each licence for a station performing a General Radio Service is subject to the condition that the operator of the licensed station shall be
(a) the licensee of the station, or
(b) a person other than the licensee, if that person is twelve years of age or more,

and notwithstanding anything in these Regulations, such operator is not required to be the holder of any certificate of proficiency in radio.

(3) Licensed stations are exempt from the radio log requirements of section 36.

Equipment

78. The radio apparatus of a licensed station shall be of a type that has been approved in accordance with the applicable Radio Standards Specification issued by the Minister and shall bear the departmental type-approval number assigned to that radio apparatus.

79. (1) A person holding a valid license issued by the Government of the United States for a Class D station in the Citizens Radio Service may operate his radio station in Canada for such period as the Minister prescribes if
(a) he submits an application in a form approved by the Minister, and
(b) he is authorized in writing by the Minister to operate the station in Canada.

(2) Any person operating a station in Canada under subsection (1) shall operate that station as if it were a licensed station subject to sections 71 to 77.

(3) For the purpose of subsection (3) of section 75 the assigned call sign of any station operated under subsection (1) of this section is the call sign assigned to it by the licensing authority of the United States followed by the name of the province where it is located and its general geographical location in the province.

Exemption from Licensing

6. (3) A radio station is exempt from the operation of section 2A of the Act: if
(a) it operates only in the frequency band 26.97 to 27.27 Mc/s;
(b) the station is used only for two-way radiotelephone communication with
  (i) other stations to which this subsection applies,
  (ii) stations performing a General Radio Service, or
  (iii) stations licensed by the Government of the United States as Class D stations in the Citizens Radio Service and operated in Canada pursuant to section 79.
(c) it is not capable of emitting Hertzian waves of a field strength greater than that produced by the radio frequency energy radiated by a single antenna element not exceeding five feet in length fed from a transmitter having a plate or collector power input of 100 milliwatts to the final radio stage; and
(d) the transmitter emissions of the station are suppressed to 20 db or more below the unmodulated carrier on frequencies below 26.97 Mc/s and above 27.27 Mc/s.

(3a) A radio station that is exempt from the operation of section 2A of the Act under subsection (3) and that communicates with a station described in subparagraph (ii) or (iii) of paragraph (b) of that subsection is subject to sections 72 to 77 as if it were a licensed station.

(4) No station referred to in this section shall be operated so as to cause interference to any licensed radio station or private receiving station.

(5) Where interference is caused by a station referred to in this section, the operator of the station may be required by a departmental radio inspector to take such steps as are necessary for the prevention of further interference or to restrict or cease operation of the station pending a satisfactory adjustment of the equipment.

INDEX